Signalers and Receivers

Signalers and Receivers

Mechanisms and Evolution of Arthropod Communication

Michael D. Greenfield

OXFORD
UNIVERSITY PRESS

2002

Oxford New York
Athens Auckland Bangkok Bogotá Buenos Aires Cape Town
Chennai Dar es Salaam Delhi Florence Hong Kong Istanbul Karachi
Kolkata Kuala Lumpur Madrid Melbourne Mexico City Mumbai Nairobi
Paris São Paulo Shanghai Singapore Taipei Tokyo Toronto Warsaw

and associated companies in
Berlin Ibadan

Published by Oxford University Press, Inc.
198 Madison Avenue, New York, New York 10016

Library of Congress Cataloging-in-Publication Data

Greenfield, Michael D.
Signalers and receivers: mechanisms and evolution of arthropod communication/
Michael D. Greenfield
p. cm.
Includes bibliographical references (p.)
ISBN 0-19-513452-4
1. Arthropoda—Behavior. 2. Animal communication. I. Title
QL434.8.G74 2001
595'.159–dc21 2001021267

2 4 6 8 9 7 5 3 1

Printed in the United States of America
on acid free paper

Preface

This book was born of several compelling factors and players, not the least of which were the arthropods themselves and the scientists who have been devoted to their study. Arthropod communication has, of course, been covered in great detail by various experts in the field, and most topics included in this book have been the subject of entire volumes. But, the field has heretofore escaped unifying treatment by a single author, and its components have largely developed, and flourished, in near isolation from each other. Thus, the various arthropod groups, the different modalities of communication, and physiological and evolutionary modes of inquiry have all acquired their own scientific subcultures, terminologies, and specialist literature. Having been a participant in several of these subcultures, I have been motivated to create a work that crosses traditional boundaries and indicates comparisons and contrasts where such revelations can further general or specific understanding. I have not intended this work to replace any of the specialized treatises on various groups and modalities. Rather, my objective is to identify unifying principles and themes in a field that is expanding steadily in multiple directions.

While I have prepared the book as a single entity, I have also recognized that many readers may wish to read or consult only certain chapters or sections thereof. To that end, I have tried to have each major section stand on its own and be intelligible without the need for constant reference to material presented in other chapters. I have taken a similar approach to the mathematical underpinnings of signaling and communication. Ultimately, signaling and perception are physical and chemical phenomena that demand some level of quantitative treatment. But, the fundamental aspects of these phenomena can all be described verbally without losing essential meaning, and I have minimized the number of equations in the text proper. For those readers seeking a more complete treatment, mathematical and physical details are presented in figures, boxes, and notes separated from the main text.

Most sections of the book include some historical information, as familiarity with the ideas of eighteenth, nineteenth, and early twentieth-century workers contributes greatly to appreciation of our current knowledge and research directions. Nonetheless, the bulk of material presented concentrates on recent findings and paradigms, especially in areas where the field has taken radically new turns. Perhaps the most important of these paradigms is the use of phylogenetic analysis to resolve questions on the evolution of signaling systems and communication, an approach that would have received scant attention as recently as ten years ago.

The writing of the book was largely completed while I was on sabbatical leave from the University of Kansas and in residence in the Division of Neurobiology at the University of Arizona. I am indebted to Nick Strausfeld, a leader in evolutionary neuroanatomy and one of its founders, for providing a most stimulating environment in the Division in which to think and write and to Terry Markow, Director of the Center for Insect Science at the University of Arizona, for involving me in the Center's activities. Various reviewers offered numerous detailed criticisms of earlier versions of the manuscript: both Rafa Rodriguez and Andy Snedden read the entire text and, in addition to pointing out errors and other inconsistencies, helped me to refine my writing and organize the disparate sections. These concerns also benefited immensely from the editorial advice of Kirk Jensen, Oxford University Press, which I relied on at all stages. Tom Christensen, John Douglass, Klaus-Gerhardt Heller, Pyotr Jablonski, Stuart Krasnoff, Kate Loudon, Mike Ritchie, Ron Rutowski, Justin Schmidt, and Liz Smith each read chapters for which they had special expertise; I am particularly grateful to John Douglass for his patience in clarifying technical aspects of visual neuroanatomy and perception. David Clayton, Rex Cocroft, Don Davis, Paul Faure, Larry Field, Paul Flook, Carl Gerhardt, Vince Gottschick, Eileen Hebets, Cesar Nufio, Johannes Schul, Hayward Spangler, Eran Tauber, Dagmar von Helversen, Otto von Helversen, and Mark Willis provided literature and manuscripts and imparted lore on various arthropod groups and their communication. Many ideas presented here evolved from discussions with colleagues and students in Lawrence, Kansas, including David Alexander, Bill Bell, LaRoy Brandt, Marc Branham, Nancy Cohen, Bob Collins, Sylvia Cremer, Yikweon Jang, Feng-You Jia, Kate Loudon, Bjorn Ludwar, Bob Minckley, Nick Prins, Klaus Reinhold, Andy Snedden, and Michael Tourtellot. At an earlier time and in a more fundamental way, my graduate mentor, Michael G. Karandinos, was instrumental in sustaining my interest in theoretical and quantitative approaches to problems in biology and in driving me to seek ever more effective means for communicating my ideas. All of the above scientists helped me to develop the book in its present form.

I was most fortunate to have had the able computer assistance of Al Brower and Terry Yuhas for keeping me wired and connected while in Tucson, Arizona, and the graphical services of three talented illustrators, Jim Busse, Megan Gannon, and Shari Hagen, all of Lawrence, Kansas, to design and produce the figures and ensure a standard format. The Department of Ecology and Evolutionary Biology, University of Kansas, helped to defray illustration costs, and the University of Kansas offered sabbatical funding during my stay in Tucson. In several sections I relied extensively on examples from my own research, and I thank the U.S. National Science Foundation for generously funding these endeavors over the past 18 years.

Throughout the writing of the book, I was sustained by my wife, Valery Terwilliger, whose love and wisdom were ever present. Her belief in this project never wavered, and her support served as a constant inspiration.

Tucson, Arizona
August 2000

Contents

List of Symbols

α_{10}	exponential damping coecient (dB·cm^{-1})
$\Delta\Phi$	angular separation (degrees)
Δp	pressure deviation (= force/area)
θ	phase angle (degrees)
θ_c	contralateral phase angle (sound)
θ_i	ipsilateral phase angle (sound)
θ_I	angle of incidence (light)
θ_ℓ	angle of reflection (light)
θ_r	angle of refraction (light)
λ	wavelength
λ_{vac}	wavelength in a vacuum (light)
μg	microgram (= 10^{-6} g)
μm	micron (= 10^{-6} m)
μs	microsecond (= 10^{-6} s)
ν	kinematic viscosity (= fluid viscosity/density; m^2·s^{-1})
ρ	density (= mass/volume)
A	longitudinal displacement (of fluid particle in sound field)
B	magnetic field or vector (of electromagnetic wave)
c	velocity of wave propagation
c_g	group velocity (vibrational waves)
cv	coecient of variation
D	coecient of diusion (m^2·sec^{-1}; olfactory channel); diameter (visual channel)
dB	decibel
E	electric field or vector (of electomagnetic wave)
E	energy
f	wave frequency (Hertz)
h	Planck's constant (= 6.626×10^{-34} Joules·Hz^{-1})
Hz	Hertz (= 1 cycle·s^{-1})
I	intensity (= power/area; mechanical signal); concentration (= quantity/volume; chemical signal); irradiance (= photons·s^{-1}·m^{-2}; light)
IL	intensity level (sound)

J	Joule (= Newton·m)
J	flux rate (= quantity/time)
kHz	kilohertz (= 10^3 cycles·s^{-1})
l	length (of object in fluid flow)
ℓ	reflectance (light)
ms	millisecond (= 10^{-3} s)
Mya	million years ago
N	Newton (= kilogram·m·s^{-2})
n	refractive index
ng	nanogram (= 10^{-9} g)
nm	nanometer (= 10^{-9} m)
p	pressure (= force/area)
Pa	Pascal (= Newton·m^{-2})
pg	picogram (= 10^{-12} g)
Q	quantity (grams or moles)
r	radial distance from the source of a stimulus
Re	Reynolds number
rms	root mean square $\{= \sqrt{(\Sigma x_i^2/n)}\}$
SD	standard deviation
SI	Système international (units)
SPL	sound pressure level
sr	steradian (degrees of solid angle)
T	period (time)
u	current velocity (fluid flow); particle velocity (sound)
v	velocity (wind)
V	volume
W	Watt (= Joule·s^{-1})
Z	specific acoustic impedance (Rayl; 1 Rayl = 1 kg·s^{-1}·m^{-2} or 1 Pa·s·m^{-1})

Signalers and Receivers

1 Communication in a Lilliputian World

In most terrestrial and freshwater or marine habitats, the vast majority of animals transmitting and receiving communicative signals are arthropods. This generalization holds whether the accounting is by species, individuals, or even the number and rate of messages transmitted. Given the attention paid to the abundance and overall importance of arthropods in the earth's ecosystems in recent years, these contentions may engender little surprise when taken at face value. Nonetheless, the ways in which the myriad of arthropods communicate are often beyond our perceptual abilities or imagination, and we may be generally unaware of these activities taking place in our midst. Many arthropod signaling mechanisms rely on channels along which we have little or no sensitivity, and even when we are aware of the signals, their functions can remain obscure. Arthropod receivers may extract and use information from these signals in ways that we cannot; hence, their responses may pass unnoticed or be difficult to interpret. Moreover, the biological contexts in which these communications occur are generally hidden from casual view. The objective of this book is to examine these strange and diverse mechanisms and functions of communication among arthropods and to present them as the means that the majority of living organisms have evolved for exchanging information.

The transmission and reception of information by arthropods would still be a vital chapter in the study of animal communication even if these organisms did not comprise more than 80% of all catalogued species. First, most arthropods are small to minute in size, and such dimensions pose actual physical problems for both signaling and perception along all channels. That is, arthropod communication is not merely a miniaturization of the vertebrate processes more familiar to us. Rather, being small limits the very nature of signals that can be generated and transmitted, the sensitivity with which they can be detected, the accuracy with which they can be localized, and the sorts of information that can be extracted from them once they are perceived.[1] Thus, the study of arthropod communication provides us with some of our best examples of the effects of scale in biology and how structural, neural, and behavioral traits adapt to limitations imposed by physical laws.

Second, arthropod communication is not merely an offshoot or predecessor of the vertebrate phenomena that we may be more accustomed to observing and analyzing. Arthropods are part of a metazoan lineage, Ecdysozoa, that is quite distinct from the lineage Deuterostoma, leading to chordates, including the vertebrates (Knoll and Carroll 1999). The most recent common ancestor of these two lineages existed in the late Neoproterozoic (Precambrian) Era, 560 Mya at the very latest (Figure 1.1). Evolutionary inference indicates that it was a bilateral, coelomate creature, and the group to which it belonged has been dubbed Urbilateria. Nevertheless, it is highly unlikely that urbilaterian sensory systems and appendages were developed beyond a rudimentary state, and any similarities observed between arthropod and vertebrate communication should reflect parallel or convergent evolution. Consequently, observed similarities—of which there are many—may imply that there exist only a very limited number of available solutions, or outcomes, for certain problems confronted by the nervous system and overt behavior (cf. Conway Morris 1998). Presumably, these limitations are imposed by the physical universe, which affects both lineages equally. On the other hand, where marked contrasts are observed, arthropod adaptations to size constraints or their very separate phylogeny may be indicated. Hence, a comparative study of communication with an emphasis on arthropods can offer an overall view of adaptation and chance in evolution.

Of the various non-vertebrate groups on which we might choose to focus, arthropods (Table 1.1) probably afford the most opportunities for making clear and meaningful comparisons. Arthropods are not only abundant, diverse, and evolutionarily distinct from vertebrates, but they comprise a phylogenetically discrete clade. While hypotheses concerning relationships among the various

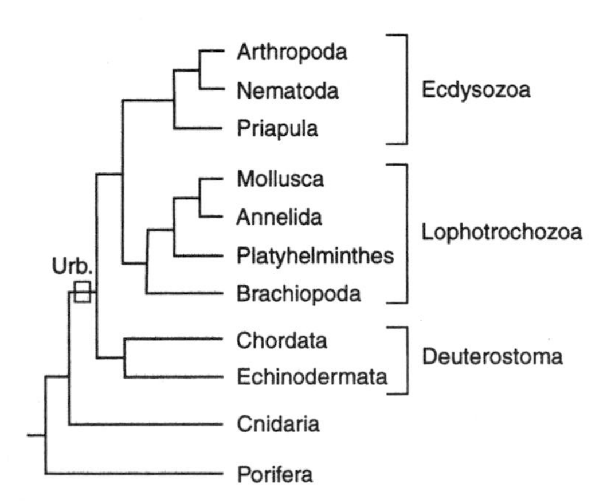

Figure 1.1 Phylogeny of animals (Metazoa), indicating grouping of major bilaterally symmetric phyla into Ecdysozoa, Lophotrochozoa, and Deuterostoma. Paleontological evidence supports divergence of these three groups before the beginning of the Cambrian period, 540 Mya; Urb. indicates Urbilateria, their hypothetical common ancestor (redrawn from Knoll and Carroll 1999).

Table 1.1 Extant Arthropod Classes, Arranged in the Four Groups (subphyla) Shown in Figure 1.2

Chelicerata	
Arachnida (spiders, scorpions, mites, etc.)[t]	Pycnogonida (sea spiders)[m]
Merostomata (horseshoe crabs)[m]	
Myriapoda	
Diplopoda (millipedes)[t]	Pauropoda[t]
Chilopoda (centipedes)[t]	Symphyla[t]
Crustacea	
Cephalocarida[m]	Remipedia[m]
Branchiopoda(fairy shrimp, etc.)[fw]	Tantulocarida[m]
Ostracoda (ostracods)[fw, m]	Mystacocarida[m]
Copepoda (copepods)[fw, m]	Branchiura[fw, m]
Cirripedia (barnacles)[m]	
Malacostraca (crabs, lobsters, shrimp, mantis shrimp, amphipods, isopods)[t, fw, m]	
Hexapoda	
Protura[t]	Diplura[t]
Collembola (springtails)[t]	Insecta (insects)[t, fw]

General habitat: t: terrestrial; fw: freshwater; m: marine.

arthropod classes remain in a state of flux among systematists (Figure 1.2), there is a growing consensus that arthropods do represent a monophyletic group within the ecdysozoan lineage (Ballard et al. 1992, Brusca 2000). And, with the possible exception of cephalopod molluscs, it is only within this group that communication has evolved to degrees of complexity rivaling those seen in vertebrates. Arthropod and vertebrate communication, though extremely dissimilar in many aspects, are evolutionary summits in the landscape of animal signaling and reception.

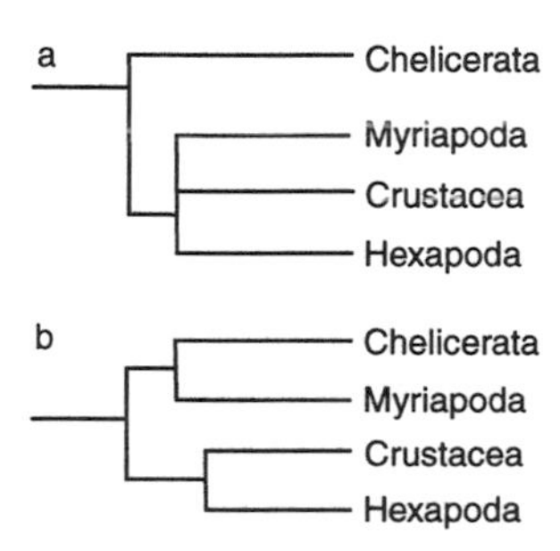

Figure 1.2 Major arthropod groups and their phylogenetic relationships. Two monophyletic schemes, determined via (a) mitochondrial gene arrangements (redrawn from Boore et al. 1995) and (b) nuclear ribosomal gene sequences (redrawn from Friedrich and Tautz 1995), are shown; the schemes differ slightly in their resolution but agree in grouping Hexapoda with Crustacea and distant from Chelicerata (see Table 1.1)

1.1 Communication, Signals, and Cues

The number of definitions of "communication" and "signals" may equal the number of books that have been written on the subject. It is not my intent to contribute yet another or to corrupt my overall objective with semantic exercises, but some clarifications are needed at this point for delimiting the scope of this book and sharpening evolutionary arguments and inferences. In keeping with several recent authors (Hauser 1996, Bradbury and Vehrencamp 1998),[2] I restrict communication to exchanges of information between individuals wherein both the signaler (transmitter) and receiver may expect to benefit in some way from the exchange. By this definition, communication would encompass not only intraspecific sexual and social signals (messages) but also certain interspecific signals that are directed toward receivers among predator, competitor, or mutualist species. Signals are taken to be modifications of the physical or chemical environment induced by the signaler's specialized behavior, and a receiver treats them as information revealing the transmitter's identity, its physiological or behavioral state, or information that the transmitter itself had received from or about other individuals or the environment. This information is perceived by, and possibly stored in, the receiver's nervous system, and it influences the receiver to respond in a manner such that it and the signaler benefit on average. Here, questions of interpretation may arise, particularly among arthropods. For example, if an animal emits a chemical compound bearing nutritive value and that compound is ingested or otherwise absorbed by another individual (Section 3.2), has communication occurred? Perhaps, insofar as the compound's chemical, and possibly physical, properties are detected and processed neurally when contacted and then elicit a change in behavior. However, any physiological response resulting strictly from assimilation of the compound would be just that: growth and development.

The stipulation that communication entails expected benefits to both the signaler and receiver raises additional questions. Certainly, many examples and models of animal communication are not completely "honest" (e.g., Hasson 1994, Krakauer and Johnstone 1995, Backwell et al. 2000), and a signaler may often reveal only partial information concerning itself (Johnstone 1997). Nonetheless, a net benefit to the receiver is expected because selection should eliminate attention toward signals when the responses are, on average, unrewarding or detrimental (see Zahavi 1977). Signal generation and transmission requires some outlay of energy, and selection would then act similarly on the signaler, eliminating those signals from the repertoire that fail to elicit any receiver responses ultimately of benefit to itself. The specific criteria for assessing net benefits and the expected energy outlay and other costs in signaling are altered when communication occurs between genetic relatives (Johnstone 1998a; see also Reeve 1997), but the same basic premises hold.

By focusing on mutual benefit to signalers and receivers, we avoid the awkwardness of having to classify, as animal communication, stimuli emitted by a prey individual and detected and used by a predator to locate and consume that prey. Few biologists today would call this information transfer "communication," and such stimuli are generally termed cues. I designate stimuli as signals when two conditions are met: (1) Individuals derive expected net benefits from emitting and receiving the stimuli; (2) The stimuli have undergone evolutionary modifications that enhance the benefits derived from providing information to receivers and influencing their behavior. The second condition is a subtle but necessary one for distinguishing signals from cues because animals cannot avoid emitting various chemical, vibratory, and optical stimuli, some of which are perceived by other individuals. In many cases, these stimuli are a liability in that they can reveal an individual to its predators or prey, and selection may generally lead to their suppression. However, inadvertent stimuli might also be perceived by conspecifics, and their emission is not necessarily detrimental. For example, males may recognize and locate conspecific females by detecting odors released at the time of molting or by contact with cuticular compounds. Females might be unable to prevent releasing or bearing these chemical stimuli, but they can benefit from the emissions by achieving pair formation. In due time, these inadvertent stimuli might evolve elaborations or modifications that shorten the time to pair formation, place the time at which mating occurs under the female's control, increase the number of males attracted, or facilitate choosing a higher "quality" mate and avoiding heterospecific ones. At that point, the stimuli become known as (sex) pheromones or pheromonal signals, but until specialized features exist indicating that such evolution has occurred, they would be classified as cues.

I speculate on the evolution of signals from cues in subsequent chapters and emphasize that a clear distinction between the two is essential for understanding the way in which many communicative signals may have originated. In practice, signals may be distinguished by the measurable amount of energy necessary for their production and by much of their display being restricted to a specific communicative activity (see Hauser 1996). Among arthropods, body and wing vibrations are often performed for thermoregulation, but they may also be perceived by conspecifics as a cue indicating where the vibrator has been or intends to go in the future (Section 4.4.3). Vibrations that are performed only in the context of information exchange, rather than during general maintenance activities, and that require an expenditure of energy exceeding that necessary for thermoregulation, however, would be considered signals. In general, the stimulus form observed merely reflects the particular evolutionary stage that we happen to be witnessing. The female cuticular hydrocarbons that elicit male copulatory responses in today's populations may become the specialized pheromonal signals of descendant generations (Sections 3.1.5 and 6.1).

1.2 Scope and Coverage

Animal communication is essentially the generation and transmission of signals and their reception and processing by receivers. I am guided by this outlook throughout the book and devote considerable portions to describing the special ways in which arthropods produce, receive, and respond to signals. While adhering in principle to an operational definition of communication, I have chosen to focus on intraspecific communication, which exists in sexual and social contexts. Unfortunately, this restriction bypasses some fascinating interspecific exchanges: aposematic communication between arthropods and their vertebrate predators (Dunning et al. 1992), mutualistic communication between flowering plants and their insect pollinators (Silberglied 1979), and functionally similar communications between lepidopteran larvae (DeVries et al. 1993) and pupae (Travassos and Pierce 2000) and their attendant ants.

All arthropod communication falls into three basic mechanistic categories: chemical (pheromonal), mechanical (acoustic and vibrational), and visual. There are no reported cases of electrical communication among terrestrial or aquatic arthropods as exist among aquatic vertebrates. Several other forms of energy, such as infrared (IR) radiation (Schmitz et al. 1997, Schmitz and Bleckmann 1998) and geomagnetism (Wiltschko and Wiltschko 1995), can be perceived by various arthropods, but as far as we know these modalities are never used for generating or transmitting signals. I have attempted to be fair in covering the three basic categories, but some lopsidedness remains. Any failures to be even-handed reflect my own biases and background (terrestrial over freshwater and marine, acoustic and chemical over vibrational and visual, long-range communication over contact pheromones and tactile signaling, sexual over eusocial) and disparities that persist in available information. For example, there exists far more knowledge of pheromonal and acoustic communication than of visual and vibratory signaling, and our understanding of communication in terrestrial species is generally much greater than in freshwater or marine forms.

Each of the three chapters devoted to chemical (Chapter 3), mechanical (Chapter 4), and visual communication (Chapter 5) treats both mechanistic aspects of signaling and perception and special topics in adaptation and evolution pertinent to these modalities. I have indicated comparisons and contrasts between the modalities by endnotes and cross-references as well as by synopses inserted at key points. The more theoretical aspects of adaptation and evolution are treated again in two concluding chapters that describe the potential influences of sexual selection on signal evolution and fundamental problems in the modification and diversification of signals.

2

Signal Theory and the Language of Communication

To many an observer, chemical, mechanical, and visual signaling may appear as vastly different phenomena, processes that share few mechanistic features. Nonetheless, all three are involved in transferring information between individuals, and to accomplish this transfer they must often be capable of transmission over substantial distances in reasonably brief time intervals. They must also avoid significant alteration by the environment in the course of their transmission. Thus, a unified body of theory that can focus on these commonalities would be most helpful. Here, principles borrowed from the branch of engineering formally known as signal, or signal detection, theory have proven useful for describing certain fundamental aspects of animal communication and comparing communication in its various modalities. In this chapter I introduce those abstract constructs of signal theory that are applicable to the signals that animals transmit and that can improve our understanding of the ways in which animal signals function. I also introduce the basic terminology, the language of animal communication, used throughout the book.

2.1 Channels, Signals, and Signal Characters

The pathway linking a signaler and receiver is known as a channel. This entity is both the actual path in space along which the signal travels and the physical means by which the signal is transmitted. Thus, multiple channels, for example, chemical, mechanical, and visual, may simultaneously exist between the same two points in space, and a given signaler may use several of these coexisting channels to communicate in a multi-modal fashion with a given receiver (Section 7.3).

Animal signals are essentially physico-chemical disturbances, that is, modifications of the background environment, that are transmitted along a channel from a signaler to a receiver. A chemical signal, for example, may consist of a distinctive chemical compound (pheromone) emitted by an individual and carried downwind or downstream to a distant receiver, who may perceive that compound against the background odor of the local environment and respond by moving toward, or away from, the emitter. While chemical signals are often emitted continuously during an individual's activity period, mechanical and visual signals can be transmitted in the form of temporally discrete packets of energy. This feature may be an artifact of the rhythmic muscular activity responsible for producing the signal, but the discrete packets could also represent an important signal feature or even separate signals. Here, we can delimit a given signal by the discrete behavioral response it predictably elicits from a receiver: An aquatic beetle generating mechanical disturbances on the water surface in the form of 5-s packets of concentrically spreading waves that recur every 15 s may be considered as transmitting a single, temporally modulated ripple wave signal. However, if it alters the fundamental nature of the waves in a packet and thereby elicits a qualitatively different response from receivers or influences the behavior of a different class of receivers, the altered wave packet(s) can be deemed a separate signal serving a distinct function (Figure 2.1). Similarly, chemically distinct pheromone compounds that elicit different responses would represent separate signals.

In many cases, individuals modify their signals without altering their fundamental nature. When an individual merely increases a continuously varying feature of a signal and elicits a different receiver response as that feature exceeds a threshold level, the signal is termed graded. Signals functioning in sexual contexts are typically graded; for example, a courted female may exhibit different levels of receptivity or acceptance commensurate with the amplitude or frequency of a male's ripple wave signal or the concentration of his courtship pheromone. Likewise, territorial males may attack or retreat from a rival neighbor depending on such features in the neighbor's display. On the other hand, the (graded) signals that males transmit when courting females and when interacting with neighboring males may differ in their basic design. Different pheromone compounds could be released in the two contexts, or surface waves may be transmitted continuously rather than temporally packaged in rhythmically repeated 5-s bursts (Figure 2.1).

The variable features of a signal are known as signal characters, which are either quantitative or qualitative. All signals are quantitatively characterized by intensity, which is a measure of the energy transmitted by a mechanical or optical signal or of the amount of matter disseminated by a chemical signal. Intensity may be recorded at a given instant or averaged over a time interval, and it may be measured at a given point in space distant from the signaler or at the signaler's location. A chemical signal, for example, may be characterized by its flux rate at the source, normally given in units of nanograms or micrograms of pheromone per minute, or by its concentration, often given in units of molecules per cubic centimeter, at a distant point where receivers may be located. In similar fashion, the energy transmitted by a sound signal may be characterized

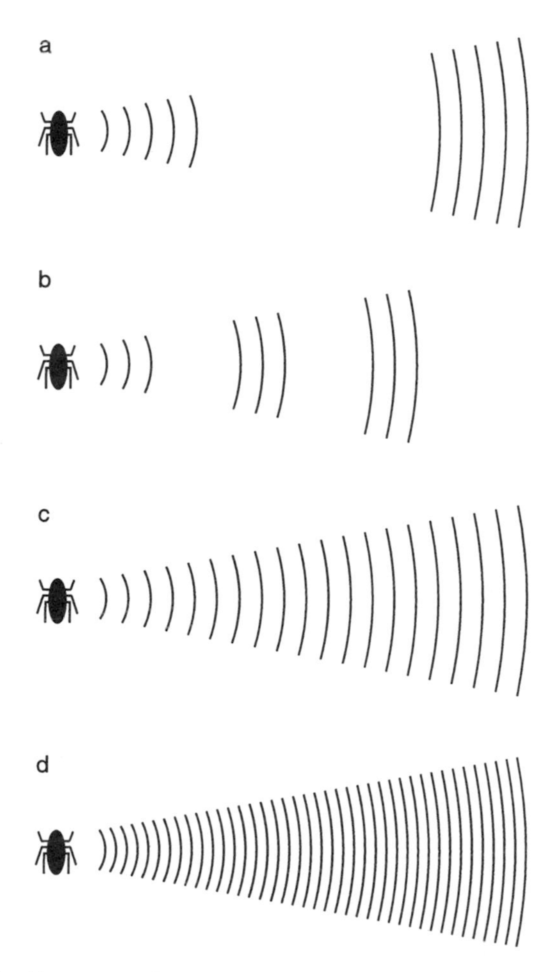

Figure 2.1 (a, b) Temporally modulated ripple wave signals spreading concentrically from the source, an aquatic beetle disturbing the water surface. Arcs depict the crests of individual waves as they exist at a given instant in a particular direction from the source. Individual waves are generated at equivalent frequencies in a and b (see Section 4.1), but the discrete packets of waves differ in duration (and number of waves) and period. (c, d) Continuously generated ripple wave signals. Individual waves are generated at a higher frequency in d than c.

by the acoustic power (watts) broadcast in all directions at the source or by the acoustic power per unit area ($W \cdot m^{-2}$, sound intensity) measured at a point far removed from the signaler.

Qualitative signal characters are intensity-invariant properties of the matter or energy transmitted by a signal. These properties are simplest in chemical signals: the chemical identity of the various compounds comprising the pheromone and the concentration ratios between those compounds. In other modalities, however, qualitative characters are potentially more complex and exist along multiple dimensions. Like chemical signals, most mechanical and optical signals are characterized by a basic "flavor," the frequencies of sound, vibrational, or light waves that impart the perceived sensations of tone to a sound and of color or hue to light (Figure 2.2). However, unlike chemical signals, many sound, vibration, and optical signals are also characterized by stereotyped tempos: rhythmic patterns of sound pulses or light flashes, which may include a nested hierarchy in which pulses of waves are themselves grouped into pulse trains. In addition to tempo, optical signals normally include spatial

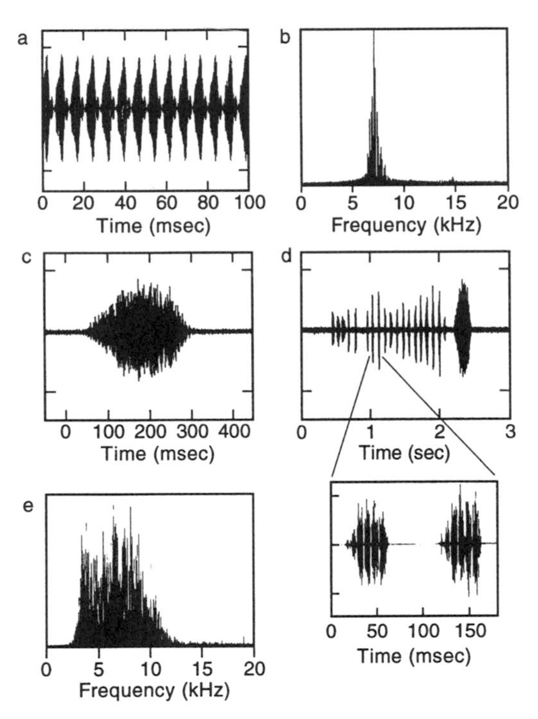

Figure 2.2 (a) Oscillogram, depicting sound pressure level (SPL, indicated by distance above or below the baseline; see Section 4.2) versus time, of male calling song of the short-tailed cricket *Anurogryllus muticus* (Orthoptera: Gryllidae: Brachytrupinae). Note the production of discrete pulses of sound, each approx. 5 ms in length and separated by 2-ms inter-pulse intervals. (b) Frequency spectrogram, depicting acoustic energy versus carrier frequency (see Section 4.2), of the *A. muticus* calling song. Note the concentration of energy within a narrow frequency range centered on 7 kHz. (c, d) Oscillograms of male calling song (c) and aggressive song (d) of the territorial grasshopper *Ligurotettix planum* (Orthoptera: Acrididae). Note the qualitatively different temporal features of the two songs and the nested hierarchy of pulses within chirps in the aggressive song. (From Greenfield 1997a; reprinted from *The Bionomics of Grasshoppers, Katydids, and their Kin*, p. 213, with permission of CAB International.) (e) Frequency spectrogram of the *L. planum* calling song. Note the relatively broad frequency range of acoustic energy. SPL and acoustic energy are presented on a linear scale in all graphs.

features that characterize the manner in which the quantitative and qualitative characters are expressed in separate parts of the signal. Assembled, these parts form the visual image perceived by a receiver.

From a biological perspective, signal characters are only meaningful insofar as their different levels can be distinguished by a receiver. In many cases, the ability of a biological receiver to distinguish these levels falls short of the precision that instrumentation is capable of evaluating. For example, an arthropod may only distinguish pheromone concentrations that differ by 50% or more, whereas a sensitive capillary gas chromatograph may offer much finer distinctions. Often, the precision with which a receiver distinguishes levels of a quantitative signal character is influenced by the time interval over which it samples the character. Longer sampling intervals generally offer greater precision and sensitivity, but this relationship can lead to tradeoffs in the precision with which other quantitative and qualitative characters are evaluated. An arthropod's ability to evaluate the intensity of a faint optical stimulus is enhanced by an extended sampling interval, but its ability to perceive and evaluate a rapid tempo of the stimulus may suffer as a result.

2.2 Peripheral and Central Filters

Biological receivers evaluate the characters of perceived signals by peripheral and central processing mechanisms in the nervous system. These processing mechanisms function as filtering devices that selectively admit certain signals or portions thereof to the subsequent step in the pathway ascending to higher neural centers in the brain. Some filters also send the signal along a distinct pathway, or line, whose identity indicates the specific nature of the signal. When an incoming signal is split among several lines, information within all of the lines may serve collectively to form a map of regional stimulation in the brain that represents the odor, sound, or visual image. Filters normally operate by transferring signals that match an internal template and blocking those that do not, but they can also operate by blocking signals that bear specific features deemed incorrect even if the internal template is otherwise matched (see Section 3.1.3; von Helversen and von Helversen 1998). The initial filter is the generally overlooked one represented simply by a sensory organ, the structural apparatus that transduces signal matter or energy to the electrical activity of a neural message. Because a sensory organ is always limited to detecting signals whose characteristics fall within a particular range, a signal is recognized as being in that range by virtue of its basic detection.

Once detected, signals are processed further at the level of the peripheral, or sensory, receptors associated with the sensory organ. In general, individual sensory receptors (neurons) will detect and respond to signals bearing certain qualitative characteristics only. The various categories of sensory receptors may have sharp ranges of response that overlap little, or they may be relatively broad and overlapping (Figure 2.3). In the former case, processing may be accomplished by a labeled line mechanism in which the central nervous system evaluates the separate neural inputs, the labeled lines, provided by each receptor category. Labeled line mechanisms are used to evaluate the qualitative characters of odor composition in olfaction (Section 3.1.3) and of frequency in acoustic and vibrational perception (Section 4.3.5). Where the ranges of response of individual receptors are broad and overlapping, processing may be accomplished by an across-fiber pattern mechanism in which the neural inputs from the various receptor categories are combined prior to evaluation by the central nervous system. Across-fiber pattern mechanisms are used to evaluate the qualitative character of color spectrum in visual perception (Section 5.3.1).

Quantitative signal characters are generally evaluated by the levels of response of the sensory receptors, which are encoded as rates of action potentials. However, peripheral filtering can supplement the evaluation of intensity in both olfaction and hearing. Here, receptors that have equivalent ranges of response to a qualitative character but that differ greatly in sensitivity may distinguish between weak and strong signals: the latter would be recognized when receptors in both sensitivity categories respond (Figure 2.4).

Additional filters exist at various levels of the central nervous system. These devices are vital for evaluating qualitative signal characters such as tempo, pulse length (Figures 2.1 and 2.2), and the spatial patterns in visual images. Other

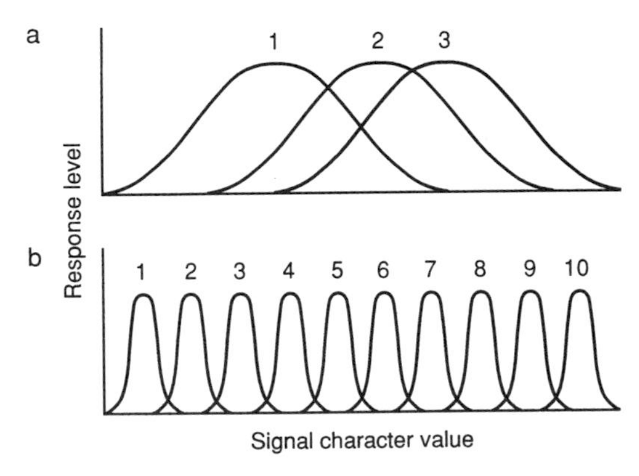

Figure 2.3 Response ranges of the various receptor categories in a sensory organ. The ranges of signal character values to which given receptor categories respond may be relatively broad and overlapping (a; three receptor categories) or narrow and separated (b; ten receptor categories). Examples of signal characters are sound frequency and wavelength of light; response level is the action potential (AP) rate elicited in a receptor of a given category.

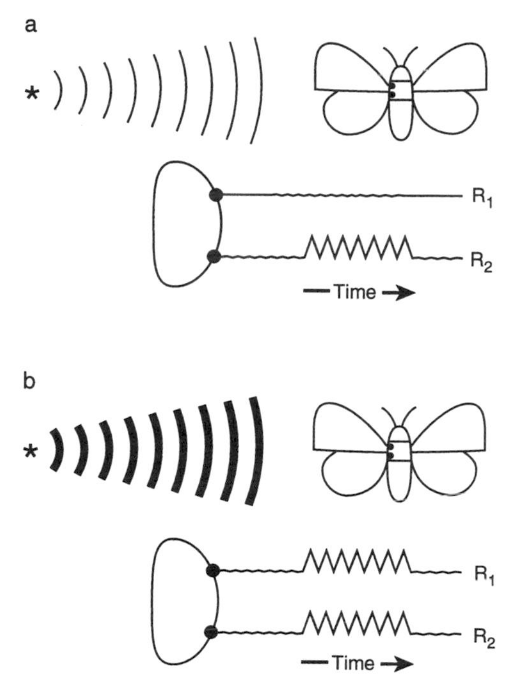

Figure 2.4 Discernment of weak (thin arcs) versus intense stimuli (thick arcs) by a sensory organ bearing sensitive (R_2) and relatively insensitive (R_1) receptors. (a) Weak stimulus recognized by firing of action potentials in R_2 only. (b) Intense stimulus recognized by firing of action potentials in both R_2 and R_1 receptors.

central processing mechanisms augment the evaluations of odor composition (Section 3.1.3) and sound frequency (Section 4.3.5) initiated at the peripheral level.

The signal processing performed by peripheral and central filters gives a receiver information with which it may recognize several attributes of the signaler: its species, sex, physiological state, motivation, and location. In some cases, the information may even reveal aspects of the signaler's familial and individual identity. But, accurate recognition of any of these attributes usually requires that the signal be transmitted with relatively little modification en route and remain conspicuous against a background of external and internal noise.

2.3 Sensory Adaptation and the Perception of Relative Intensity

Quantitative signal characters can be described and evaluated by a receiver either by absolute measurements or by measurements relative to general

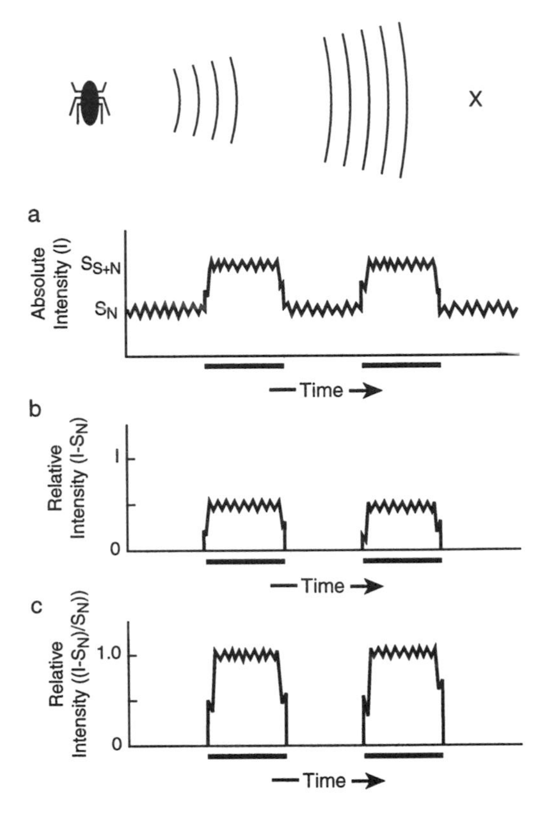

Figure 2.5 Absolute versus relative measurements of intensity. (a) Absolute intensity during times when signal is absent (background noise; $I \cong S_N$) and when present (signal + background noise; $I \cong S_{S+N}$) (b) Relative intensity expressed as the difference between absolute intensity (signal + background noise) and background noise only, $S_{S+N} - S_N$. (c) Relative intensity expressed as scope, $(S_{S+N} - S_N)/S_N$, the proportional increase in absolute intensity when signal is present. Intensity measurements are shown as evaluated at point X (top illustration; see Figure 2.1). Horizontal bars below *x*-axes indicate times of signal presence.

background intensity or a reference signal (Figure 2.5). For biological receivers, relative evaluations of signal intensity may often be more meaningful than absolute ones, and both peripheral and central mechanisms can provide them with this information. Sensory neurons subjected to a continuous stimulus often adapt, or fatigue, to the level of that stimulus by greatly reducing their rate of firing action potentials. Thus, only the transients that occur when intensity suddenly exceeds continuous or background levels may elicit a peripheral neural response sufficient to evoke a central potential. While sensory adaptation suppresses the ability to evaluate absolute intensity, it may be quite valuable for discerning chemical, acoustic, or optical signals under many circumstances. For example, a cricket may readily perceive, and localize, a nearby neighbor calling in a chorus because that neighbor's calls are evaluated as slightly more intense than the continuous din produced collectively by more distant individuals (Sections 4.3.5 and 4.4.2). Although the absolute intensities of the chorus and the chorus + neighbor may be quite similar, the receiver adjusts its basal level of peripheral neural activity to the chorus intensity and can therefore perceive the small intensity transients resulting from the neighbor's calls.

Central mechanisms may also enhance a receiver's perception of relative intensity and ability to discern signals against background noise. A common mechanism, variations of which are found in each modality, is the inhibition of response in one receptor or sensory organ by stimulation of adjacent receptors or the opposite organ (Sections 4.3.5 and 5.3.1). Such lateral inhibition magnifies spatial or temporal contrasts in perceived intensity. In hearing, lateral inhibition may allow a receiver to discern a marginal difference in intensity between sound signals located to the left and right, and in vision it can aid a receiver in detecting an object only slightly brighter or dimmer than the optical background, such as the silhouette of a flying insect seen against the blue sky or green vegetation.

2.4 Noise and the Signal : Noise Ratio

Phenomena that prevent a receiver from detecting or accurately evaluating a signal that would otherwise be perceived clearly are called noise. For biological receivers, noise may originate externally from biotic or abiotic environmental sources or internally from activities such as locomotion or respiration. Some internal noise also occurs due to spontaneous electrical activity in the nervous system. Any of these sources can generate action potentials that would obscure the potentials stimulated by incoming signals. At the neural level, the signal would be imbedded in a continuous stream of potentials, and the receiver's central filtering mechanisms may be unable to extract it. Along the chemical channel, for example, a receiver may be prevented from detecting a pheromonal signal because other environmental odors stimulate the same olfactory receptors or brain interneurons that are tuned to the pheromone components.

The problem imposed by noise is evaluated by the signal : noise ratio, which is formally calculated as:

$$(S_{S+N} - S_N)/\mathrm{SD}(S_N) \qquad (2.1)$$

where S_{S+N} and S_N are the mean stimulus intensities, most accurately measured by an observer as the receiver's levels of neural activation, when a signal is present (signal + noise) and when it is not (noise only), respectively, and $\mathrm{SD}(S_N)$ is the standard deviation of noise intensity (Figure 2.6). Thus, signal : noise

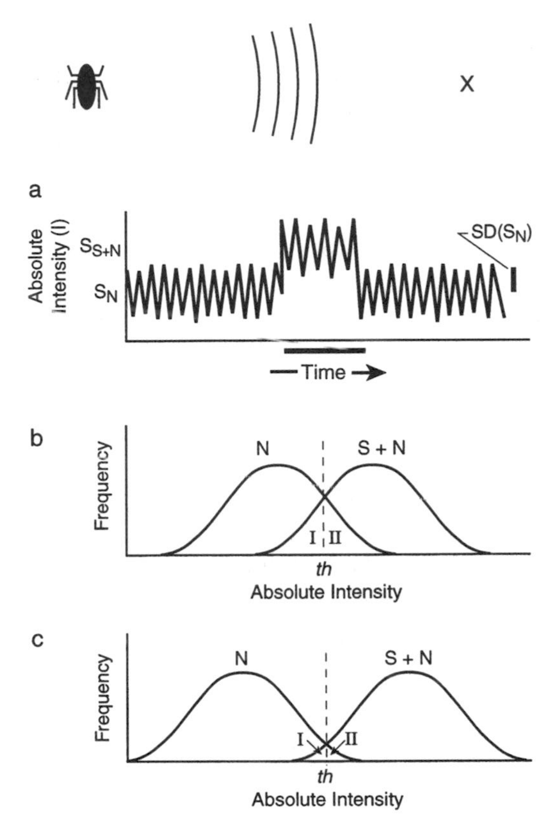

Figure 2.6 (a) Relationship between signal and noise expressed as the signal : noise ratio, $(S_{S+N} - S_N)/\text{SD}(S_N)$ (see Figure 2.5). (b) Relatively low signal : noise ratio indicated by large overlap between N (noise) and S+N (signal + noise) distributions. Assuming that the receiver sets *th* as the threshold intensity level at which signals are discriminated from noise, areas I and II represent the probabilities of missed signals and false alarms, respectively; the sum of these probabilities is minimized when the threshold is set at the intersection of the N and S+N curves (see Dusenbery 1992). (c) High signal : noise ratio indicated by small overlap between N and S+N distributions.

ratio may be viewed as a measure of the opportunity for an "ideal" receiver to make a "correct" behavioral response upon detecting and processing a stimulus: assuming that the ideal receiver appropriately sets its threshold level for responding to a signal, minimizing both false responses to noise and missed responses to actual signals (both errors being equally disadvantageous), the responses will be most accurate when mean signal intensity is much greater than mean noise intensity and noise level varies little. Otherwise, the noise level may sometimes exceed the level of signal + noise, and the chance of false alarms in response to noise and missed responses to actual signals could be substantial regardless of where the receiver sets its threshold level.

2.5 Information

When signal : noise ratios are low, we describe a signaler's message as conveying relatively little information to a receiver. As used in this statement, "information" is both a familiar term and a specialized one borrowed from signal or information theory. In the familiar sense, little information is conveyed because the receiver may be unable to recognize the signaler's state or identity from its message. In the specialized sense, information is defined as reduction in uncertainty, and in this case reception of the signal has reduced the receiver's uncertainty about the signaler by only a small amount. That amount is a quantifiable measure expressed on a binary scale in units of bits. Thus, measures of information sensu signal theory have the potential to afford objective comparisons between different signaling systems and communication modalities.

One information bit represents the amount of uncertainty reduction that results when a signal allows a receiver to distinguish between two equiprobable alternatives. In a population with a 50 : 50 sex ratio, a signal that accurately reveals a signaler's sex would contain one bit, provided that receivers can always discern the signal's two alternative states. When such discernment is obscured or restricted, as by a low signal : noise ratio or other limitations in receiver perception, an amount less than one bit would be contained. The signal would also contain less than one bit if the sex ratio is normally skewed, because a priori the signaler is much more likely to be one sex than the other. That is, the signal offers less reduction in uncertainty in this situation than when both sexes are equally common. If the signal allows a receiver to distinguish between two equiprobable conditions, for example, reproductive maturity versus immaturity, as well as sex (and the conditions are independent of sex), two bits are then contained. In general form, we calculate the information content, H, of a signal as

$$H = -\sum_{i=1}^{n} (p_i \cdot \log_2 p_i) \tag{2.2}$$

where the p_i values are a priori probabilities of the n states, conditions and identities of the signaler revealed to a receiver by the signal.[1]

From another perspective, the above methodology may also be used to estimate the transmission capacity of a communication channel. If calling songs vary in intensity, frequency, and tempo, and receivers can distinguish among y ($= 4$) different levels of each of these x ($= 3$) signal characters, the acoustic channel has the capacity to transmit 64 ($= y^x$) unique signals (see Section 4.5). These different songs may indicate the signaler's behavioral state or its individual, familial, or species identity. When indicating identity, the signals represent a signature system (Beecher 1989), and each contains much less information than the amount predicted by equation 2.2. In many cases, a given receiver only makes binary responses to a signal as if it is produced by a mate versus a non-mate, kin versus non-kin, or a conspecific versus a heterospecific. Nonetheless, in a system including *n* individuals, families, or species, the channel must have the capacity for transmitting at least *n* unique signals if errors resulting from multiple occurrences of the identical signature are to be avoided. Moreover, when signatures are adopted randomly, with no mechanism for preventing a given signature from being used more than once (random sampling with replacement), many more than *n* possibilities would be needed.

Signals typically vary in several independent characters, or dimensions, but such variation is only meaningful for providing information or expanding the transmission capacity of a channel to the extent that receivers can discern the variation. Among acoustic arthropods, for example, sound frequency of signals may vary and accurately reflect a signaler's size or age, whereas receivers often possess little facility for discriminating between frequency levels. Under such circumstances, the information content of signals and the transmission capacity of the channel would be quite limited—unless signal variation and corresponding receiver discrimination exist along other dimensions.

2.6 Reliability, Repeatability, and Redundancy

The accuracy with which a signal indicates the state, condition, or identity of the signaler is called reliability. As long as these features of the signaler do not change, a reliable signal is expected to be repeatable: values of one or more signal characters should remain relatively constant over time. The various characters of a signal may differ markedly in their reliability, and receivers are expected to attend only to the more reliable ones. While signal intensity as measured at a standard distance from the signaler may be a reliable indicator of the signaler's size, motivation, or prowess, a receiver normally cannot judge its range from the signaler and may be unable to extract much information from intensity. On the other hand, certain qualitative signal characters, such as tempo, may be range-invariant and could indicate the signaler's features regardless of its location.

Most animal communication involves some conflict of interest between the signaler and receiver (Endler 1993a). When a male transmits a sexual advertisement or courtship signal, it may be in his interest to mate with any given female receiver, but the female may benefit by being discriminating and awaiting a future suitor. For a signal to be reliable in the presence of conflict, it must generally bear a cost (see Arak and Enquist 1995) such that signalers for whom it is not in the receiver's interest to respond to cannot produce it. However, where conflicts of interest are greatly reduced or absent, as in social communication among genetic relatives or other cooperating members of a colony, costly signals may be unnecessary (Maynard Smith 1991a, 1994). Here, we may expect "whispers" rather than "proclamations," signals that are exaggerated only to the extent necessary for conveying information to the receiver (Johnstone 1998b).

A thorough analysis of an animal's communication often reveals elements of redundancy in the signal transmission process. Signaling usually continues for much longer than would appear necessary to convey information, and the information may contain more bits than needed to elicit an appropriate response from the receiver. Such apparent redundancy may be influenced by receiver motivation and perceptual ability, especially where conflicts of interest between the signaler and receiver occur. A signal of extended duration can reliably indicate the costs incurred by the signaler and thereby overcome the receiver's resistance to respond. Superfluous information bits may simply represent artifacts of signal generation, but in some circumstances designed redundancy may reduce the probability of inappropriate responses resulting from receiver error or enhance the efficiency of a receiver's response (see Sections 3.1.3 and 3.1.4).

3

Chemical Signaling and the Olfactory Channel

Signaling, sensory perception, and communication among arthropods probably invokes in most observers of natural history the image of chemical messages and the olfactory channel. By various accounts, this image is an accurate representation. Whether measured by the percentage of species or clades that rely heavily on chemical messages and olfaction, the proportions of these species' behaviors that are facilitated by olfaction, or the complexity and exaggeration of structures and behaviors responsible for chemical signaling and perception, the olfactory channel is the predominant one among arthropods. Olfactory communication is found in all of the major insect orders, the minor ones in which it has been specifically investigated, and the major classes of chelicerates, crustaceans, and myriapods. Moreover, the olfactory channel serves a variety of communicative functions in these groups, and it is exploited by aerial, terrestrial, and freshwater and marine forms.

The small to minute body sizes of arthropods may be largely responsible for their general reliance on the olfactory channel. At typical arthropod dimensions, the opportunities for communicating along the mechanical and visual channels are surprisingly limited. For all but relatively large, robust species, vibrational signaling would be restricted to a narrow set of substrates, and acoustic signaling beyond the signaler's immediate vicinity would be most difficult: small organisms cannot efficiently generate low-frequency sound signals, and the high frequencies that they can generate are not transmitted effectively over long distances (Section 4.2). Additionally, morphological and physiological constraints may have prevented most arthropods from developing sensitive hearing devices, particularly for lower sound frequencies (Section 4.5). Perceptual limitations, ultimately traceable to small body size, are also particularly severe in arthropod vision, and most receivers would be unable to extract much information from optical signals regardless of the level of detail contained in the various signal characters (Section 5.4). This constraint prevents visual communication from operating effectively over any but the shortest of distances.

Chemical signaling and perception, however, may not suffer such debilitating effects of scale. The olfactory channel was probably the earliest one in evolutionary history to be exploited for communication, and it is of critical importance among unicellular organisms. Pehaps this use reflects that communication along the mechanical and visual channels requires multicellular structures for signal generation and perception. At arthropod body sizes and levels of structural complexity, chemical communication can function most efficiently and offers several advantages over alternative modalities. An individual organism may generate and transmit a variety of chemical signals across considerable ranges and around barriers, regulate their emission, and perceive and discriminate these compounds with a high level of sensitivity. The major drawbacks to chemical communication are that signals cannot be sent rapidly over long distances and a signaler's control over the direction in which it transmits its messages is quite limited. Additionally, it may be difficult for a receiver to localize the source of a distant chemical signal. Nonetheless, arthropods have evolved some amazingly effective sensorimotor mechanisms for olfactory localization.

I have organized this chapter on a functional basis and cover seven major sexual and social processes in which olfactory communication plays a dominant role among arthropods. Each of these processes is introduced with a vignette describing a communicative feat facilitated by chemical signals. These examples are followed by analyses of the signaling and sensory mechanisms used and discussions of the evolution and adaptiveness of olfactory communication in the particular context. Lastly, it should be recognized that while olfactory communication is obviously critical for these processes, all of the examples described here also involve other modalities at various junctures.

3.1 Sexual Advertisement: Organic Lures and Beacons

Advertising one's sexual identity and status for the purpose of mate attraction may be the most basic function of chemical communication. Chemical signals and olfaction are responsible for the sexual pairing of unicellular organisms, including prokaryotes (Adler 1975, Stephens 1986). Similar intercellular communication also operates in gamete fusion in both external (e.g., Boland 1995) and internal fertilization.

Among metazoans, the most spectacular examples of sexual advertisement via the chemical channel are arguably the long-range pair-formations of animals navigating within fluids—air and water. In arthropods, this achievement has evolved to an extreme degree in Lepidoptera, namely moths. While this perfection led to initial skepticism,[1] it also generated a wealth of later investigation. Thus, sex pheromone communication among moths is a model system (Karlson and Schneider 1973), and it can serve as the standard of comparison when discussing chemical signaling in other arthropod groups.

In the vast majority of moth species that have been observed, mating occurs after a stationary female emits a scent, a sexual advertisement pheromone, from terminal abdominal glands. The scent molecules evaporate from the gland surface, diffuse away from the female, and are carried downwind. Males who happen to perceive this scent are stimulated to initiate or continue flight, and they may eventually locate the female via integration of chemical, mechanical, and visual inputs that elicit a complex series of motor activities. The spatial scale over which such pair-formation occurs and the signal's intensity make the event all the more impressive: a male moth with a body length of 1 cm may arrive at a female over 100 m distant within half an hour, and the flux (emission rate) of her chemical signal may be less than $1\ \text{ng}\cdot\text{min}^{-1}$, equivalent to approximately $10^{10}\ \text{mol}\cdot\text{s}^{-1}$. Consequently, the male may be exposed to signal intensities (concentrations) lower than $10^{4}\ \text{mol}\cdot\text{cm}^{-3}$ at the initiation of his journey and at various points along the way.[2] A geometrically proportional human feat might be represented by someone at the south end of Manhattan Island (over 20 km long) detecting the odor emanating from a person at the island's north end and then localizing and arriving at that odor source a short while thereafter. It is no wonder that the United States Department of Defense, and presumably corresponding agencies of other nations, has become most interested in the means by which arthropods accomplish this detection and localization (Malakoff 1999). Possibly, its simulation via mechanical devices would prove quite valuable in espionage and on the battlefield.

Our current knowledge of how moths and certain other animals living within fluid media use the chemical channel for sexual advertisement and pair-formation is built upon behavioral observations that began in previous centuries. Natural historians had long suspected the existence of the chemical channel, as many had observed the attraction of male moths via some invisible signal to caged females appropriately positioned in their habitat. But the distance over which such attraction and orientation seemed to be occurring, combined with the inability of humans to detect any chemical signal from receptive females, also generated strong doubts (e.g., Fabre 1879). These doubts were not fully eliminated until 1900, when Alfred Mayer reported the first experimental evidence confirming the chemical channel in pair-formation. He demonstrated that male promethea moths (*Callosamia promethea*; Saturniidae) were attracted to caged females, or their freshly detached abdomens, as long as the males had contact with the cage air via a small screened opening. On the other hand, no amount of visual stimuli from the females was sufficient to stimulate the males provided the cage was sealed (Mayer 1900).

Some early reports tended to exaggerate the perceptual aspects of the chemical channel in moths. Noting that male saturniid moths released several kilometers distant from females were attracted and arrived at those females later that night, Rau and Rau (1929) emphasized the apparently great range at which the chemical message could be discerned.[3] We now know that male moths move widely through the habitat in their search for receptive females. This activity, appetitive flight, would have brought some males within a much closer range (<100–200 m) of the females by chance alone. Importantly, appetitive flight would have begun prior to any perception of the female scent (Section 3.1.4). Other reports claimed that male moths responded to as little as one molecule of pheromone. This notion of extraordinary sensitivity confused the response of a single olfactory neuron with a behavioral response, which requires a sufficient level of neuronal activity for the pheromonal signal to be distinguished from spontaneous background noise in the neurons (Section 3.1.3).

The modern era of investigating the chemical channel in sexual advertisement is often considered to have begun with the study reported by Butenandt et al. in 1959 on the silkworm moth (*Bombyx mori*; Bombycidae) sex pheromone, "den Sexual-lockstoff des Seidenspinners." Having isolated and chemically identified the endocrine hormone ecdysone, regulating molting in insects (Butenandt and Karlson 1954), Butenandt and his coworkers applied their expertise in natural-products chemistry to deciphering external chemical messages. By extracting and analyzing material from the abdomens of approximately 500 000 female *B. mori*, they determined that its sexual advertisement pheromone was a 16-carbon aliphatic compound with two double bonds (diunsaturated) in the chain of carbon atoms and an alcohol functional moiety at the terminal ω-position (Butenandt et al. 1959). Subsequent refinements of their initial analysis revealed the geometric configuration (specific diastereomer) of the molecule at the double bonds (Butenandt 1963), that an additional compound (a related aldehyde; Kasang et al. 1978; see Figure 3.1a) was also part of the sex pheromone, and that the two compounds synergistically influenced the pair-forming sequence. These compounds, either extracted from the female or synthesized in the laboratory, attracted males like a magnet, and the findings soon created a stir in the scientific world.[4]

The initial characterization of the *B. mori* sex pheromone was a monumental effort spanning several years, but its success quickly gave birth to a field of research that continues to expand today (Cardé and Minks 1997, Hansson 1999). Two different sources stimulated this activity. First, Rachel Carson's seminal book *Silent Spring*, first printed in 1962, showed both the perils and shortcomings of chlorinated hydrocarbons and other organic compounds as pest control agents, and scientists were keen to explore alternative means that neither threatened the environment nor led to evolution of resistance in the target pests. The characterization of sex pheromones of pest species, their synthesis in the laboratory, and their deployment in the field promised escape from both pitfalls above. As will be noted in a later section (3.1.7), however, applied scientists have not realized their early hopes of exploiting our knowledge of the chemical channel.

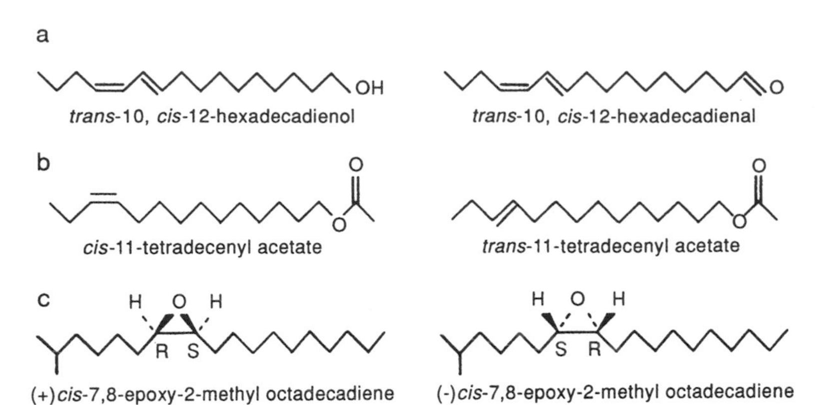

Figure 3.1 Female advertisement pheromones of Lepidoptera. (a) Two different compounds that comprise the pheromone of the silkworm moth, *Bombyx mori* (Bombycidae). (b) Two different geometric isomers (diastereomers) that comprise the pheromone of the European corn borer, *Ostrinia nubilalis* (Pyralidae). Populations use either 97:3 or 3:97 ratios of these isomers (Klun and Maini 1979). (c) (+) optical isomer (enantiomer) that serves as the pheromone of the gypsy moth, *Lymantria dispar* (Lymantriidae). The (−) enantiomer is inactive.

Second, the ready responses of male *B. mori* to either extracts of female pheromone glands or to synthetic pheromone suggested that many aspects of the chemical channel regulating moth pair-formation could be bioassayed and that copious and meaningful data would be forthcoming. To this promise of information return, we may add the fascination with a channel largely exotic to direct human experience. Some of these expectations have been fulfilled. The following sections show that we have learned a great deal about the nature of chemical signaling, perception, and associated orientation behaviors by concentrating on sexual advertisement in moths. This success has been achieved largely because most adult moths are short-lived and become sexually receptive shortly after eclosion and remain so until morbidity sets in. Moreover, because it is males—who are wont to be less discriminating than females—that are the receivers, the behavioral responses to the chemical signal are conspicuous, immediate, and occur reliably in a very high proportion of tested individuals (Section 3.1.4). Thus, researchers have not contended with the vagaries of mate choice in designing their bioassays and interpreting their results.

3.1.1 The Signal

A 1991 compendium (Mayer and McLaughlin 1991) lists identified female advertisement pheromones in several hundred moth species spanning more than a dozen lepidopteran families; extensive sets of additional identifications in seven other insect orders are also listed.[5] Most of these chemical identifications owe their existence to advances in instrumentation that allow researchers to work with exceedingly small quantities of material (Heath and Tumlinson 1984). Using a capillary gas chromatograph and coupled gas chromatography–mass spectrometry (GC–MS), a worker can now identify the chemical structure of organic compounds given as little as several nanograms of material. For pheromones, the material is generally collected with a vessel designed to entrap or suck a small, defined volume of air, "head space," surrounding an individual (female) or her pheromone glands, rinsed with an organic solvent, concentrated, and then injected into the gas chromatograph. Thus, the scent that a single individual emits into the air over a short period of time can be sampled and characterized. This technology represents a significant improvement over earlier techniques wherein solvent extracts were made from many entire females or their pheromone glands. Such methods could not distinguish material actually emitted and serving as the chemical message from that sequestered within the female, which may have included precursors of the pheromone. Additionally, by performing both qualitative and quantitative chemical identifications from a mere nanogram of material, we are afforded the opportunity to study variation in the chemical signal between individual females, times of day, physiological states, and ages. The sensitivity of current instrumentation also allows us to characterize fully the various compounds forming the chemical signal, some of which typically represent only small percentages of the entire pheromonal bouquet.

As used here, chemical signal, pheromone, and pheromone component refer to a compound(s) that elicits neurophysiological and behavioral responses in the conspecific (male) receiver. Some of the compounds that can be collected from a female's emission into the air may not elicit orientation or any other responses and should not be considered as part of the signal or as pheromone components.

Qualitative and Quantitative Characters. Three decades of chemical identifications and biological assays provide the following profile of female advertisement pheromones in arthropods. Chemically, these pheromones are relatively simple organic compounds, with molecular weights ranging from 150 to 350 (in terrestrial forms). A basic hydrocarbon structure with attached functional moieties is often found. In most cases the pheromone is composed of several compounds, some or all of which are usually related stereoisomers. Normally, the pheromone of a given species (or population) is qualitatively consistent among the female individuals and specific for that species: As a rule, it represents a mate recognition signal and is chemically distinguished from the pheromones of all other species active at that time and place[6] (Cardé 1986; Chapter 7).

Among moths, in which the majority of the chemical identifications and biological studies have been made, the advertisement pheromones are mostly straight-chain, mono- or diunsaturated aliphatic molecules with acetate, alcohol, or aldehyde functional moieties at the terminal ω-position. Chain lengths range from 10 to 21 carbon atoms, an even number of carbons prevails, and the double bonds in diunsaturated compounds may or may not be conjugated (separated by two positions along the carbon chain; Figure 3.1a). But, in some moth species, the pheromones are tri- or even tetraunsaturated, terminal moieties are lacking, or various branch groups occur along the chain (Baker 1989). The last category would include the pheromones of gypsy moths (*Lymantria dispar*; Lymantriidae), in which both epoxy (see Millar 2000) and methyl branch groups occur (Figure 3.1c). While most moth species produce pheromones that are either unbranched or branched molecules, some arctiid moths produce both classes of compounds.

Because most advertisement pheromone molecules are unsaturated compounds, they exist in specific geometric configurations (diastereomers), and those molecules that include branch groups exist in specific chiral forms (+/− optical isomers or enantiomers) as well (Figure 3.1c). In most cases, an individual's pheromone includes several stereoisomers of the same or different compounds; the latter represent chemicals having different molecular formulae. The stereoisomers and additional compounds seldom occur in equivalent proportions. Rather, one is usually much more abundant than are the other (minor) components (Figure 3.1b). These biased ratios per se may enhance the ability of (male) receivers to orient toward (female) signalers (Section 3.1.4).

Flux (emission) rates of female advertisement pheromones tend to be extremely low, even when measured at the peak of the advertisement period. Reported values in moths range from $\ll 1$ to $10\,\text{ng}\cdot\text{min}^{-1}$ at the source (Sower and Fish 1975, Bjostad et al. 1980, Charlton and Cardé 1982). In some species, flux rates may increase over the several days following eclosion (Greenfield 1981, Schal et al. 1987).[7]

Biosynthesis and Emission. In moths, female advertisement pheromones are normally synthesized in and emitted from modified epidermal cells, an exocrine gland, at the base of the ovipositor; a common location is the membrane between abdominal segments 8 and 9 (Percy-Cunningham and MacDonald 1987, Itagaki and Conner 1988; cf. Butenandt et al. 1963).[8] The abdomen may be elevated and the terminal segments extruded to expose the glands (Figure 3.2). Such signaling is often performed at a relatively conspicuous perch on or near a host plant and restricted to particular periods of the day or night when the males are active and receptive. The biosynthesis of pheromone via enzymatic transformation of its precursors, certain fatty acid compounds, may coincide with the signaling period (Raina 1993) and be subject to both neural and endocrine regulation (Rafaeli et al. 1993, 1997, Tillman et al. 1999). In various species, the final biosynthetic steps are under the control of a peptide hormone (pheromone biosynthesis activating neuropeptide, PBAN) that is manufactured in the subesophageal ganglion of the brain.[9] PBANs reach the pheromone glands either through release, from brain neurohemal sites, into the hemolymph or by

Figure 3.2 Female clearwing moth (*Synanthedon pictipes*; Sesiidae) transmitting sexual advertisement pheromone. The elevated abdomen with extruded terminal segments is the typical advertisement posture among female moths. Arrow indicates location of pheromone gland (photograph courtesy of James W. Mertins).

neurosecretory transport to the terminal abdominal ganglion (TAG) via descending fibers of the ventral nerve cord. From the TAG, PBANs then stimulate local innervation of the glands (Christensen and Hildebrand 1995).

As with male moths, females in most species are sexually mature at the time of adult eclosion and begin their initial signaling period within 24 hours. However, in species that undergo predictable or unpredictable migrations in relation to severe habitat decline, females normally complete a sustained long-distance flight first and may forgo synthesizing and emitting pheromone for several days. Additionally, they may require exposure to host-plant odors (Riddiford and Williams 1967, Raina et al. 1992, 1997). In the true armyworm moth, *Pseudoaletia unipuncta* (Noctuidae), which migrates predictably each year, females do not eclose with a supply of mature eggs, and the post-migration increase in juvenile hormone (JH) titer appears to influence concomitantly egg development and pheromone biosynthesis via release of PBAN or increased sensitivity of the pheromone glands to PBAN (Cusson and McNeil 1989). The reasons for delaying pheromone synthesis and emission in migratory species are not fully clear. We may expect females to mate, or at least attempt to do so, prior to dispersal, presumably as a precaution against the possibility that no males will be found in the new habitat. Whereas egg maturation might demand resources that are needed to fuel the migratory flight (i.e., food reserves which could be replenished later), signaling is unlikely to make such demands: the precursors of female advertisement pheromones are minute quantities of fatty acid byproducts of secondary metabolism (Section 3.1.5). Nevertheless, time as well as energy may represent a factor when budgeting effort among these activities. Thus, a delay in female signaling might arise because an early onset would occur when all nocturnal hours must be devoted to migration.

3.1.2 Signal Transmission: Plumes and Filaments

Sexual advertisement signaling is generally performed when the locations of potential mates are both dispersed in space and unknown to the signaler. Should these locations be clustered and known, advertisement signaling might not be a necessary or worthwhile mating strategy: moving toward those potential mates and initiating or eliciting courtship could be a viable alternative for the signaler. Therefore, when organisms do broadcast sexual advertisement signals to potential mates spread among distant locations, their signals are expected to be relatively incessant and transmitted over long distances. They may also be dispersed in a wide range of directions in order to reach as many individuals as possible. For chemical signals, transmission over such spatial scales poses special problems, and again scientists have studied these problems most intensively in the female sex pheromones of moths. Here, the signals are largely transmitted by wind, and a precise description of how the transmitted signals vary over space and time has proved far more difficult to obtain than initially imagined. We approach this problem via a sequence of physical models of increasing complexity and reality: a chemical signal continuously emitted from a point source in still air, in air subject to uniform flow, and in air subject to turbulence. Whereas the last model is the only one relevant to actual field conditions, it is instructive to view it in relation to the first two.

In moths and other arthropods relying on an aerial chemical channel, pheromone molecules evaporate and diffuse away from the signaler in all directions (Box 3.1). Were simple molecular diffusion the only process available for dispersion of the pheromone, though, sexual advertisement via the chemical channel would not be possible: 30 minutes following onset of emission would be required for a concentration of pheromone detectable to receivers to spread spherically to only 1 m, and the time would increase exponentially at longer radii (Dusenbery

Box 3.1 Diffusion and Convection in Olfactory Signaling

We approach the problem of long-range olfactory communication by examining how diffusion and convection disperse chemical compounds in a fluid (Bossert and Wilson 1963). To begin, chemicals move through a stationary fluid medium in accordance with Fick's law of diffusion,

$$J_r = -D \cdot \frac{dI}{dr}$$

where J_r is the flux rate of the chemical along an axis extending radially outward from its source and across a unit area perpendicular to that axis, D is the diffusion coefficient of the chemical in a given medium, and dI/dr is the concentration gradient of the chemical along the axis. In air at 25°C and a pressure of 1 atmosphere, D for the silkworm moth pheromone component bombykol is approximately $2.5 \times 10^{-6}\ m^2 \cdot s^{-1}$ (see Schneider et al. 1998b), and we shall use this value in the calculations that follow.

If a quantity Q of a chemical with diffusion coefficient D is instantaneously released into a stationary fluid at a point source, the concentration $I(r,t)$ of the chemical at a distance r from the source and time t after release is determined by

$$I(r, t) = \frac{Q}{(4\pi \cdot D \cdot t)^{3/2}} e^{-r^2/4D \cdot t}$$

This determination (Roberts 1923) assumes that the chemical is free to spread spherically in all directions. When a reflecting or partially absorbing boundary impedes movement of the chemical, as would occur when an arthropod releases a sudden burst of pheromone while on the ground, the concentration at a given point and time is much greater. For a perfectly reflecting boundary, it is effectively doubled. Substituting values typical of arthropod pheromones for Q (10^{10} molecules; this amount may be released by a moth in 1 s assuming that $J = 10^{10}$ molecules$\cdot s^{-1}$, and it is treated here as an instantaneous release) and D ($2.5 \times 10^{-6}\ m^2 \cdot s^{-1}$) shows that threshold concentrations typical of receivers ($I_{th} = 10^4$ molecules$\cdot cm^{-3}$ for moths) can be attained by simple diffusion in less than 1 s at a distance of 1 cm from the signaler, but approximately 10 min would be needed to attain this concentration at 20 cm. Following the initial spherical spreading, chemical concentrations decline and eventually dip below threshold values. At time t after pheromone release, the maximum distance from the source at which concentration still exceeds a given threshold value is determined by

$$r_{th}(t) = \sqrt{4D \cdot t \cdot \ln\left(\frac{Q}{I_{th}(4\pi \cdot D \cdot t)^{3/2}}\right)}$$

This distance represents the radius of the active space at time t. Setting $r_{th} = 0$, the time required for concentrations to fade below I_{th} at all points (at the source, and hence everywhere else) is determined by

continued

Box 3.1 (*continued*)

$$t_{\text{fade-out}} = \frac{1}{4\pi \cdot D}\left(\frac{Q}{I_{\text{th}}}\right)^{2/3}$$

Volatile chemicals, which have relatively high diffusion coefficients, have shorter fade-out times than less evaporative substances.

With the possible exception of alarm pheromones (Section 3.6), though, most chemical signals of arthropods are transmitted for extended time intervals. The concentration $I(r,t)$ of a chemical released continuously at a point source beginning at time 0 is determined by

$$I(r,t) = \int_0^t \frac{J(t^*)}{(4\pi \cdot D(t-t^*))^{3/2}} e^{-r^2/4D(t-t^*)} dt^*$$

where $J(t^*)$ is the flux, or release rate, at time t^*, given in molecules·s^{-1}. After a long time interval has elapsed following the onset of a constant release rate J, the concentration stabilizes and its determination may be simplified as

$$I(r) = \frac{J}{4\pi \cdot D \cdot r}$$

Substituting I_{th} for $I(r)$ and solving for r allows us to determine the maximum distance from the source at which concentrations exceed the receiver threshold,

$$r_{\text{max}} = \frac{J}{4\pi \cdot D \cdot I_{\text{th}}}$$

For values typical of moth pheromone release and perception (see above), an $r_{\text{max}} \cong 32$ km is obtained! It should be noted, though, that expansion to this incredible range would require a time interval much longer than the signaler's lifespan.

The presence of wind alters the shape of the active space, the volume within which concentrations exceed I_{th}, and it reduces the time required to attain a given concentration at a point within the active space. However, wind does not increase the maximum dimension of the active space, r_{max}. In air flow that is perfectly laminar, a chemical released continuously at rate J from a point source creates a long, narrow plume in which concentration at a given point can be approximated by

$$I(r,\theta) = \frac{J}{4\pi \cdot D \cdot r} e^{\left(\frac{-(1-\cos\theta) r \cdot v}{2D}\right)}$$

where r is the straight-line distance from the source to the point of interest, θ is the angle between that straight line and the wind direction, and v is the wind velocity. Along the wind axis, the exponential term = 1 at all wind velocities, and r_{max} would again $\cong$ 32 km using the above values for I_{th}, J, and D. Away from the wind axis, however, wind reduces the transverse spreading of the chemical, and higher wind velocities yield increasingly narrower plumes. But even in a 1 m·s^{-1} wind, the 32-km plume is only 50 cm across at its greatest width. When turbulence

Box 3.1 (*continued*)

> is accounted for (Sutton 1953), models of eddy diffusion predict that the active space becomes shorter but wider, an effect that is more pronounced in higher wind velocities (Bossert and Wilson 1963).

1992). Therefore, some amount of air convection would be absolutely essential for advertisement via chemical signaling as long as signalers and receivers are separated by a meter or more. Because perfectly still conditions seldom if ever occur, advertisement communication is not constrained to such close ranges.

Convection begins to operate strongly outside of the boundary layer that surrounds any solid surface within a moving fluid or itself moving through a stationary fluid (Figure 3.3). But within the viscous sublayer of this boundary layer, fluid movement relative to the solid surface is impeded by a "no-slip" condition, and substances in the fluid largely move through it via diffusion or self-propulsion; as most arthropods do not actively propel pheromone away from their bodies,[10] diffusion is the primary means. The thickness of the viscous sublayer may be estimated from the dimension of the surface measured from its leading edge to the point of interest, the unimpeded (mainstream) velocity of local flow, and the coefficient of diffusion of a substance in the fluid (Denny 1993). For arthropods emitting pheromone into air subject to a light wind ($1\ m{\cdot}s^{-1}$), this layer would be less than 3 mm thick, and a detectable concentration of pheromone molecules would begin passing through the viscous sublayer within 1 second following onset of the emission. At greater distances from the source, the relative influence of diffusion becomes negligible, and convection creates an elongated plume, an "active space" wherein pheromone concentration exceeds a threshold level for eliciting a behavioral response in (male) recei-

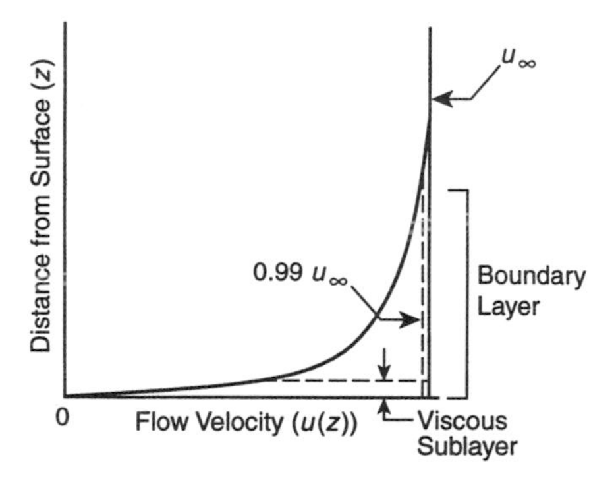

Figure 3.3 Velocity profile within the turbulent boundary layer adjacent to a (smooth) surface of a solid object surrounded by moving fluid. The velocity gradient (du/dz) is extremely steep within the viscous sublayer immediately adjacent to the object's surface; flow in this sublayer is laminar. In the outer portion of the boundary layer, fluid velocity ($u(z)$) approaches that found in mainstream flow (u_∞); flow in this outer layer is turbulent (from Denny 1988; cf. Weissburg 1997).

vers. Because diffusion continues to exert some influence outside the boundary layer, the beginning of the plume is located slightly upwind of the signaler if wind velocity is low (Figure 3.4a).

Under conditions of laminar flow, assuming that convection is uniform in velocity throughout a given region downwind from the source, a continuously emitted chemical would create a rather long, narrow plume (Murlis et al. 1992; Box 3.1). Here, the velocity of the wind affects the shape of the plume and the rapidity with which it is established after emission begins. However, wind velocity has no influence on the length of the plume, the maximum distance downwind at which receivers can detect the pheromone. At higher wind velocities convection exerts an increasingly stronger influence relative to diffusion, and a narrower plume is created that rapidly expands to its eventual dimensions. Counterintuitively, a plume created by a stronger wind will have a smaller active space.

What are the expected volumes of pheromone plumes subject to simple laminar flow? For female moth pheromones, calculations predict an exceedingly long

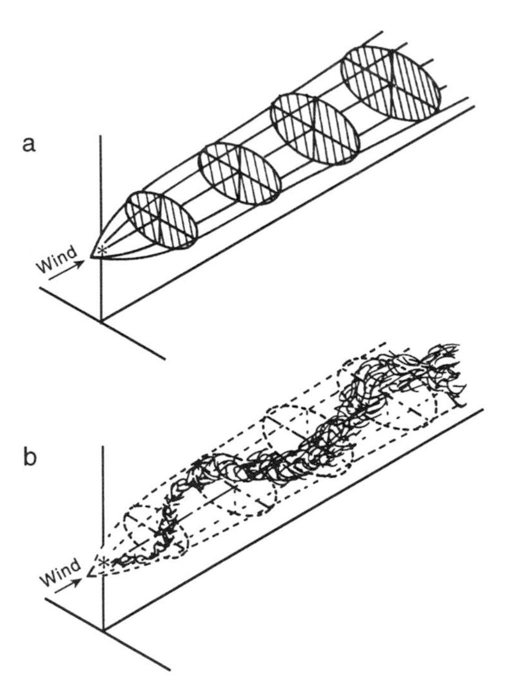

Figure 3.4 Active space of the time-averaged pheromone plume maintained by laminar flow (a) contrasted with the filamentous plume structure found in turbulent flow (b). Short wavy lines represent individual filaments of odor, and the meandering volume enclosing these lines is an approximate snapshot of the overall plume at a given instant. Asterisk indicates location of the pheromone source. (From Murlis et al. 1992; reprinted with permission from the *Annual Review of Entomology*, vol. 37 © 1992 by Annual Reviews www.AnnualReviews.org.)

and narrow plume (32 km × 50 cm) in even a light wind ($1\,m{\cdot}s^{-1}$). These dimensions are calculated based on realistic values for signal flux ($10^{10}\,mol{\cdot}s^{-1}$), threshold intensity (concentration) for male receivers ($10^4\,mol{\cdot}cm^{-3}$), and coefficient of diffusion ($2.5 \times 10^{-6}\,m^2{\cdot}s^{-1}$, typical of moth pheromone compounds; see Schneider et al. 1998b). Nonetheless, all of the available data on moth behavior in the field indicate that the 32-km predicted length is at least two orders of magnitude greater than actual communication distances. The discrepancy arises because airflow in even the simplest environments is never laminar over an extended range. Rather, turbulent flow is the norm and must be incorporated in any model of chemical signal transmission in air.

Turbulent Transport. Turbulence arises when airflow is obstructed by solid surfaces or when local differences in pressure occur, as induced by thermal layers. When these factors are present, inertial forces exceed the viscous forces that promote fluid cohesion, and a torque will act on a volume of air drifting along in a current. Air subject to such torque will move irregularly and shed whirls, eddies, and vortices. Thus, a pheromone plume formed under turbulent flow conditions will be repeatedly disrupted into separated scent-laden parcels whose dimensions and locations cannot be predicted with accuracy.

Early modelers of chemical signal transmission were aware of the inevitability of turbulence, and they incorporated it in calculating the predicted dimensions of pheromone plumes. However, they assumed that the vortices of odor molecules formed in turbulent flow would disperse according to the principles of diffusion, and they simply replaced the molecular diffusion coefficient with an estimated coefficient for vortex diffusion in their calculations. Thus, their predicted lengths of moth pheromone plumes (1.8–4.5 km, which assumed a higher value for D than I have above; Bossert and Wilson 1963) were still far greater than what we know the actual values to be. This assumption of vortex diffusion also led to misjudging the overall volume of the plume as an elongated hemi-ovoid shape. As discussed below, pheromone dispersion within the active space is far from continuous or even regular. Moreover, pheromone concentration at any given point within the overall active space changes constantly over time. The simplified model above estimates concentration at a point in space as an average over time and therefore generates values that are not biologically meaningful.

Atmospheric scientists and others working with fluid dynamics concede that transport via turbulent flow remains one of the least tractable phenomena in physics. It is largely chaotic on a fine scale, and only a few workers have attempted to apply recently developed Lagrangian techniques to construct precise models of odor dispersion (e.g., Suckling et al. 1999a,b). Consequently, most of our understanding of the dispersion of chemicals in air and the structure of plumes has been acquired through various empirical approaches. Pheromone plumes have been estimated indirectly by filming the visible track left by smoke-producing compounds (e.g., titanium tetrachloride) emitted in laboratory wind tunnels (Baker and Linn 1984) or by deploying an ion generator in the field and detecting the emitted ions at various locations downwind and measuring their concentration (Murlis and Jones 1981). Ions are used because they can be detected nearly instantaneously following capture, thereby allowing very precise temporal resolution of the plume. In both cases, the tracers were assumed to behave in a manner physically similar to pheromone compounds because of comparable diffusion coefficients and responses to convection (Murlis et al. 1992). The latter method confirmed the suspicion that plumes are not regular hemi-ovoidal volumes within which chemical concentration is everywhere greater than a threshold value, the concentration at the volume's upper and outer boundaries, but rather collections of meandering filaments within a sea of largely clean air (Figure 3.4b). Moreover, the filaments are thin wisps of odor and do not provide continuous tracks back to the source. Owing to turbulence, they are repeatedly broken into separate segments.

Individual filaments tend to move outward from the source in the direction of the wind at the time they are formed. Thus, a snapshot of the overall plume at any moment would reveal the history of changes in wind direction, and older, downwind filaments would not necessarily be aligned with the present wind direction at their location (Figure 3.4b depicts flow that remained largely unidirectional). Parallel filaments are packed more densely at the center of the plume than at the edges, but odor concentrations within center and edge filaments would be comparable.

According to this picture, pheromone concentration at any given point within the plume, fixed with respect to the ground, would fluctuate constantly, remaining near zero at most times but occasionally rising to values far in excess of the receiver threshold as filaments form and drift by. Because (male) receivers constantly update their neurophysiological tracking of pheromone concentration, it is these momentary peak values rather than an average taken over an extended time period that would be critical.

Experiments testing the behavior of live gypsy moths (*Lymantria dispar*; Lymantriidae) have corroborated the filament model of pheromone plumes described above and added some quantitative refinement. By holding a large number of (male) receivers in small-screen cages at gridded locations in the field and bioassaying their response (pre-flight wing-fanning) to a dispenser emitting synthetic pheromone, Cardé and Charlton (1984) were able to evaluate in a precise manner the signal's active space and its alteration over time in response to changing wind direction. Their observations agreed with the expectations that pheromone concentration is not uniformly distributed and that it changes constantly at a given point, with values passing above and below the receiver threshold many times. The time-averaged plume inferred from their observations was only 80–100 m long, but it spread considerably away from its longitudinal axis and extended across approximately 40° of azimuth. Similar findings were later obtained in the Oriental fruit moth (*Grapholita molesta*;

Tortricidae) by replacing the live insects with whole antenna preparations and making neural recordings (electroantennograms, EAG; see Bjostad and Roelofs 1980, Roelofs 1984, Bjostad 1988) of the antennal responses in the field (Baker and Haynes 1989). These methods underscore the point that the very best chemical detector to date is a biological one. We have yet to develop synthetic detectors that can reliably mimic the accuracy or sensitivity of the live organism or in vitro preparations of its receptor organs (cf. Malakoff 1999, Webb 2000).

Pheromone Tempo and Beaming? Can (female) signalers adjust their transmitted chemical signals by behavioral or physiological mechanisms? The raised and extruded abdominal posture typically assumed during pheromone emission has been noted above and certainly enhances evaporation rates (Conner and Best 1988). In some species females appear to fine-tune this posture by pumping the abdomen such that the terminal segments bearing the pheromone glands are rhythmically exposed and covered by the adjacent anterior segment at a rate as high as 3 times per second (Itagaki and Conner 1987). This regular movement has been interpreted as an adaptation that either imparts an additional defining character, temporal pattern, to the chemical signal (Conner et al. 1980) or that increases the peak intensities to which (male) receivers are exposed (Schal and Cardé 1985, Dusenbery 1989).

In general, constituent compounds and the intensity (concentration) of those compounds define a chemical signal. If intensity varied over time in a regular fashion, though, tempo might characterize an individual's signal along a third dimension (Sections 2.1 and 4.2.7; Figure 3.5). Recognizing this possibility gave rise to both theoretical and experimental work on the pulsing of advertisement pheromones. After deriving the formulae predicting plume dimensions in laminar and turbulent airflow, Bossert (1968) modified the basic diffusion model to accommodate periodic as opposed to continuous pheromone emission. His modification predicted that the emission rhythm would be maintained close to the source as a regular modulation of signal intensity. The depth of modulation (i.e., relative difference between maximum and minimum concentration during a pulse period) would decline with increasing distance from the source and with an increasing pulse rate. In both cases, diffusion would smooth out the peaks and valleys of pheromone concentration. With airflow, however, slow to moderate pulsing was predicted to extend much farther, and Bossert postulated that tempo could function as a chemical signal character within an elongate plume. But, this hypothesis, like the time-averaged model for pheromone dispersion, was made prior to our current knowledge of plume fine structure and how it is influenced by turbulent flow. Actual pheromone plumes already include pulses, filaments of odor that pass a given point at irregular intervals, and a modulated emission is unlikely to be reflected in a regular appearance of these pulses anywhere except quite close to the source, and only for unrealistically slow modulation rates. Consequently, the alternative explanation that modulated emission is

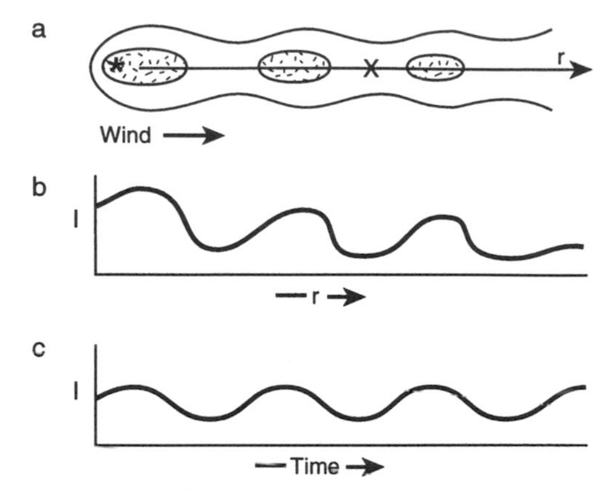

Figure 3.5 (a) Plume of a temporally modulated pheromonal signal subject to laminar flow. Higher odor concentrations as occurring at a given instant are represented by stippling (see Dusenbery 1992). (b) Change in pheromone intensity (*I*) along the plume axis (*r*) at a given instant, as depicted in (a), above. (c) Change in pheromone intensity over time, as measured at point X. Asterisk indicates location of the pheromone source.

an adaptation for increasing the peak pheromone concentration within filaments may be more valid. And, it is possible that modulated emission simply represents an artifact of the rhythmic abdominal movements necessary to maintain the hydrostatic pressure everting the pheromone gland surfaces.

Because transmission of an aerial chemical signal over any but the shortest distance occurs at the mercy of local airflow, a female is unlikely to possess any means of actively beaming the signal in a particular direction should that be desirable. However, she may be able to control the horizontal spread of the signal by altering pheromone diffusion and thwarting or enhancing the impact of turbulence on the plume. If diffusion were reduced, the plume would continue as a single, narrow filament for a much longer distance away from the source. This effect follows from a minimum threshold for vortex diameter (= l in the formula determining a fluid's Reynolds number; see equation 3.1 below), which is approximately 3 cm in air flowing at $1\ \text{m}\cdot\text{s}^{-1}$. Until a plume expands laterally to this diameter, turbulence is unlikely to disrupt it. Females of some arctiid moths release their advertisement pheromones in the form of liquid aerosol droplets (Krasnoff and Roelofs 1988), which inhibits diffusion greatly. Can the males more easily track the resulting ribbon-like plume—once they have found it? Or, does unusually low sensitivity in the males require females to emit pheromone at a very high concentration, one that could only be attained by hindering the compounds from entering the vapor phase and dispersing?

Verdicts on these hypotheses, and on any proposed functions of signals, must consider receiver perception and response. I therefore defer further speculation on the active modulation and expected intensities of female advertisement pheromones to the sections on receivers and signal reception (Section 3.1.3) and evolutionary origins and adaptations (Section 3.1.5).

Pheromone Transmission in Water. The principles of diffusion, convection, and turbulence that influence aerial pheromone plumes operate in any fluid medium. Many freshwater and marine arthropods rely extensively on olfaction, and various species emit chemical advertisement signals. How would their pheromone plumes compare with aerial plumes given the properties of water?

Because diffusion coefficients of small molecules in water are approximately 10^{-4} times as great as in air, a pheromone continuously emitted in still water would expand, ceteris paribus, to fill an active space roughly 10^4 times the radius of that produced in air. But, the advantage of this increased—astronomical—size would be more than offset by an exponential increase in the time required for the active space's leading edge to reach a given radius. Thus, sexual advertisement via chemical signals would be as reliant on current in water as it is in air.

Turbulence is also as prevalent in water as in air. The likelihood of turbulent flow in a fluid can be predicted by the Reynolds number (*Re*), a dimensionless parameter calculated as

$$Re = (u \cdot l)/\nu \qquad (3.1)$$

where u is the unimpeded current velocity, l is the distance separating two points in a fluid between which the likelihood of turbulence is to be assessed, and ν is the kinematic viscosity (ratio of a fluid's viscosity to its density; ν is measured in units of $m^2 \cdot s^{-1}$). Turbulent flow is more likely to occur at higher current velocities, over an extended distance within the fluid, and in less viscous fluids. A rule of thumb is that turbulence can be expected with certainty at *Re* values $>10\,000$ and that laminar flow would prevail at *Re* values <2000. Because the kinematic viscosity of water is roughly 15 times greater than that of air but current velocities in water are typically 15 times slower, *Re* values are similar in the two media. Thus, over equivalent distances, turbulence is as likely to occur in freshwater and marine environments as in air, and pheromone plumes of arthropods inhabiting such environments are also expected to be represented by collections of meandering filaments of high odor concentration. This predicted plume structure has been substantiated by various tracer detection methods both in flow tanks and in the ocean (Moore and Atema 1991, Finelli et al. 1999, Weissburg 2000).

Unfortunately, far less is known about the mating systems, sexual advertisements, and generation of actual pheromone plumes of arthropods in water (see Zimmer and Butman 2000 for review). In some marine crustaceans, females mix their chemical advertisements with urine before releasing them into the water via the gills (Bushmann and Atema 2000). Unlike terrestrial arthropods, water conservation is not critical for them, and the urine current represents a pre-existing means of propelling the advertisement compounds away from the body (Atema 1995). Because freshwater and marine environments are typically stratified and males and females may inhabit several strata and move among them, chemical signals may have to be transmitted vertically as well as (or rather than) horizontally to reach receivers (cf. Hamner and Hamner 1977).[11] Can such transmission be accomplished by using substances of altered buoyancies which thereby rise or sink from the signaler?

3.1.3 The Receiver and Signal Reception: Molecular Sieves and Olfactory Transduction

Chemoreception is unique among perceptual functions in that the sensory neurons must be directly exposed to the external stimuli. The neurons may be enclosed within receptacles and protected by various mucous substances, but there is no escape from the necessity of direct contact between the chemical serving as the signal and binding sites on the neuronal membrane (Schneider 1971, Krieger and Breer 1999). Advertisement pheromones as used by moths are no exception.[12] In this section we examine the route and steps by which pheromone molecules in a filamentous aerial plume reach the contact points on sensory neuronal membranes in (male) receivers, are processed by the peripheral and central nervous systems, and are distinguished from the various sorts of external and internal noise that may occur along the chemical channel.

Chemoreceptor Location and Structure. Among non-chelicerate arthropods, the universal structures for olfactory reception are the paired antennae on the head (Schneider 1964). This location contrasts with that of gustatory (taste) receptors, which are normally on structures that contact surfaces potentially coated with relatively high concentrations of chemicals of interest. For examples, gustatory receptors for food are usually on the mouthparts (palps) and legs (tarsi), and those for oviposition substrates may be located on abdominal segments or appendages that probe the ground or vegetation. But olfactory receptors often need to detect exceptionally low concentrations of material in the surrounding fluid, and enlarged or elongated appendages such as antennae can offer an increased surface area specialized for intercepting as much chemical signal as possible. Moreover, the anterior location of the antennae allows the receiver to detect the chemical signal (or its absence) before progressing through the region harboring it. Thus, appropriate orientation maneuvers may be made as soon as possible. The anterior location may also allow chemical signals to be intercepted in the absence of turbulence that could impede diffusion toward the receptors:

large, rapidly flying arthropods are likely to generate considerable turbulent flow and a boundary layer posteriorly around the body and wings. The dimensions (cylindrical shaft with diameter $\leq$1 mm) and posture (often held transverse to the body axis) of antennae would also minimize the thickness of the viscous sublayer through which chemicals must pass. Consequently, a moth flying into a plume would not have to await diffusion to carry pheromone molecules across the final 1–3 mm to the chemoreceptors. Rather, it should be able to detect the molecules almost immediately upon entering the pheromone's active space (but, see Schneider et al. 1998b).

Both the gross anatomy and fine structure of moth antennae may represent engineering features that enhance interception of the chemical signal still further. Males and females alike may respond to various habitat odors such as those emitted by larval host plants, but typically only males respond strongly to female advertisement pheromones (Hansson 1995).[13] Correspondingly, the sexual dimorphism of antennae is quite pronounced in some species, particularly those that may regularly orient over extended distances. Male antennae in these dimorphic species bear pectinations, comb-like projections arranged in one or more rows running the length of the antennal shaft (flagellum) (Figure 3.6), and each pectination is fitted with numerous trichoid chemosensilla, thin-walled, multiporous sensory hairs roughly 1 μm in diameter and 30–600 μm long. Pores that perforate these sensilla serve as the points through which pheromone molecules penetrate the omnipresent cuticle en route to the dendrites of the sensory neurons within. Thus, the total number of sensilla may be very great without compromising either flight aerodynamics or the diffusion of intercepted molecules into the pores: the pectinate structure is a molecular sieve that provides an expanded surface area on which more than 100 000 sensilla can be borne while allowing airflow to pass through. Even among the majority of moth species that display little or no sexual dimorphism in their antennae, the number of sensilla on the entire antennal flagellum is very great.

The fine structure of antennae may augment the interception of pheromone in a number of additional ways (see Loudon 1995), most of which have not been fully investigated. First, certain positions and orientations of pectinations on the antennae, and of sensilla on the pectinations, might increase the proportion of pheromone molecules in the airstream that contact the sensilla (Vogel 1983). Second, adjacent pectinations, and adjacent sensilla on those pectinations, may be arranged in relative positions and orientations such that they effectively capture molecules filtering through or deflected around upstream pectinations and sensilla, just as tiers of leaves can be arranged on a tree for the trapping of light (Loudon 1995, Koehl 1996). It is also possible that individual pectinations and sensilla can be shaped such that relatively little oncoming air is deflected around them and that an optimal placement of pores on the sensilla exists. Each of these microstructural receiver features may be amplified by antennal move-

Figure 3.6 Pectinate antennae of male cecropia moth, *Hyalophora cecropia* (Saturniidae) (photograph courtesy of James W. Mertins).

ment, which can increase the rate at which pheromone molecules are intercepted (Koehl 1996; cf. Section 5.2.3, Signal Amplifiers).

Once adsorbed on a sensillum, a small proportion of pheromone molecules diffuse into the pores and may then bind to (protein) receptor sites on the sensory neurons.[14] The proportion actually diffusing into these entryways may be slightly assisted by grooming. Arthropods regularly clean their antennae, and in some moths this activity is aided by a specialized spur, the epiphysis, on each front leg. By pulling an antenna through the gap between the spur and the leg, a moth can rid the antennal surface of debris that would otherwise clog the pores and obstruct the entry of pheromone molecules (see Callahan and Carlysle 1971).

Labeled Lines to the Brain. The receptors on antennal neurons are highly specific and under natural conditions will generally not bind effectively to any but the correct compound, diastereomer, and enantiomer: that found in the conspecific female pheromone.[15] A related category of neuronal receptors bind specifically to

pheromone antagonists, compounds that inhibit the receiver's response to its conspecific pheromone (Roelofs and Comeau 1968); in general, antagonist compounds are components of the pheromones of related species. In the various species that have been studied, each component of the pheromone and its antagonists are associated with a different category of olfactory receptor neuron (ORN) whose binding sites are specific for that compound,[16] but several ORN categories are normally co-compartmentalized within a single sensillum (Todd and Baker 1999), and different categories of sensilla may be recognized that are innervated by certain combinations of ORN types. Thus, pheromone appears to be evaluated via a labeled line process (analogous to how the ear evaluates the frequency spectrum of a sound; Section 4.3.5) as opposed to an across-fiber pattern (analogous to how the eye evaluates the spectrum of light; Section 5.3.1) of neuronal activity: Each component and antagonist has a separate (labeled) line to the brain, wherein higher neural centers evaluate pheromone composition by recognizing the input from certain ORNs as the presence of a specific component and then integrate the input from the various ORN categories (Boeckh 1984, Hansson 1995; Figure 3.7).

Pheromone compounds and antagonists were originally believed to reach the sensory neurons (ORNs) by diffusing along canals leading inward from the pores on the sensillum surface. Our current understanding is quite different. The interior of a sensillum is filled with a hydrophilic lymph similar to the mucous coatings in vertebrate olfactory organs, and pheromone molecules, which are hydrophobic, need assistance in penetrating to the ORN. Specialized peptides (pheromone-binding proteins, PBPs) serve to shuttle pheromone through to the ORN, and they may also form a pheromone–PBP complex that actually binds to the receptor site on the neuronal membrane (Breer 1997). Binding at the receptor site then releases an intracellular cascade of secondary messengers that is transduced to membrane hyperpolarization and the transmission of an action potential (AP). Single-cell neurophysiological recordings from antennal ORNs demonstrate that pheromone binding at the receptor site or at the PBP is highly specific: a slight change in a compound's structure (e.g. altering a diastereomer from the *cis*- (Z-) to the *trans*- (E-) configuration or moving the position of the double bond) will reduce the rate of action potentials greatly (Hansson 1995, Todd and Baker 1999; Figure 3.7).

Pheromone Degradation and the Avoidance of False Alarms. Pheromone-receptor site binding is only momentary. As typically occurs in olfaction, a bound (pheromone) molecule is released within milliseconds, and the site is then free to bind with another molecule shuttled in subsequently. Once released, a given pheromone molecule is unlikely to rebind with the ORN because PBP, which is oxidized following its interaction with the ORN-binding site, radically switches its function and acts as a pheromone inactivator (Hansson 1995). Further inactivation results from specific enzymes (esterases) that aggressively degrade

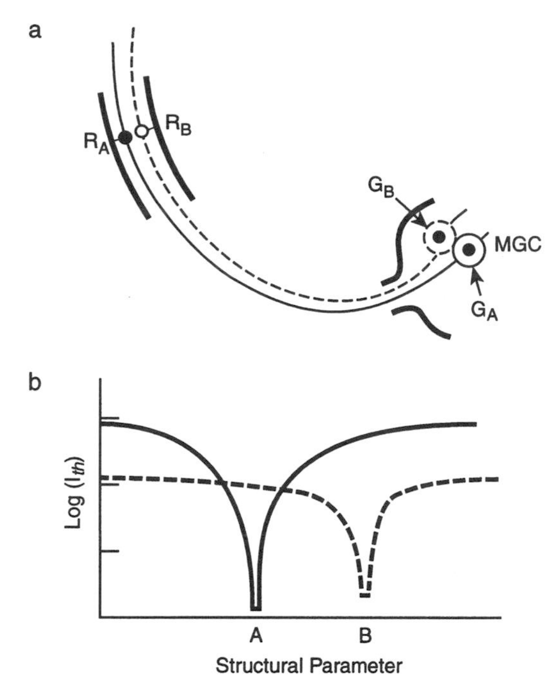

Figure 3.7 Labeled line model of pheromone perception. (a) Receptor categories R_A and R_B in an antennal sensillum have direct lines to glomeruli G_A and G_B, respectively, in the macroglomerular complex (MGC) of the brain. (b) Sharply tuned responses of receptors R_A and R_B to compounds A and B, respectively. Small changes in the structure of A or B (e.g., modification of the carbon chain length or position of double bond(s)), greatly increase threshold intensities (I_{th}) for response.

unbound pheromone within the sensillum lymph. Similarly, the intracellular cascade of secondary messengers that is transduced to an action potential is inhibited very quickly. In principle, these inactivation, degradation, and inhibition processes would allow external pheromone intensity to be coded accurately by the rate of action potentials. Additionally, pheromone intensity that is temporally modulated at rates as high as 10 times per second may be faithfully tracked by the sensory neurons.

Unlike mechanical and electromagnetic signals, chemical messages may linger in the environment long after their onset. Unless properly inactivated or degraded, these residues could threaten the signaler or receiver with confusion in many circumstances. In our present context, pheromone might adhere to the antennae and body of a male moth who had earlier flown through the plume emitted by a female. Should some of this pheromone eventually diffuse into the chemosensillum pores, the residue could potentially elicit an inappropriate response. But, such false alarms evidently do not occur, and their absence may largely reflect efficient activity of the pheromone-degradation enzymes:

one hypothesis maintains that only pheromone molecules intercepted directly by chemosensillum pores at the instant a male receiver enters a plume filament have a reasonable chance of binding to PBP and being carried in to the sensory neuron (Futrelle 1984). Those molecules diffusing into the pores later due to contamination of the body do so at a slower rate and are much more vulnerable to enzymatic attack. In addition, some pheromone residue may be degraded externally by specialized esterases even before entering the pores (Vogt and Riddiford 1981, 1986, Ferkovich et al. 1982). Consequently, males can track a rapid modulation of pheromone as would occur when flying through the filaments comprising a plume (Baker 1989, Vickers et al. 2001) or when perceiving a pulsed pheromone source (Conner et al. 1980; but see Section 3.1.2). The elimination of residual odor molecules is a general problem in chemoreception, and these pheromone-degradation enzymes have counterparts among vertebrate olfactory systems (Lazard et al. 1991).

Efficient action aside, degradation enzymes cannot suppress a more insidious form of interference, that which would arise from pheromone adsorption to and re-emission from foliage and other surfaces in the vicinity of a signaling female (Noldus et al. 1991a). Would males occasionally track these residual odors, and would it be in a female's best interest to avert creating such confusion? I address the second question tangentially in a later section, Evolutionary origins and adaptations (Section 3.1.5). At this point, though, it is critical to recall that females generally signal from relatively open locations in their microhabitat. Thus, convection would carry most of the emitted pheromone far downwind before it would be adsorbed, and concentrations on foliage at these distant points may be sufficiently reduced that males are seldom misled.

Central Processing. The sensory neurons lead from the antennal chemosensilla to the deutocerebrum of the brain (antennal lobes), where they terminate in specialized glomeruli grouped within a macroglomerular complex (MGC). At a glomerulus, the inputs from a great many sensory neurons of a given ORN type converge and are passed to a central ascending neuron (projection neuron, PN) which conveys the message to the protocerebrum; this specific neural architecture is distinctly sexually dimorphic, being found only in males. Integration of inputs from different categories of labeled lines occurs both in the MGC (via interglomerular synapses) and in the protocerebrum (see Mustaparta 1997 for variations) and is responsible for evaluating the pheromone's composition. Both behavioral and neurophysiological studies demonstrate that central evaluation of pheromone composition is also highly specific: Removal of a minor component or even a change in its relative concentration normally has an adverse effect on the response level (Mustaparta 1997).

Input from ORNs specialized for pheromone antagonists is also integrated with inputs from the ORNs specialized for pheromone components, and this integration also occurs in the MGC and protocerebrum of the brain.

Antagonists inhibit behavioral responses to the conspecific pheromone, but they do not do so by inhibiting or blocking activity in peripheral ORNs in the antennae. Probably, peripheral inhibition of antagonist ORNs would be an inefficient mechanism, as the rate at which action potentials are generated spontaneously in these neurons is already quite low: any inhibitory response at this level would not represent a significantly lower AP rate (see Boeckh 1984). Similarly, mechanisms in which antagonists inhibit responses to conspecific pheromone by competing with and blocking pheromone components at the binding sites on pheromone ORNs would normally be ineffective, as only a relatively high antagonist : pheromone component ratio would lead to thorough inhibition. However, central mechanisms for evaluating antagonists can be very sensitive and generate inhibition at even rather low antagonist : pheromone component ratios. Sensitive inhibitory responses to antagonists have probably been selected because receivers are thereby prevented from orienting toward the signals of those other species whose pheromones include both the antagonists and components of the receiver's conspecific pheromone (Table 3.1; cf. von Helversen and von Helversen 1998). The potential for such heterospecific interference is actually quite high among moths and other insects (e.g., Leal 1996), and it represents a curious problem in signal evolution (Section 3.1.5).

External and Internal Noise. In a provocative article, Freeland (1980) suggested that additional interference along the chemical channel would arise from environmental odors that block or otherwise impede pheromone recognition. Given our knowledge of olfactory advertisement signals and their perception by insect receivers, is this concern a valid one and, if so, how do receivers deal with it? In part, the answer to the first question depends on the extent to which the labeled line model (Figure 3.7) characterizes the pheromone reception and recognition process. A minimum signal : noise ratio is a basic necessity for effective communication along any channel (Section 2.4), but background environmental noise should generally be absent from the chemical channel if labeled lines are used exclusively. That is, volatile compounds from vegetation and other organic and

Table 3.1 Pheromone Components and Antagonists of Three Sympatric Species

Species	Female advertisement pheromone (major component(s))	Antagonist inhibiting male response to conspecific female pheromone
1	A	B
2	B	A
3	A + B	

Inhibitory responses of species 1 and 2 to compounds B and A, respectively, prevent inappropriate orientation toward species 3.

inorganic sources are unlikely to mimic the precise chemical properties of female pheromone components or antagonists either to compete with signals or to inhibit a receiver's responses to them. Moreover, such compounds are generally hydrophobic and, being unable to bind to the specialized PBPs, would not readily penetrate the lymph of antennal sensilla where they could block or otherwise interfere with pheromone reception. These predictions are in agreement with a literature survey (Cardé 1983) that uncovered no evidence of general chemical noise from the environment reducing the efficacy of sexual advertisement communication via pheromones in arthropods.

Nonetheless, recent information on olfactory processing suggests that adopting a view based entirely on the classical labeled line model and the supposition of a firewall separating the pheromonal and plant-odor olfactory systems could lead to some oversights. Pheromone receptor neurons may be very sharply tuned, but, as with other channels, their responses to chemical stimuli ultimately depend on intensity (see Roelofs 1978). An antennal ORN specialized for a particular pheromone component will respond to certain other compounds that are structurally related in various ways,[15] and these ORNs may even respond to some plant odors presented at supernormal intensities (Hansson and Christensen 1999). When such stimuli are presented at natural intensities, a greatly reduced yet potentially interfering level of response might remain. More realistically, however, it is at the level of central processing that volatile emissions of plants and other biotic or abiotic sources could pose a serious threat in certain conditions. Recognition of pheromone blends may require evaluation of across-fiber patterns in addition to the labeled lines (e.g., Christensen et al. 1991), and antennal lobe interneurons that respond to both pheromone components and low concentrations of plant volatiles are found among moths (Anton and Hansson 1995, Dickens 1997). Conceivably, volatile compounds emitted by some plant species might elicit false orientation (cf. Schiestl et al. 1999; Section 3.1.5), while others might inhibit receiver attraction to conspecific pheromone. The original assessment of the notion that environmental odors could interfere with pheromonal communication was made before the parallel, and interlocking, natures of plant odor and pheromone perception were appreciated to the extent they are today. In light of our current understanding of odor processing, reopening the question of environmental interference along the chemical channel is likely to yield some interesting results.

Under most conditions, labeled lines and central filtering mechanisms may safeguard pheromone perception from much external noise. These neural features, however, cannot help a receiver to avoid the more pervasive problem of internal noise from spontaneously generated action potentials in the ORNs. Nonetheless, receivers can and do detect exceedingly low concentrations of pheromone (Kaissling and Priesner 1970), so they must somehow discern the barely elevated rate of action potentials resulting from reception of a few hundred or thousand pheromone molecules. In principle, the discernment could be achieved probabilistically by (1) summing action potentials from a given ORN over a lengthy time interval or (2) summing potentials over a short interval but from a great many ORNs (Figure 3.8). In either case, the spontaneous background rate from receptor noise would stabilize, and that barely elevated rate resulting from pheromone reception would be predictably higher (Boeckh 1984, Dusenbery 1992). Thus, excitatory post-synaptic potential (EPSP) thresholds required for stimulation of projection neurons ascending from the antennal lobes of the brain could be set such that they reliably fire in response to pheromone and are seldom triggered by false alarms originating in spontaneous neural activity. Neuroanatomy, specifically the convergence of inputs from a great number of ORNs of a given category within antennal lobe glomeruli, shows that the second option is used. The complex and rapid maneuvering by receivers tracking aerial odor plumes, to be described in the next section (3.1.4), indicates the adaptiveness of this option: Receivers must update their instantaneous evaluation of pheromone several times per second in order to orient within a plume and localize its source (see Vickers et al. 2001).

Intensity Discrimination. The neural mechanisms by which receivers discern low concentrations of pheromone may also afford them some ability to differentiate pheromone concentrations at levels well above their threshold. As with other sensory channels, signal intensity (concentration) would be coded by the rate of action potentials. But, detectable pheromone concentrations may vary over several orders of magnitude, and a receiver's dynamic range for differentiating concentrations might be limited if it relied solely on this coding. Additionally, many olfactory neurons respond to pheromone in a phasic manner and quickly adapt to the level of stimulation, which would obscure any relationship between

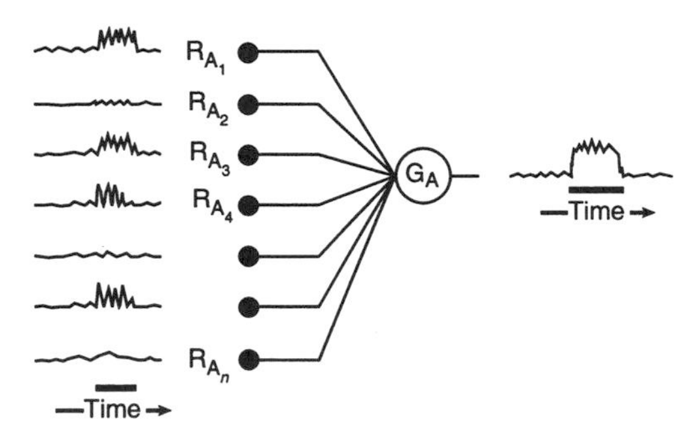

Figure 3.8 All n category A receptors converge at glomerulus G_A in the macroglomerular complex (MGC) of the brain. Whereas many individual receptors do not respond to a given signal of pheromonal component A, there are a sufficient number of receptors such that their summed potential in a projection neuron ascending from G_A is significantly higher than background activity, even when evaluated over a brief time interval. Horizontal bar indicates time when A is present.

action potential rate and absolute signal intensity (Section 2.3). Instead, findings in tobacco budworm moths (*Heliothis virescens*; Noctuidae) showing that the various neurons of a given ORN category have markedly different sensitivities, suggest that receivers may differentiate absolute intensities of a compound by the same type of labeled line process used for differentiating between compounds (Almaas and Mustaparta 1991, Mustaparta 1997). That is, a high concentration of a pheromone component could be recognized by input from the least sensitive neurons of the category specialized for that compound (cf. Figure 2.4). This neural solution is also prominent in acoustic perception (Section 4.3.5).

A Cautionary Note Revisited. Before leaving the subject of signal reception, the distinction between those compounds emitted by a (female) signaler and those actually detected by and stimulating a (male) receiver must be re-emphasized. In general, the number of identified ORN types in the receiver is less than the number of compounds emitted by the signaler. Moreover, the receiver may only require stimulation by some of the minor pheromone components (Linn et al. 1984, Todd et al. 1992; cf. Fadamiro et al. 1996). For example, a signaler may emit a major amount of compound A and minor amounts of B, C, D, and E, the receiver may have ORN types specialized for only A, B, and C, and it may exhibit higher level responses following stimulation by either A and B or A and C. In other words, either minor component is adequate, and the signaling system may be considered to include an element of redundancy (Section 2.6). Neuroanatomical tracing via specialized staining techniques indicates that redundancy arises when a single minor component cross-stimulates two or more glomeruli owing to the arborescent patterns of ORN axons (Todd et al. 1995).

3.1.4 Orientation and Signal Localization: Fluid Current, Optical Cues, and Internal Programming

Localizing and arriving at the source of a distant chemical signal in a moving fluid is no mean feat for a receiver, even one equipped with a vast array of sensilla on pectinate antennae. A bacterium or protozoan detecting a nutrient compound in its aqueous medium 20 μm away can find that nutrient rather easily via movement responses that are inversely proportional to the instantaneous concentration perceived (Berg and Brown 1972), and an ant walking along a terrestrial odor trail, laid down with sharp boundaries, need only rely on its paired antennae tapping the ground and now and then turn in the direction opposite from an antenna that loses the trail's scent (Bell 1984; Section 3.7). But how can a small animal faced with drifting wisps of barely detectable odor ever trace their source?

The first issue to clarify is that arthropods within moving fluids do not localize chemical signals by following a concentration gradient. Such tracking would not be feasible even if pheromone dispersed solely via diffusion and expanded to fill a spherical active space: at points farther than several meters from the source, concentration of the emitted pheromone would simply change too little over a radial distance to be detectable by a small organism, whose paired chemoreceptors would be separated by a few centimeters at most. For example, in air, pheromone emitted by a signaler at 10^{10} mol·s^{-1} would change in concentration at a 50-m distance by only 159 mol·cm^{-3} as a receiver moves either 1 cm toward or 1 cm away from the source (cf. Bossert and Wilson 1963). This change would represent a minute fraction of the estimated threshold concentration (approximately 10^4 mol·cm^{-3}) for male receivers in various moth species.[17]

Unlike the idealized diffusing cloud above, real pheromone plumes are bordered by steep changes in concentration, but, most of these changes occur at the lateral edges of filaments and would not point toward the source. Moreover, the discontinuous nature of the filaments implies that pheromone would decrease in concentration—to near zero—as a receiver reaches the end of the filament toward the source. Consequently, a plume of meandering filaments would be no easier to track in this hypothetical fashion.

Anemotaxes, Appetitive Flight, and Perching. The initial clue to odor tracking in moving fluids came not from studies of moth sex pheromones but from the responses of mosquitoes to carbon dioxide. Kennedy's landmark experiments showed that *Aedes aegypti* mosquitoes respond to odor (CO_2) associated with their vertebrate hosts by flying upwind (Kennedy 1939). Other studies found that salmonid fish tracking habitat or sexual odors in water currents essentially do the same by swimming upstream (see Arnold 1974 for review). Thus, investigators testing the silkworm moth, *Bombyx mori*, after determination of its sex pheromone were prepared, and their behavioral bioassays focused on upwind movement (positive anemotaxis) in males. But, *B. mori* is flightless, having been bred intensively for efficient silk production over the past 4000 years, and these bioassays of running over limited distances offered little insight to the complex maneuvering necessary in a three-dimensional aerial plume replete with the unpredictable patterns of turbulent flow. Biologists began to appreciate this complexity as soon as they turned their attention to flying moths. *B. mori* was not a suitable "laboratory rat" with which to probe this particular aspect of the chemical channel (but, see Kanzaki 1998).

Laboratory and field investigations elucidating the mechanisms by which male moths track aerial chemical signals began over 30 years ago, and various experiments continue today. By studying the flight of moths in several families, biologists have found that successful orientation and localization normally relies on perception of visual cues as well as the chemical ones. Additionally, internal, or self-directed, programs of movement are critical (but, see Witzgall 1997, who challenges the importance of internal programming). Our current understanding of aerial tracking (Vickers 2000), assembled in a sequence progressing from initial odor detection and orientation over a large spatial scale to finely tuned maneuvers, is discussed next.

Because some amount of air flow is nearly always present in natural habitats and is responsible for establishing a chemical signal's plume, much of the laboratory work has been conducted in wind tunnels (Baker and Linn 1984). Flow in these chambers can be adjusted to a low velocity and nearly laminar conditions. From studies of several species of moths, it appears as if the initial responses to detectable pheromone concentrations in the airstream are wing-fanning, which may be a form of "sniffing" serving to enhance the flux and intensity of intercepted pheromone (Schneider 1964, Loudon and Koehl 2000), and taking flight if stationary or heading upwind if already airborne. Thus, sensitive perception of the direction and magnitude of air flow and course correction in response to this information are integral parts of orienting to a chemical signal.

A receiver's behavior prior to even contacting a pheromone plume is equally important, however, in its ultimate localization of the source. Males are generally believed to search their habitat via appetitive flight, an activity that typically exhibits a daily periodicity starting before the onset of female signaling and continuing after most females have stopped. This pre-stimulus behavior cannot be examined fully in laboratory wind tunnels, and, as with most phenomena that can only be approached in the field, it remains poorly understood. Several early studies suggested that males fly at a right angle to local winds (reports in Cardé and Charlton 1984), presumably to increase their chance of intersecting a plume, but a later study (Elkinton and Cardé 1983) did not confirm this flight pattern. Moreover, it is not clear how the right angle heading would be established, a point we return to shortly. Generally, we do not know whether appetitive flight represents searching for receptive females only directly via their pheromone plumes or also indirectly via sites likely to harbor them. In many cases such sites would be suitable host plants, which might be recognized by a combination of volatile chemical[18] and visual cues: silhouette, and spectral and polarization reflectance patterns (cf. Harris and Miller 1983). Energetics must also be an important factor influencing the pattern of appetitive flight, particularly given the restricted (liquid) diet of adult moths: might it be less costly, and perhaps equally profitable as measured by the rate of encountering females, for a male to drift downwind until entering the broad expanse of a pheromone plume (Murlis et al. 1992)?

A variation on the theme of appetitive flight occurs in cockroaches and offers a valuable perspective from which to view the searching behavior of moths. Instead of ranging widely through their habitat, males in various arboreal cockroach species perch on foliage in open, elevated locations during the daily period of female pheromone emission (Schal 1982, Bell et al. 1995). These perches may offer males increased opportunities to detect the pheromone filaments from signaling females as the filaments form and sail past. Thus, a male awaits an indication that a receptive female is nearby, and only then does he take flight and orient upwind. Does this reversal of the moth protocol reflect the more expensive locomotion energetics of cockroaches? Or do these cockroach species occur in population densities that are high enough to make this strategy for contacting pheromone plumes more rewarding?

Optomotor Responses. After a receiver arrives at a plume, even a simple one established by laminar flow, positive responses to odor and wind alone would be insufficient to allow it to track the chemical signal reliably (Figure 3.9). A receiver moving upwind must maintain a positive velocity with respect to the ground as well as to the air if it is eventually to attain the source of an odor plume; otherwise it would risk being swept downwind and away from the source even while maintaining a positive air velocity. Animals solve this problem by using visual feedback—longitudinal movement of patterns from the ground flowing posteriorly across their field of vision—to adjust their air speed to values greater than the wind speed and thereby progress toward their goal (cf. Section

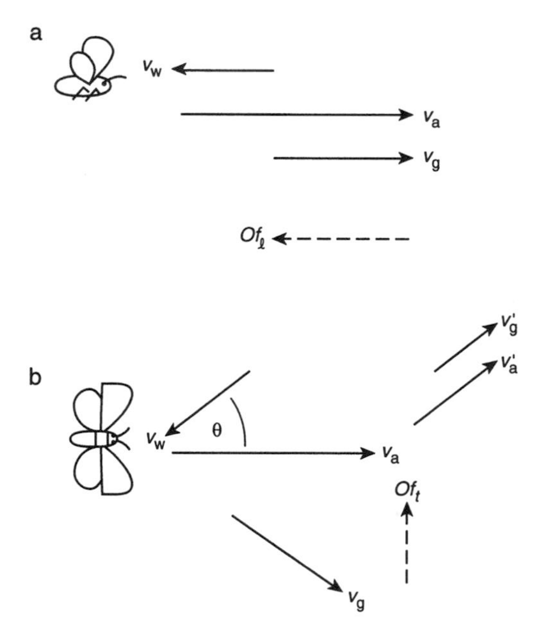

Figure 3.9 (a) Wind (v_w), air (v_a), and ground velocity (v_g) for an insect flying directly upwind within a pheromone plume. Of_ℓ represents the longitudinal optical flow seen by the insect. (b) Wind, air, and ground velocity for an insect flying at angle θ with respect to v_w (overhead view). Of_t represents the transverse optical flow seen by the insect. v_a' and v_g' are the air and ground velocities, respectively, at which the insect travels if it corrects its course until all transverse optical flow is eliminated.

5.3.1; Kennedy 1939). In addition, they may simply refrain from tracking odor plumes in strong wind; similarly, signalers may refrain from emitting odor under such conditions.

The process of relying on visual feedback during upwind flight, known as an optomotor response or optomotor anemotaxis, has been documented in wind tunnels with various moth species (cf. Fadamiro et al. 1998). By placing a conveyor belt on the floor of the tunnel and moving a pattern of transverse black and white stripes in the same direction as the wind, a male can be kept flying at a ground velocity of zero, or even backwards, if the pattern is moved fast enough (Miller and Roelofs 1978). Presumably, the male interprets the pattern flowing longitudinally and posteriorly beneath him as an indication that he is progressing toward the chemical signal emitted at the upwind end of the tunnel. Evidence from several species indicates that males may adjust their effort (air speed) to fly at a preferred ground speed (Willis and Arbas 1991, Baker and Haynes 1996). Visual patterns from the ground need not be as sharply defined and conspicuous as the transverse stripes used in the first conveyor belt experiments. In the field, nocturnal species probably only discern faint contrasts from objects that are illuminated by the moon or stars.

Animals flying within moving fluids also depend on visual feedback and the optomotor response to adjust their course azimuth to align with the upwind direction (Figure 3.9), but the reason for this reliance is less obvious. One would think that wind direction would be detectable via mechanoreceptors at the bases of the antennae (Johnston's organs; see Section 4.3.1) or in fields of sensory hairs on the head, but in fact, once aloft, an animal will experience few aerodynamic forces or pressure gradients and differences that indicate wind direction with respect to the ground, or even its own frame of reference. By monitoring the transverse flow of ground patterns across its field of vision, however, a flying animal can detect that it is being blown laterally. In response, it can adjust its course until that transverse component of visual flow disappears, at which point it is aligned with the wind axis. Proper course adjustment would require the animal to turn initially in the same direction as the transverse flow across its visual field. To do otherwise would result in perfectly aligned but downwind flight! This elaborate sensorimotor process, which involves olfactory, visual, and possibly some mechanical inputs, must be conducted with considerable finesse and speed. Via in-flight electromyographic (EMG) recordings from moths maneuvering within wind tunnel plumes, we are beginning to understand how this feat is accomplished (see Willis and Arbas 1997).

Finally, visual feedback is probably the mechanism by which a receiver maintains a relatively consistent altitude during flight (cf. Baker and Haynes 1996). Ground patterns in a local area may have a particular grain that is registered by the compound eyes. Should this grain be suddenly perceived as much coarser or finer, a receiver may respond by raising or lowering its flight path, respectively. Olfactory information and the perceived grain of ground patterns might conflict over certain topographies, though. How should receivers resolve a dilemma wherein visual information recommends one altitude whereas olfactory information indicates another? We might predict that the olfactory information would have priority in this hierarchy and that the receiver would follow the plume.

Tracking Plumes and Filaments. Were pheromone plumes the regular hemi-ovoidal active spaces predicted from models of laminar or time-averaged turbulent flow, the simple combination of a positive response to odor and optomotor anemotaxis would suffice for orientation. Because wind is a collimating stimulus that would unambiguously point toward the source of the chemical signal, a (male) receiver would not be faced with the problem of determining trail polarity that besets animals tracking terrestrial odor trails (e.g., Stirling and Hamilton 1986; cf. Section 3.7.1). But, as shown above (Section 3.1.2), odor plumes reflect a history of changing wind directions and turbulence, and receivers therefore require an additional mechanism(s) to track them. Otherwise, they would rapidly fly out of the plume and never arrive at its source.

Observations of the zig-zagging of moths flying toward pheromone (Figure 3.10) were the first indication that an additional mechanism existed. In a filamentous plume created in a wind tunnel, moths do not fly straight into the wind at all times but instead regularly break into a transverse casting flight. These maneuvers are considered to be a means for regaining the plume should it be lost: the male breaks into horizontal casting back and forth, with some vertical excursions, that soon returns him via trial-and-error to within the plume, at which point he resumes upwind flight. Close to the source of the plume, casting tracks are made at a greater angle (nearly perpendicular) to the upwind direction, reversals (counterturns) of the tracks are more frequent, and the tracks are

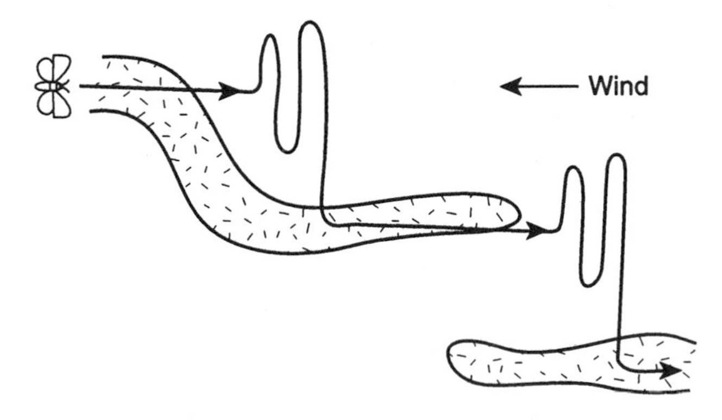

Figure 3.10 Flight path of an insect orienting upwind within a pheromone plume (overhead view). The insect surges directly upwind when exposed to pheromone but casts transversely if it leaves a filament of the pheromone plume. Stippled areas represent pheromone filaments.

therefore shorter (Baker 1986, Vickers and Baker 1997). What features of the plume might effect this change in flight pattern? The excitation of low-sensitivity neurons may indicate a higher pheromone concentration should it be present, but several workers have suggested that concentrations within filaments are not necessarily higher at the source end of the plume than at the downwind end. In addition, differences in plume structure that would be detectable by a receiver are not obvious (Todd and Baker 1999). Nonetheless, the shorter and more transverse casting tracks may prevent the male from flying out of the upwind end of the plume and also help him to pinpoint the female's location.

Why do male receivers undertake the elaborate exercise of casting flight when they are equipped with paired receptor organs? After all, many animals can track and orient toward terrestrial odors (Section 3.7.1), and sound sources (Section 4.3.5), via a simultaneous comparison mechanism in which they monitor relative intensities perceived by each receptor and simply turn toward the greater stimulus. A bilateral comparison process, however, is unlikely to work for a small organism flying within an aerial odor plume. Here, paired receptors per se may be of little value because (1) concentration gradients at filament edges are too shallow for the animal to detect by simultaneously comparing left and right antennal inputs, (2) edges are normally intersected rapidly and at nearly right angles, or (3) the brain cannot integrate inputs from the chemoreceptors quickly enough to correct the flight path.[19] We cannot evaluate the relative importance of these possibilities in the absence of information on the specific chemical inputs received by the antennae at the precise moment that maneuvers are being made in flight. Recent efforts to develop and apply techniques for in-flight electroantennographic (EAG) recording of male moths in wind tunnel pheromone plumes (see Vickers and Baker 1994, Baker and Vickers 1997, Vickers et al. 2001)[20] promise to reveal the nature of this chemical input and why it may be insufficient for more direct tracking mechanisms.

Parallel studies on the orientation of benthic crustaceans within aqueous odor plumes (Atema 1995, 1996, 1998, Zimmer-Faust et al. 1995) and of beetles, cockroaches, and silkworm moths walking within terrestrial ones (Bell et al. 1995) offer some insight into to the special problems encountered in aerial plumes. Lobsters, which may inhabit fairly turbulent waters, apparently do not make direct use of current or an optomotor mechanism in order to localize the source of an odor plume from a food item or a sexual advertisement. Rather, they sequentially evaluate the concentration profiles of individual filaments and then head in the direction in which profile slope, measured from the onset of a filament to its peak intensity, increases maximally between successive filaments. In vivo neural recordings of orienting lobsters indicate that they process the information received from individual filaments with sufficient rapidity (see Gomez and Atema 1996, Gomez et al. 1999) to make the appropriate course corrections, leading them reliably to the plume's source. Crabs orienting in odor plumes in calmer water do make direct use of current as well as chemical cues (Weissburg and Zimmer-Faust 1994), but they are able to remain within the plume without resorting to casting. Evidently, departures from the plume are avoided because the ratio of the organism's antennule separation to filament dimensions is high, plume structure in some aqueous environments is more stable and regular than in air, and orientation movements can be relatively slow.

Insects walking within terrestrial pheromone plumes may similarly reach the source without casting, and some may do so by sequential evaluation of concentration (Bell et al. 1995). Here, a simple orientation algorithm would suffice: reverse or change direction if a concentration decrease or constancy over time is perceived, but otherwise maintain the same course. Further experiments with the American cockroach (*Periplaneta americana*; Blattidae) demonstrated that they rely on anemotaxis and simultaneous comparison of inputs to their left and right antennae to remain within a plume if the requisite information (wind and a steep lateral concentration gradient) is available. In the absence of these cues, however, sequential comparison of input, to even one antenna, serves as a backup mechanism (Bell and Tobin 1981).

Moths flying in an aerial plume may have none of the advantages or options available to walking arthropods, and casting may be their only viable means of regaining an odor once lost. Nonetheless, moths might retain alternative orientation mechanisms in reserve and rely on these mechanisms under certain conditions. For example, some moth species can apparently locate a pheromone source in nearly still air (Mankin and Hagstrum 1995), albeit over a limited distance, and it is tempting to propose that they too process and compare antennal inputs in a sequential fashion to accomplish this orientation (cf. Section 4.3.5). Various planktonic crustaceans may confront analogous conditions in freshwater and marine environments, and studies of pair-formation in copepods suggest that males in some species do locate females emitting advertisement chemicals by the sequential comparison of olfactory inputs (Yen et al. 1998).[21]

Endogenous Counterturning. Further analyses in wind tunnels revealed that the casting flight performed by moth receivers is not simply a cross-wind movement made in response to the loss of an odor plume. Rather, several behavioral observations can only be reconciled with the existence of other controlling factors. First, casting may continue inside a homogeneous pheromone cloud filling a tunnel (Kuenen and Baker 1983). Second, males that begin casting after losing a pheromone plume may continue to cast after the wind is eliminated (David and Kennedy 1987).[22] Moreover, motor activity indicative of casting has been witnessed in tethered animals that had no access to visual feedback, pheromone, or wind (Baker et al. 1984, Arbas 1997). Taken together, these findings indicate that counterturning is an internal, self-directed program of movement whose only spatial reference may be the animal's last direction prior to casting. This internal program, however, would be modified by external stimuli such as wind.

Neurophysiological recordings suggest that the casting rhythm, approximately 2 counterturns$\cdot$s^{-1} in some species, may be controlled by certain identified neurons, aptly named "flip-flop" interneurons, descending from the brain to the prothoracic ganglion (Olberg 1983, Kanzaki and Mishima 1996). The flip-flop interneurons comprise a pair of left and right connectives, each of whose activity is influenced by the onset of a pheromone pulse and by the other connective. Thus, the activity of each interneuron occurs out-of-phase from the other.

Armed with the understanding that counterturning is a rhythmic, internal program and that odor plumes are composed of meandering filaments, several laboratories advanced the above studies a critical step by testing the responses of male receivers to individual filaments of pheromone (Mafra-Neto and Cardé 1994, 1995, 1996, Vickers and Baker 1994, Cardé and Mafra-Neto 1997). In these experiments, an encounter with a single filament was represented by a brief (0.02–0.25 s) pulse of pheromone puffed past a male receiver flying in a wind tunnel. The intermittent nature of a plume, multiple filaments separated by clean air, that a male would experience while flying through it was simulated by pulsing the pheromone periodically. Analyses of videotapes of the flight tracks showed that after a short latency, a motor delay $\cong$ 0.3 s, males surged directly ahead for a fraction of a second upon encountering a pheromone pulse and then began casting (Figure 3.10). If successive pulses were encountered with sufficient rapidity, the male flew in a relatively straight track, because he began to surge again before his casting program was triggered. At slower pulse rates males showed a higher incidence of casting, and in a homogeneous cloud of pheromone they casted continuously. From these observations, the following phasic–tonic model of orientation is inferred: initiate a straight, upwind surge (phasic response) upon detecting a sudden increase in pheromone concentration and then lapse into an indefinite period of casting (tonic response) unless another sudden increase in concentration is detected. Given the filamentous structure of odor plumes, this basic sensorimotor process, with timing parameters modified among species due to constraints imposed by their different flight abilities and habitats, should be an efficient means of orienting toward distant odor sources. Here, the phasic nature of the response to pheromone onset would be particularly important because it lets a receiver advance quickly toward the source upon contacting a plume filament.

The accuracy with which the phasic–tonic model describes what live receivers actually do is currently being explored via computer simulation (Belanger and Willis 1996, Belanger and Arbas 1998) and "robots" (Ishida et al. 1995, Kanzaki 1996; cf. Consi et al. 1995). By distilling surging and casting into a set of algorithms controlling motor activities that respond to various sensory inputs and internal states, modelers can compare the tracks of simulated male receivers with those of live ones. Similarly, the tracks of mechanical devices that detect ions or odorants in a wind tunnel plume and then orient, in accordance with the phasic–tonic model, can also be compared with the tracks of live males. Both approaches have been invaluable in forcing biologists to consider previously overlooked yet crucial elements of the orientation and localization process.

Localization versus Recognition: Processing Pheromone Components and Antagonists. Our examination of orientation and signal localization has thus far viewed the chemical signal as an indivisible entity. But, advertisement pheromones in arthropods are generally blends of several related compounds, and specialized inhibitory responses to pheromone antagonists are also common (Table 3.1). How do receivers perceive and respond to the various pheromone antagonists and components while en route to signalers? Are the antagonists and components processed before or during localization, and is perception of all components necessary throughout localization?

Receivers may discriminate antagonists from pheromone components with the utmost spatial precision. That is, male receivers appear not to be inhibited from tracking the conspecific signal unless the sources of that signal and of an antagonist are emitted from virtually the same point in space (Baker et al. 1998, Fadamiro et al. 1999). Probably, the filaments of spatially separated chemical signals do not mix, and the time intervals over which pheromone and antagonists are evaluated neurally are very brief. Thus, a male receiver would continue to track conspecific females in the presence of other species whose signals include both the conspecific pheromone and the antagonist. There are no reported findings suggesting that a species' geographic range or temporal activity pattern are restricted by heterospecific signals that include both the pheromone components and antagonists of the focal species.

Individual pheromone components may be processed in a variety of ways. Some behavioral studies indicate that the entire blend is evaluated as one entity and that all components must be perceived from the onset of orientation within the plume until the signaler is contacted (Linn and Roelofs 1983, Cardé and Charlton 1984). In other words, the various individual components are not responsible for eliciting different behavioral steps en route. This view, however, is at odds with neurophysiological data from certain species. The pheromone of the tobacco budworm moth (*Heliothis virescens*; Noctuidae) includes a major and minor component in a 16 : 1 ratio, and the sensory (and antennal lobe) neurons for the major component are both much more sensitive and more numerous than those for the minor component (Mustaparta 1997). Consequently, a male receiver would detect only the major component at the onset of its orientation and immediately upon entering a filament as well. Could this differential neural response accentuate the rapidity of the phasic response—an upwind surge following the motor delay—when entering a filament? If yes, at what specific stages does the minor component(s) then play a role in orientation?[23] Would accelerated phasic responses be more likely if only one component is evaluated, "computational time" in the moth brain being thereby reduced, and conversely, would this likelihood explain why the component ratios within pheromones are normally highly biased?

Neurophysiological data from the tobacco hornworm moth (*Manduca sexta*; Sphingidae) provide yet another view of odor processing during signal localization. Here, the major pheromone component elicits an excitatory response (EPSP) in certain projection neurons (PNs) ascending from the MGC, while the minor component elicits an inhibitory response (IPSP) that appears after a very brief latency in those same PNs and continues after the stimulus has ended (Christensen and Hildebrand 1987). When both components are present simultaneously, as in pheromone emitted by a female, these PNs are initially inhibited for approximately 100 ms and then excited. Thus, an effect analogous to the forward masking known in auditory processing and precedence effects (Section 4.4.2) accentuates the neural representation of a pheromone pulse's onset, its leading edge, and the moth may thereby track modulation of its pheromone at frequencies as high as 5–10 pulses per second (Figure 3.11).

Altogether, these behavioral and neurophysiological observations reveal that a diversity of schemes organize the processes of pheromone recognition and localization. Such diversity could have a phylogenetic basis, but it may also reflect the various habitats in which different species typically occur as well as their flight speeds and maneuverabilities (cf. Section 4.3.5). For example, rapidly

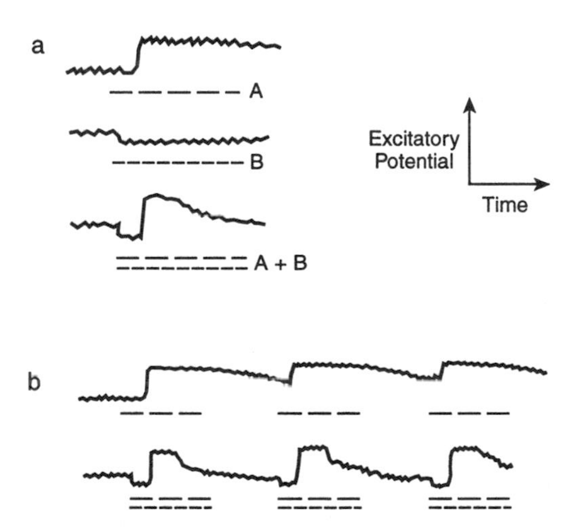

Figure 3.11 Parallel processing of pheromone components. (a) Pheromone components A and B elicit excitatory and inhibitory potentials, respectively, in the same projection neuron ascending from the macroglomerular complex (MGC) in the brain. When presented simultaneously, A and B elicit an initial inhibitory response followed by a brief excitatory one. Long and short horizontal dashes indicate time intervals when A and B, respectively, are present. (b) Sensory adaptation reduces the relative excitatory potential in response to pulses of A. The relative excitatory potential is restored in responses to pulses of A+B (see Christensen and Hildebrand 1987).

flying species might be neurally incapable of processing all pheromone components each time they encounter a plume filament, which could favor processing schemes that focus attention on the major component.

Final Approach and Contact. While appetitive flight and the phasic–tonic process of surge casting should allow a male receiver to find a plume and orient within it, are these mechanisms sufficient for approaching and making final contact with the signaler? Here, supplementary visual stimuli from the female might guide the male in some species and under certain circumstances. Nonetheless, such information is probably unavailable to most nocturnal species, and other mechanisms would have to be used over the final centimeters. These mechanisms could include evaluation of pheromone concentration, but the detection of subtle changes in plume composition and structure that occur close to the signaler are more likely candidates (cf. Section 4.3.5, Ranging). Because most sex pheromones are blends of several components, blend ratios might change predictably over time or distance from the receiver if the components differ markedly in their diffusion and behavior in turbulent flow.[24] However, specialized mechanisms of pheromone release may maintain blend ratio constancy in some species,[25] and in many others the various pheromone components are chemically very similar and would not differ in their dispersive properties. On the other hand, the spacing of neighboring filaments and the dimensions of individual ones may be reliable indicators to a receiver that he is located near the plume's source, and these features could be assessed via feedback from the tempo of surge casting. Precise measurements of plume fine structure close to the source and of receiver orientation within the region may help to resolve this problem.

Receiver Extravagance. The inescapable conclusion from the preceding sections is that the searching, perception, and orientation performed by moths flying within aerial odor plumes are highly exaggerated features (Section 6.2). What makes them all the more remarkable is that they are receiver features. There exist few cases in other animal signaling systems where the receivers even approach this level of extravagance. Without venturing into the minefield of optimality, it is probably safe to claim that moth receivers have evolved an extremely efficient means of exploiting the chemical channel given the constraints of their body size and the distances and medium across which communication and pair-formation must occur. Receiver perfection and sophistication notwithstanding, chemical advertisement signaling by moths appears to be a rather lackluster affair, so low key that one eminent scientist[26] was even prompted to doubt whether the odors are worthy of being designated signals (Williams 1992). Signal flux rates are minimal, the compounds used are relatively simple and are not synthesized via elaborate biochemical pathways,[27] and the associated morphologies and behaviors are modest. This odd juxtaposition will serve as a focal point in considering the origin of their chemical signaling system.

3.1.5 Evolutionary Origins and Adaptations: Chemical Leaks and General Olfactory Sensitivity

Many animal signals have descended from activities other than the communication they currently effect (Alcock 1998, Bradbury and Vehrencamp 1998; Chapter 6), and we begin our evolutionary inquiry by considering whether chemical advertisement signals may represent modifications of substances and behaviors that originally served defensive, competitive, or other functions. For examples, the sex pheromones of female Lepidoptera are conceivably derived from compounds originally used in thwarting natural enemies, in aggressive encounters with conspecific competitors, or in subduing prey. Here, the evolutionary shift may have occurred because the quantity and quality of the chemical reveals to the receiver the signaler's prowess in these endeavors. Chemical advertisement signals might also be derived directly from dietary intake, in which case the substances could have originally indicated the quality of food resources at the signaler's location. Over evolutionary time, the substances may have acquired their current advertisement status because they reveal the quality, in terms of foraging ability or nutrient storage, of the signaler itself. However attractive these hypotheses may be as explanations for various chemical signals used in courtship and social interactions (Sections 3.2 and 3.3), though, none are plausible for female sexual advertisements and particularly not those of Lepidoptera. Because defense and predation are likely to be expressed in both sexes, and conspecific aggression in males, at least some vestiges of the signals and their associated structures would be expected in males. But, not even a hint of these signals or their exocrine glands is found in male Lepidoptera. Additionally, neither the aliphatic compounds serving as female sexual advertisements nor their biochemical precursors bear suspected noxious or antibiotic properties. To be sure, the female advertisement pheromones are ultimately derived from larval dietary intake, but minor alterations from one suitable diet to another normally have little or no influence on pheromone quality and emission (see McNeil and Delisle 1989).[28]

In light of the above information, have lepidopteran advertisement pheromones and the behaviors associated with their emission evolved specifically within the context of pair-formation and mating? A phylogenetic approach suggests yes, insofar as their history within the Lepidoptera is concerned. Long-range female advertisement pheromones are not only widespread among moth species, but they very likely represent the ancestral (plesiomorphic) character state for the Lepidoptera and its sister order, Trichoptera (caddisflies) (Wagner and Rosovsky 1991, Löfstedt and Kozlov 1997; Figure 3.12). Moreover, several signal characters on which we have repeatedly dwelled, 10–21 carbon aliphatic compounds that are emitted from glands at the end of the abdomen, are ancestral for the ditrysian, or "higher," Lepidoptera, that branch of the order which accounts for 87 of 107 families and ≈99% of all extant species. All evidence suggests that the communication systems in those Lepidoptera which do not follow the pair-forming scheme of female advertisement pheromones and male searching, such as butterflies[29] (Sections 3.2 and 5.2) and certain acoustic moths (Section 4.5), are secondarily derived (apomorphic). Hence, the common ancestor of modern Lepidoptera and Trichoptera most probably used long-range female advertisement pheromones.

Finding that female advertisement pheromones are an ancestral trait tells us much about their radiation and diversification within the Lepidoptera, but the initial evolution of advertisement pheromones is another matter. To approach the more difficult question of pheromonal origin, we might examine general mechanisms of olfactory orientation and perception in Lepidoptera and other insects. In moths, females exhibit the same basic orientation mechanisms for locating host

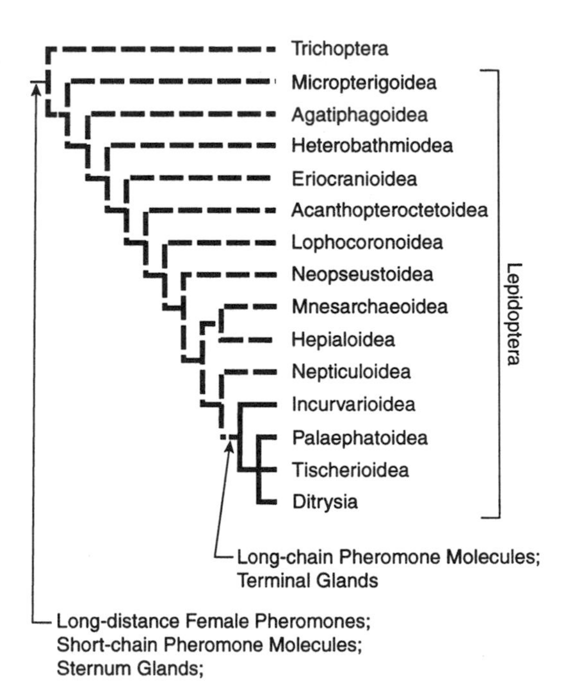

Figure 3.12 Phylogeny of the Lepidoptera + Trichoptera, showing distribution of long-distance female advertisement pheromones (Lepidoptera + Trichoptera), pheromone glands on terminal abdominal segments, and long-chain (≥10 carbons) pheromone molecules (Ditrysia + three related superfamilies). Ditrysia include the vast majority of extant species, and families (87 of 107), in the Lepidoptera (redrawn from Löfstedt and Kozlov 1997).

plants on which to oviposit as males exhibit when locating pheromone-emitting females (e.g., Haynes and Baker 1989, Willis and Arbas 1991, Hansson 1995). This point was elegantly demonstrated in an experiment in which imaginal disks for male antennae were implanted into female *Manduca sexta* larvae: once adult, the females tracked female pheromone as if they were males (Schneiderman et al. 1986). Evidently, all of the motor commands and apparatus for orientation were present and merely needed to be wired to antennal ORNs whose binding sites were specialized for pheromone rather than for volatile emissions of host plants. In some moth species, males are also attracted by host plant volatiles and will respond more readily toward female pheromone when those volatiles are present (Dickens 1997, Landolt and Phillips 1997). These observations suggest that the general sensory and orientation mechanisms for localizing habitat, food, and oviposition sites also underlie the means by which males localize mates.

A review of other insects indicates that olfactory localization mechanisms like those found in Lepidoptera are widespread, particularly among phytophagous species (Miller and Strickler 1984) and certain other guilds such as parasitoids (Godfray 1994). The general reliance on olfaction for resource localization may reflect both the shortcomings of insect vision and the olfactory cues that living organisms invariably leak to the environment. A comparison of insect with human perception illustrates the first point: supplied with a field guide and some botanical training, many human observers are perfectly capable of making plant identifications visually, but this ability reflects our high visual resolution and still normally depends on close inspection of the plant. Operating under the handicap of vastly inferior visual resolution (Chapter 5), the majority of insects would not be expected to rely solely on the visual channel for identifying host plants. Additionally, time and energy constraints may deter them from using a long-range channel that would also require close inspection of host plants and concomitant trial-and-error searching. Conversely, the chemical channel could work rather efficiently. Inadvertently or not, plants emit volatile chemicals, many of which are quite specific and indicative of physiological state, and they may do so during both day and night. And, as elaborated above, insects are certainly capable of evolving the perceptual apparatus and motor programs necessary for detecting and localizing faint odors at a relatively great distance.

Because a non-sexual context in which female advertisement pheromones might have originated cannot be readily identified and olfactory perception and orientation mechanisms are so widespread, a parsimonious explanation for these signals is that they evolved from simple inadvertent cues. Males, like females, would have been fitted with efficient olfactory mechanisms for localizing resources, and it is not difficult to imagine that they also evolved reliance on these acute sensitivities for finding mates, who must have necessarily emitted some odors if only unintentionally: all that males would have needed to acquire were olfactory receptor neurons (ORNs), binding sites, and binding proteins (PBPs) with high affinities for female scent. Thus, female advertisement pheromones in the Lepidoptera and Trichoptera, and in various groups of Blattodea, Mantodea, Hemiptera, Homoptera, Coleoptera, Diptera, and Hymenoptera with similar pair-forming systems, may have originated by coevolving with male olfactory abilities (see Section 6.1.2). In the following section, we analyze the specific modifications by which female sex pheromone signals may have become differentiated from ancestral cues and the ways in which male perception and localization behaviors may have selected for these modifications.[30]

Advertisement Pheromones and Female Choice: Minimal Energy and Passive Filtering. Inferring how the scents used by females in sexual advertisement acquired their specific properties is best done by applying sexual selection theory (Thornhill and Alcock 1983, Andersson 1994) to pair-formation. Because females confer more parental investment per offspring than males, they are predicted to be more discriminating in their choice of mates. In lieu of mate discrimination, males should behave in a manner that maximizes the number of females with which they mate. Thus, in a population at a 50 : 50 sex ratio, the mean numbers of matings of females and males must be identical, but the variance in the male number is expected to be much higher: Almost all females mate one to several times, whereas some males mate many times and others not at all. In a pair-forming system wherein the female is the advertising sex and the male responds to and locates the female advertisement, the male searching protocol and its associated apparatus assume extravagant proportions and become the instrument with which males compete to mate with as many females as possible. Searching assumes the function of the stag's antlers, the peacock's tail, and the male cricket's chirp, and those males most sensitive to female advertisements and best at localizing them quickly will encounter the most mates.

Female advertisements are subject to rather different selection pressures. To enjoy any significant level of fitness, females in a non-social species must mate at least once before they die! Additionally, females who attract a greater number of males may have an expanded opportunity to mate with a higher quality one(s), provided that they later screen the quality of attracted males in a close-range courtship activity of some sort. These pressures would demand that sexually receptive females do slightly more than inadvertently emit odors that they cannot prevent from leaking into the environment. Thus, the raised abdominal postures and rhythmic pumping of females during pheromone emission and the exocrine glands[31] that synthesize and emit pheromone may have been selected. Such features would enhance signal intensity, thereby increasing the likelihood that mating occurs within the limited time available and that a sufficient cadre of males are encountered to ensure that at least one is likely to be of high quality. In this regard, the female odors may not be exaggerated displays, but they are true signals (cf. Williams 1992).[32]

Opposing these pressures, other factors would select for a minimum level of extravagance in female advertisements. Because the stakes of winning at sexual competition are much lower for females than for males, females are expected to incur little risk in their pair-forming activities. Therefore, female sexual advertisements should be modest lest the advertiser expend energy needed in another arena or attract predators, parasitoids, or parasites. Several features of female moth pheromones fit this prediction. Signal flux (emission rate) is extremely low, the chemicals merely evaporate into the surrounding medium with a minimum of physical activity by the female, and they are derived from the byproducts of secondary metabolism. Acting in concert, a passive mechanism of female choice may enhance the selection for a minimum level of signal extravagance. In pair-forming systems with male searchers and female signalers, a male's sensitivity to the female signal and his ability to locate it quickly may represent his quality in a Fisherian (arbitrary) or good-genes (viability indicator) sense (Section 6.2). A female might improve her chance of mating with a high-quality male, then, by producing a barely detectable signal: thus, she would filter out many of the males with low sensitivity and poor localization ability. Evidence consistent with this hypothesis comes from various moth species in which a female's fecundity is maximum if she begins mating immediately after eclosion (Greenfield 1981) yet her pheromone flux or length of daily activity do not peak until several days later.[7] This incongruence suggests that the female initially exhibits passive choice or filtering of males but abandons the practice after several days, if she has not yet mated.

The intensity of the female advertisement signal should reflect the combined strengths of the various counteracting selection pressures described above. Unfortunately, this expectation is most difficult to evaluate. Presently, there exists no information on the genetics of male sensitivity to pheromone concentration and searching ability and whether females would actually mate with superior receivers and searchers by virtue of a minimally extravagant signal. Similarly, we lack information on the energetics of these female chemical advertisements and whether a higher flux rate might be compensated by reduced egg development and fecundity.

Has sexual selection influenced other characters of female advertisement pheromones beside their intensities? As reviewed earlier, the compounds serving as female sex pheromones probably do not indicate any individual attributes of the signaler, and it is not apparent that males, who in most moths contribute little or no parental investment (e.g., Greenfield 1982),[33] would be concerned about such attributes anyway. These compounds may simply be used because they issue from secondary metabolism and are therefore very, very cheap. But, the biosynthesis of pheromone components normally entails specific steps of chain-shortening, desaturation, and conversion of terminal moieties to acetates, alcohols, or aldehydes (Roelofs and Bjostad 1984). The pheromone compounds resulting from these steps may possess molecular weights and other properties that improve their dispersion in atmospheric plumes or that are more readily detected by receivers selected originally to respond to volatile emissions of host plants. Nonetheless, if passive choice does operate and impart female sexual advertisements with minimum extravagance, selection for compounds bearing ideal chemical properties would be tempered.

Selection for Dedicated Channels? At another level of characterization, female sex pheromones, forming part of a species' mate recognition system (Chapter 7), are generally chemically specific for the species and in some cases, where "pheromone dialects" occur, for the population as well (Löfstedt et al. 1986, Löfstedt 1993). The few instances in which sympatric species share the same pheromone and would be susceptible to cross-attraction involve species active at different times of the day or season (e.g., Haynes and Birch 1986). Chance events in biochemical evolution during speciation may have contributed to some of this pheromonal distinctiveness among species, but various sources of evidence suggest that sexual selection sensu lato has played a major role (see Ryan and Rand 1993). Females who produce an advertisement identical to that of all other females in the population and different from that of all other species may attract only conspecific males and thereby encounter an appropriate mate more quickly: specificity in the male response would be expected to coevolve with the female advertisement (Section 6.2), and males would spend little time distracted by heterospecific females. While no firm evidence exists that pheromonal specificity in the signaler or receiver originated via reinforcement of pre-mating isolation at the initiation of speciation events (see Chapter 7),[34] in many cases specificity may have been selected for afterward as outlined above. For example, among sesiid moths the pairwise divergence of species' pheromones is inversely related to their divergence in daily or seasonal activity periods (Greenfield and Karandinos 1979; see also Phelan 1992, Löfstedt 1993, Butlin and Trickett 1997). The general multi-component nature of pheromones may have arisen as part of the selection for specificity (see also Section 7.3), but, alternatively, it may merely represent the chemical form in which the end-products of the pheromone biosynthetic pathways are produced.

On the receiver side, inhibitory responses to antagonists and redundancy in multi-component pheromones also suggest that dedicated channels lacking heterospecific interference have been selected in many species (see Gwynne and Morris 1986 for an example of antagonistic signals in acoustic communication). Moreover, in some sesiid moths the specialized inhibitory responses to antagonist compounds are found during seasons when other species whose pheromones include the antagonists are active but not when they are absent (Greenfield and Karandinos 1979; see Table 3.1).[35] Cases in which male responses to the female pheromone require only detecting the major component plus any one of several minor components suggest that male receivers may use the minimum amount of information necessary to identify a conspecific female reliably. The male's flexibility and minimum requirements may ensure a rapid response to the pheromone and more efficient orientation within the plume (Figure 3.10).

Predilection for the Olfactory Channel: Economy and Enemy-free Space. Pair-forming systems in which males search and females advertise are known in a variety of animals other than Lepidoptera. Among arthropods, notable examples include certain groups of Crustacea (Copepoda; Yen et al. 1998), Arachnida (Araneae; Miyashita and Hayashi 1996), Blattodea (Schal 1982), Mantodea (Robinson and Robinson 1979), Hemiptera (Miridae; Aldrich 1995), Homoptera (Aphidoidea; Pickett et al. 1992, 1997), Coleoptera (Scarabaeidae; Leal 1998), and Hymenoptera (Symphyta; Anderbrant 1993). In nearly all cases, the long-range female advertisement signal is a chemical one.[36] While this reliance on the chemical channel by female signalers may have originated in inadvertent female cues coevolving with male sensory (olfactory) abilities (Section 6.1.2), subsequent evolution of olfactory communication per se could have been driven by specific advantages conferred to female signalers, who are expected to value conservation of energy and avoidance of risk. That is, advantages intrinsic to chemical signaling may have maintained and bolstered female reliance on this channel.

Unlike acoustic or visual displays, chemical signals may require only minimum energy during development and actual emission, particularly if derived from the byproducts of secondary metabolism and merely evaporated to the surrounding medium (but, see Section 4.2.9). Chemical signals may also be comparatively immune to the danger of attracting illegitimate eavesdroppers, natural enemies who would locate the signaler by means of her advertisement (see review in Zuk and Kolluru 1998). Among Lepidoptera, there are no known cases of sex pheromones acting as kairomones and attracting predators, but several exist among Coleoptera (e.g., Vité and Williamson 1970). Predators, who are seldom specialists, usually need to recognize the cues from a variety of prey, and this broad recognition may be inherently difficult to accomplish for chemical cues: Can a receptor be engineered that responds to a wide range of 10–21 carbon acetates, alcohols, and aldehydes and yet is not distracted by chemically related compounds emitted from vegetation and other environmental sources? Significantly, the Coleoptera (bark beetles; Curculionidae: Scolytinae) that do attract predators to their sex pheromones are typically found in dense aggregations, and these predators (checkered beetles; Cleridae) are specialists. Probably, the dense aggregations of prey make specialized predation worthwhile. Parasitoids are typically specialists, however, and the general paucity of sex pheromones acting as kairomones for them[37] is puzzling at first. But parasitoids, being generally unrelated to their hosts, may be unlikely to benefit from an ancestral perceptual ability (Section 6.1) with which host signals might be detected. Additionally, parasitoids must locate hosts that will survive long enough for their offspring to complete development, and many mature insects, and nearly all mature Lepidoptera, probably will not. In the few known situations wherein parasitoids are attracted to sex pheromones of adult female Lepidoptera, they localize nearby vegetation to which pheromone adhered and do so in order to search for eggs deposited by the signaler (Noldus 1988, Noldus et al. 1991a,b). The parasitoids then oviposit in the host eggs. However, in many host species signaling and oviposition may be separated in time and space or too little pheromone adheres to or otherwise remains near local vegetation to make this strategy profitable for an egg parasitoid.

The relative safety in which Lepidoptera use the chemical channel for sexual advertisement is perhaps best appreciated by noting that only one case is known in which adults themselves are attacked specifically because of their communicative activity. In what may be the most spectacular example of aggressive mimicry among animals, female bolas spiders (*Mastophora*; Araneae: Araneidae) attract males of various moth species with a lure that includes compounds identical to major components of the female sex pheromones of those moths (Figure 3.13; Eberhard 1977, Stowe et al. 1987, Yeargan 1994, Gemeno et al. 2000). The male receivers are deceived[38] (and eaten), whereas the female signalers are not directly involved and remain untouched. This extreme case calls attention to the generally overlooked risks that may be experienced by searchers in pair-forming systems (Section 4.4.1).

3.1.6 Territorial Male Advertisements: Tending the Chemical Trapline

Sex normally determines roles in a pair-forming system, and it is far more common for males to advertise and for females to search and orient toward males. Reversals of this scheme, as in moths, could arise when ecological circumstances render searching the more costly role. One such circumstance might be a wide dispersal of the sites at which males and females encounter each other, as would occur for species associated with widely dispersed host plants.[39] In the more typical scheme of male signaling and female searching, male advertisements are generally mechanical (sound or vibration; Section 4.2) or visual. When males do use long-range pheromones, the scents are usually supplementary to signaling along another channel[40] or emitted at a focal resource to which females must come.[41] Exceptions to this generalization occur among Hemiptera,[42] Coleoptera (Cerambycidae), and Hymenoptera (Anthophoridae, Apidae, Sphecidae). The last mentioned demonstrate how chemical advertisement signals can function in territorial mating systems.

In various species of bumblebees (*Bombus*; Apidae) (Bergman and Bergstrom 1997) and solitary hunting wasps (bee wolves; *Philanthus*; Sphecidae) (Evans and O'Neill 1988), males regularly mark vegetation with scents secreted from labial and mandibular glands, respectively (Figure 3.14). The scents are deposited daily at various sites within or surrounding a defended territory, which is normally established in the vicinity of female nest locations. Bumblebees may apply the scents several times in the morning and then repeatedly fly along a circuit connecting the scented locations. This behavior gave rise to the notion that males scent-mark because it guides them along their own territorial circuit before they have fully memorized its visual landmarks, a form of autocommunication. Later findings, however, showed that the scents, fatty acid derivatives as well as sesqui- and di-terpenes, attract females and that males attempt to mate with females encountered only at their scent marks. The quantity of the scent compounds within the glands may be high (40 µg in one species), and our ability to extract these compounds from the head space surrounding marked vegetation indicates that signal flux and intensity may be relatively high as well. A male's daily renewal of scent compounds within his glands is unknown but can be expected to influence the frequency with which he reapplies scent throughout the day and on successive days. Do males select surfaces bearing particular absorptive qualities on which to apply their scent? Non-absorptive surfaces would offer a high signal flux over a short time interval, but this lack of porosity could be disadvantageous if the male cannot reapply scent frequently. Male scents, either fatty acid derivatives or isoprenoids, secreted from thoracic glands, that serve as long-range attractants for females have also been found in carpenter bees (*Xylocopa varipuncta*; Anthophoridae) (Minckley et al. 1991). Unlike female sex pheromones, the quantity and quality of long-range male scents may be expected to indicate reliably to the (female) receiver critical features of the signaler (Sections 3.2 and 6.2).

What compels these social and solitary Hymenoptera to break the pattern of long-range mechanical or visual signals by advertising males? A major distinction of chemical signals is that they remain in the environment long after the onset of their transmission. This lingering may be a liability in some cases, but it may be advantageous for territorial males in sparse populations: by establishing transmitting signals—continuous beacons—at various sites spread over a wide area and then revisiting the sites on a regular basis to check for attracted females, a male may increase his encounter rate with potential mates.[43] Thus, the male tends a trapline, which is likely more efficient than advertising at a single site, particularly if females do not forage widely in their habitat. In higher population density, though, the rewards of traplining should diminish, because more females would be encountered at select locations and a male's scent marks would be more vulnerable to cheaters, non-marking individuals who merely wait for females to arrive at the marks. These potential influences of population density and dispersion on signaling systems await further study.

3.1.7 Prospects for Control?

A primary impetus for research on insect pheromones, particularly sexual advertisement pheromones, has always been the development of environmentally sound pest management. Because advertisement pheromones are typically

a

b

c

d

e

Figure 3.13 Aggressive mimicry by the bolas spider *Mastophora bisaccata* (Araneidae). (a) Female *M. bisaccata* luring male moths with an odor mimicking the female moth pheromone. The odor is emitted from glands on the spider's body and is only attractive to certain species of moths. Arrow indicates sticky bolas used to ensnare prey moths. (b) Lured male moth (*Acrolophus plumifrontinellus*; Tineidae) approaching female *M. bisaccata*, (c) ensnared by the bolas, (d) reeled in, and (e) wrapped while a second moth approaches the spider, who remains attractive (photographs courtesy of Mark K. Stowe; photographs taken in Gainesville, Florida, July 1981).

Figure 3.14 Territorial scent-marking by the male bee wolf *Philanthus bicinctus* (Sphecidae). (Redrawn from illustration by Byron A. Alexander. Reprinted from Howard Ensign Evans and Kevin M. O'Neill, *The Natural History and Behavior of North American Beewolves*. Copyright © 1988 by Cornell University. Used by permission of the publisher, Cornell University Press.)

species-specific and have no known adverse effects on non-target organisms, they would appear most desirable for such management programs. This goal has certainly influenced the preponderance of sex pheromone identifications among the Lepidoptera: Moth larvae comprise a significant proportion of our major agricultural pests.

Given over 30 years in which to develop pheromone-based pest-management practices, what strategies are now used and how successful are they? Four basic pest-management strategies have been considered and attempted: (1) simple trapping of the (male) population using synthetic sex pheromones as baits; (2) trapping with baits as above, followed by killing the baited individuals with another, usually chemical, means; (3) using sex pheromone baits to monitor the presence and abundance of a pest population; (4) dispersion of synthetic sex pheromone evenly over an agricultural area to interfere with pair-formation and mating in a pest population (Wyatt 1997). Of these strategies, only (3) and (4) have met with any measure of economic success. The first two strategies fail because of the impossibility of trapping all, or even nearly all, males in a population: remaining males simply enjoy the expanded mating opportunities. The two latter strategies are hindered by commercial resistance to adopt practices that do not immediately yield dead bugs and by deficiencies in our knowledge of the ecology of most pest species. For example, we generally cannot accurately extrapolate a given sample of males baited in a monitoring station to the pest's population size, let alone age structure, in the field. Here, human aesthetic sensibilities and academic pressures may be to blame. Behavioral ecologists and evolutionary biologists are drawn to the exotic and extravagant, and the dispersal, mating behavior, and other features of most species accorded pest status may simply be unappealing for study.

Dispersion of synthetic sex pheromone as interference with pair-formation and mating has been moderately successful in suppressing populations of lepidopteran pests in several agroecosystems. The interference may result from various mechanisms: (1) males are distracted to follow false pheromone trails emanating from the devices releasing synthetic pheromone; (2) the dispersed synthetic pheromone forms a cloud that masks the pheromone plumes of local females; (3) continuous exposure to synthetic pheromone leads to sensory adaptation or central habituation in males. This technique is currently limited by technical difficulties in maintaining a continuous release of synthetic pheromone over the entire area where pair-formations occur. Additionally, some concern exists over the potential for pest resistance. Target species might evolve to form their sexual liaisons in areas outside the agroecosystem where synthetic pheromone is dispersed[44] or to shift the component ratios in their pheromones.[45] Presumably, practitioners of pest management could stay a step ahead of these evolutionary events, particularly if they anticipate them. Findings of (unintentional) selection for novel sex pheromone component ratios and a coevolved shift in the (male) receiver response in a laboratory population of cabbage looper moths (*Trichoplusia ni*; Noctuidae) (Haynes and Hunt 1990) indicate that these concerns are justified.

3.2 Courtship: Volatile Aromas as Reliable Indicators

When mating signals are produced only in the presence of a member of the opposite sex, the signals may be considered to function in the context of courtship rather than advertisement. The individual being courted may have already been attracted by the suitor's advertisement signal (or, as in many of the Lepidoptera, the suitor has already been attracted by the courtee's advertisement), and the suitor behaves in a manner indicating that he is responding to the other individual's presence. Other than in exceptional cases of sexual role reversals in parental investment, courtship signaling is initiated by males and serves to influence their acceptability to females.

3.2.1 Courtship Behavior in the Lepidoptera

Chemical signals functioning in a courtship context occur widely among arthropods, but again they are best known and perhaps most highly developed in the Lepidoptera. Such signals are found both in species wherein pair-formation is effected by female advertisement pheromones and in species relying on other means, such as butterflies, which generally use visual advertisement signals. Unlike advertisement pheromones, lepidopteran courtship signals are normally associated with a conspicuous anatomical feature, their emission is often accompanied by rather overt behavior, and the scent may be perceived by a human observer sniffing at close range. Thus, their existence was far less of a mystery than long-range advertisement pheromones were, and lepidopteran courtship pheromones figured prominently in various nineteenth century treatises, and descriptive notes (see Boppré 1984 for review). But, the deployment of synthetic courtship pheromones in agroecosystems never appeared as if it would be effective pest management, and far less research was devoted to these signals. Most of the current research efforts are made within the contexts of sexual selection and chemical ecology.

A typical courtship encounter involving chemical signals may proceed as follows. Having been attracted to a signaling female by her long-range advertisement pheromone, a male moth lands and approaches the female while wing-fanning and elevating his abdomen, which bears specialized scent scales on the terminal segments (Phelan and Baker 1990). These scales may be arranged on a hair pencil structure, which is extruded only during courtship (Figure 3.15). If duly aroused, the female will turn to face the male while he repeatedly thrusts his abdomen toward her, either laterally or directly over his head. The female, being generally larger than the male, retains control of the encounter and either accepts or rejects the male's attempts to copulate. Acceptance typically includes arrestment of motion and raising the abdomen above the wings.

Courtship pheromone is normally released after the male lands and begins wing-fanning and extruding his hair pencils. Its absence may result in uniform rejection of the male (e.g., Royer and McNeil 1992), and its presence may represent a graded signal whose seductive message is commensurate with intensity and quality. Visual, acoustic, or other mechanical signals, the latter including either tactile contact or airflow generated by wing-fanning, often co-occur with presentation of the courtship pheromone and may exert significant influences on acceptance versus rejection of the male (e.g., Baker and Cardé 1979). Butterflies differ slightly from this scheme in that much courtship occurs in flight, and visual signals, of course, predominate in both the advertisement and the courtship stages of pair-formation (Vane-Wright and Boppré 1993, Rutowski 2002; Section 5.2). In some moth species the female may move the final few centimeters to the male, attracted by his courtship pheromone and other signals.

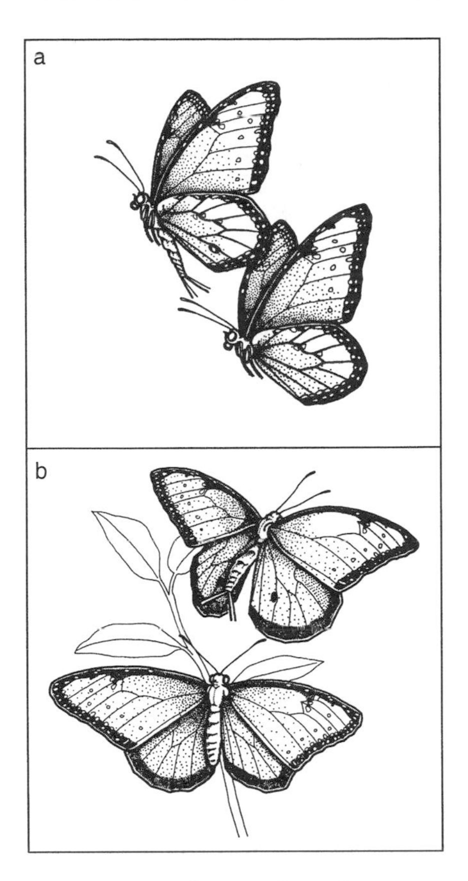

Figure 3.15 Male presentation of courtship pheromone via abdominal hair-penciling in the Queen butterfly, *Danaus gilippus* (Danaidae), (a) during flight and (b) after the female has alighted. (From Brower et al. 1965; drawings by Lincoln Brower. Used with permission of the Wildlife Conservation Society/Bronx Zoo.)

Chemistry and Emission of Courtship Pheromones. Because courtship, being close-range and interactive, is a fundamentally different activity from advertisement, we might expect courtship pheromones and their emission to differ significantly from advertisement ones. In the Lepidoptera, this expectation is upheld in many ways. Courtship pheromones are often lower in molecular weight (<200) than advertisement pheromones, and they are considerably more volatile as a result. Volatility would be a decided advantage when forming a long-range odor plume is neither important nor desirable but rapid transmission of a chemical signal over a short distance is. Many of these substances also have chemical structures very different from those of advertisement pheromones: aromatic structures and amine moieties, as in alkaloids, are particularly common (Figure 3.16). When straight-chain organic compounds occur, they are often saturated. Larval diet does exert a major influence on the quantity and quality of courtship phero-

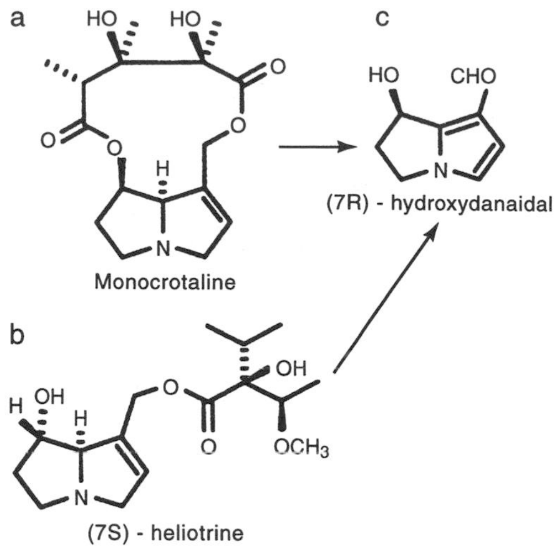

Figure 3.16 Plant alkaloid precursors and male courtship pheromone in arctiid moths. Male *Utetheisa ornatrix* convert the 7R alkaloid monocrotaline (a) to their pheromone, (7R)-hydroxydanaidal (c). Male *Creatonotus transiens* also use (7R)-hydroxydanaidal as their courtship pheromone, but they can derive it from the 7S alkaloid (7S)-heliotrine (b) as well as monocrotaline (Schulz et al. 1993).

mones (e.g., Krasnoff and Roelofs 1989, Löfstedt et al. 1989b), and in some species a shift from one host plant to another, or to another acceptable diet, alters the signal markedly. Lack of a critical host plant in larval nutrition may even result in the adult failing to develop the pheromone-disseminating structure (e.g., Schneider et al. 1982, Boppré and Schneider 1985). Adult diet can also influence the courtship pheromone in some species. This phenomenon is best known among certain danaid butterflies and arctiid moths in which males visit plants (Asteraceae, Boraginaceae, Fabaceae, and several other families) containing a class of secondary compounds known as pyrrolizidine alkaloids (PAs), imbibe fluid from the (decaying) vegetation, and use imbibed PAs to form various of their courtship pheromone components (Schneider et al. 1975, Boppré 1978, 1986, 1990, Edgar et al. 1979, Krasnoff and Dussourd 1989). Quantities of courtship pheromones sequestered by males are relatively high, and as much as 500 μg has been recorded in some species. No attempt to translate these quantities to signal flux or intensity has been made, however. Finally, courtship pheromones do not have the degree of species specificity seen in female advertisement pheromones. For example, benzaldehyde is a major component of the male scent of numerous Lepidoptera, especially noctuid moths, and identical dihydropyrrolizine compounds are used by several arctiid moths (Krasnoff et al. 1987).

Phylogenetic inference suggests that courtship pheromones are not an ancestral feature of the Lepidoptera, but rather originated independently many times in the various moth clades and in the butterflies (Papilionoidea). The structures in which courtship pheromones are sequestered and from which they are released vary extensively (Boppré 1984, Birch et al. 1990) and cannot be homologized to a single form (see Schneider 1992). Unlike the exocrine glands secreting female advertisement pheromones, courtship pheromone glands may be found on various abdominal segments of the male or on different locations on his legs or wings. The male's glands (androconial organs; see Boppré and Vane-Wright 1989) may be associated with highly modified scales or eversible tubular structures (coremata) that telescope out to full extension only during courtship events (Figure 3.17). Males may actively position their courtship pheromone structures close to the female, and females may actively contact these structures with their antennae. In some species, the male uses his hair pencils to dust the female with scent.

3.2.2 Mate Recognition or Mate Choice?

Much of the original work on animal signals focused on their evolution and function in the contexts of species and sex recognition. Courtship pheromones, like advertisement ones, fell under this policy, and they were often interpreted as a form of insurance against cross-attraction between species, a secondary or back-up system should long-range advertisement fail to provide species recognition. That is, if a male mistakenly pursued and courted a female of another species, the female could discern his error by the incorrect or missing courtship pheromone

Figure 3.17 Extruded coremata, eversible scent structures, of male saltmarsh moth, *Estigmene acrea* (Arctiidae: Arctiinae). (Photograph courtesy of Mark A. Willis; reprinted with permission from "Male lek formation and female calling in a population of the arctiid moth *Estigmene acrea*," M.A. Willis and M.C. Birch, *Science*, vol. 218, p. 168, Copyright 1982 American Association for the Advancement of Science.)

and then break off the engagement. Thus, butterflies in Müllerian mimicry complexes were assumed to rely on male courtship pheromones for species recognition because their visual signals had converged to a single aposematic form across many species, genera, and even families. Specific, non-visual mate-recognition signals would also be needed in these groups to distinguish, and avoid, their Batesian mimics (Boppré 1978). In a study specifically addressing the evolution of courtship pheromones, the incidence of these substances and associated structures among moth taxa was found to be related to the sharing of host plants by closely related species (Phelan and Baker 1987). Presumably, the species recognition afforded by long-range female advertisement pheromones and antagonists might be insufficient if females of two or more different species signal in close proximity because they are found on the same hosts.

Findings on the nature of visual signaling in Lepidoptera and more thorough phylogenetic analyses make this interpretation of courtship pheromones doubtful for most groups, however. First, reflectance patterns of ultraviolet (UV) and polarized light are common on the wings of many diurnal Lepidoptera (Section 5.2) and may not have been subject to convergent evolution as patterns in the visible spectrum have: although some of the vertebrate (avian) predators whose visual perception and cognition are assumed to have selected for aposematic patterns may perceive ultraviolet light (Cuthill et al. 2000), no studies suggest that such perception is critical for locating prey or that UV aposematic patterns have exploited it (Lyytinen et al. 2001). Thus, specialized reflectance patterns might provide the necessary species-recognition signals where coloration, as perceived by humans, is otherwise similar. Second, the study purporting to demonstrate a relationship between shared host plants and the incidence of courtship pheromones within a taxon may not be as robust as claimed, because it tabulated individual species that used courtship pheromones rather than independent evolutionary origins of courtship pheromones as data points. And, multiple species often share some or all of the major components of their courtship pheromones.

If not vital for species recognition, in what contexts might courtship pheromones in the Lepidoptera and other arthropod groups have evolved and function? Because male courtship pheromones and associated organs often exhibit exaggerated features, it seemed likely that females might rely on these olfactory signals to assess the quality of potential mates who were already recognized as conspecific (Sections 4.4, 5.2 and 6.2). Several studies of moth courtship pheromones set out to test this notion, and evidence for female choice based on male pheromone characters was found in some species. In the tobacco moth (*Ephestia elutella*; Pyralidae: Phycitinac), Phelan and Baker (1986) found that a male's pheromone titer is correlated with his size and that females discriminate against small males during courtship. This discrimination is likely made via olfaction, as it continues in darkness and after most of a male's wings have been removed. Both fecundity and offspring survival of females who mate with larger males are enhanced, supporting the hypothesis that courtship allows a female to assess some aspect of mate quality in this species (Section 6.2).

Further insight to the nature of the quality that might be assessed during courtship has been obtained from investigations of the arctiid moth *Utetheisa ornatrix* (Conner et al. 1981, Eisner and Meinwald 1995). Larval *U. ornatrix* feed on leguminous host plants (*Crotalaria* spp.) rich in PAs, substances which adult males later convert to their courtship pheromone, hydroxydanaidal (Culvenor and Edgar 1972). A male's pheromone titer is proportional to his acceptance during courtship, to his body size and larval diet (Conner et al. 1990), to his amount of systemic PA, and to the amount of PA in the spermatophore he transfers to the female during copulation (Dussourd et al. 1991).[46] Females then shunt some of the PA received to their ovaries, while retaining the remainder in somatic tissue. Because PAs are toxic to many animals, the deposited eggs are better defended against predators as a result of the male's transfer (Eisner et al. 2000). Thus, the female receives both direct protection (Gonzalez et al. 1999) against natural enemies as well as parental investment from her mate. A reasonable hypothesis would be that courtship pheromone is a reliable indication of the direct, material benefits that a female may expect to obtain from her suitor (Sections 4.4 and 6.2): there may simply be no means by which a male deficient in systemic PAs can produce a sufficient titer of hydroxydanaidal. Similar findings have been obtained in the queen butterfly, *Danaus gilippus* (Danaiidae) (Dussourd et al. 1989).

In *Utetheisa* and *Danaus* and other Lepidoptera, the male courtship pheromones, while chemically related to host plant precursors, have undergone specialized biochemical transformations (von Nickisch-Rosenegk and Wink 1993). These transformations may reflect selection for exaggerated signal features, but pre-existing receiver properties and biases probably restrict the effectiveness of signal exaggeration to a limited range of compounds. In many cases, some chemical alteration would be advantageous because the host plant precursors could be unsuitable for rapid dissemination by signalers and immediate detection and evaluation by receivers in a courtship interaction. Unlike female advertisement pheromones, high signal intensity and ease of detection by the (female) receiver may be favored in male courtship pheromones. Such olfactory conspicuousness would be offered most readily by volatile compounds for which a receiver bias already exists (Section 6.1) and that could be easily distinguished from other environmental odors. Because (female) receivers may be expected to have pre-existing biases for the classes of compounds found in host plant odors, courtship pheromones from these same classes may be selected (see Boppré 1978). But, a courtship pheromone identical to or resembling a host plant odor too closely may not present a sufficient signal : noise ratio to a receiver if mating occurs on or near the host. Thus, some divergence from compounds found in host plant odors might be selected even if these substances possess ideal properties for signal dissemination and receiver detection.

Indicators of Genetic Quality? A critical, but easily overlooked, point in the *U. ornatrix* signaling system is that both female and male larvae feed pharmacophagously on *Crotalaria* and sequester PAs. Consequently, the major portion of a female's chemical defense for herself and her offspring will ordinarily be derived from her own systemic PAs (see Dussourd et al. 1988). In light of the female's ability to provide for her own defense, are male-contributed PAs mainly important under certain environmental circumstances when these materials or the host plants are scarce? Or are females also obtaining genetic benefits—alleles associated with a high fitness level—for their offspring when choosing males based on their courtship pheromone?

Additional experiments with *U. ornatrix* have pointed toward the latter explanation: the opportunity to procure genetic benefits influences females to be choosy. Sexual selection does not end at courtship, as females who mate with multiple males apparently assess spermatophore mass and only use sperm from the largest spermatophore(s) to fertilize their eggs (LaMunyon and Eisner 1993, 1994). Material benefits, PAs, are not so differentiated, though, and a female will accept these donations from any male with whom she has mated (see LaMunyon 1997). Because spermatophore mass is proportional to male size, a female's post-copulatory mate choice (see Eberhard 1996) would lead to larger males siring most of her offspring. Should size be heritable (Iyengar and Eisner 1999a) and genetically correlated with foraging proficiency or the capacity for sequestering PAs without suffering any toxic or other debilitating consequences, female choice may then ensure that offspring inherit these desirable traits (see Iyengar and Eisner 1999b; Section 6.2).

Both *U. ornatrix* and *D. gilippus* are aposematically colored lepidopterans that possess chemical deterrence derived from toxic host plant materials consumed and sequestered by larvae. Therefore, a general connection might exist between aposematism and courtship pheromones, although not in the species-recognition context initially suggested. The general connection would be founded on the possibility that a female can reliably ascertain a mate's defensive prowess, a potentially critical quality, from his chemical signal. Outside of the Lepidoptera, a similar connection may be proposed, albeit with far less support, for male pheromones in various terrestrial Hemiptera. Many hemipterans emit volatile compounds from multi-cellular exocrine glands on the abdomen or thorax (Aldrich 1995). These compounds may deter natural enemies and are typically stored prior to emission within a reservoir associated with the glands. Thus, their flux can be quite high. A marked sexual dimorphism in both the gland and reservoir exists in some species, the male structures being enlarged. In other species, the function of a given gland is known to change with age, as nymphs emit compounds possessing deterrent properties and adult males emit sex pheromones. Alternatively, glands may not exhibit any sexual dimorphism in structure, but females emit deterrent compounds whereas males emit sexual ones. In general, courtship and advertisement functions[47] of these chemical signals of male hemipterans cannot be distinguished because the distances and circumstances in which they act in natural populations are poorly known. Nonetheless, the volatile nature of the compounds suggests that they function primarily over a close range and in courtship. From the above observations, we may infer that male sexual signals are derived from more widespread chemical deterrents and that females might assess some aspect of mate quality via these signals. But, again, far less information on chemical signaling is available for Hemiptera than Lepidoptera, and the above points must be treated as entirely speculative.

3.2.3 Intrasexual Responses

Males as well as females normally perceive and respond to the signals of conspecific males, and both behavioral and neural (EAG) responses (e.g., Jacquin et al. 1991) to male courtship pheromones seem to be found in males whenever they are investigated with sufficient thoroughness. In the Lepidoptera, these responses range from simply curtailing wing-fanning (Lecomte et al. 1998) to interrupting upwind flight toward a female emitting her sexual advertisement pheromone (Hirai et al. 1978). Such receiver responses have probably been selected for because they prevent the male from wasting time and energy pursuing a female who is already being courted by another male (but, see Fitzpatrick et al. 1988). Many female lepidopterans typically mate but once or enter an extended refractory period following mating (Raina et al. 1994, Kingan et al. 1995), and a male's efforts would be better spent seeking another advertising female than one about to mate.

More elaborate inter-male responses to courtship pheromones may be expected to occur in arthropods found in higher population density. For example, males of the arctiid moth *Estigmene acrea* gather in lekking aggregations on certain occasions (Willis and Birch 1982; see Figure 3.17), and these aggregations may be facilitated, in part, by orientation toward courtship pheromones. Would such orientation afford males a "shortcut" to sites at which numerous females may be encountered? In the cockroach *Nauphoeta cinerea*, which may cluster in the vicinity of food resources and suitable habitat, males exhibit a dominance hierarchy that is maintained by their sex pheromone. The males emit a three-component scent that serves as a courtship pheromone influencing female acceptance and that determines a male's status (Moore et al. 1997). Males that have high titers of two of the components and a low titer of the third both behave as dominant individuals and are treated as such by other males. But female choice, as based on pheromone composition, does not perfectly coincide with the inter-male evaluation, and subordinate males do enjoy some mating success. Pheromone composition is genetically controlled, and inconsistencies between female and male evaluation may maintain, in part, the (genetic) variation in composition (Moore 1997; Section 6.2).

3.3 Nascent Sociality: Conspecific Cuing, Mass Attacks, and Swarms

Behavioral interactions between courting males and potential responses of females to the advertisement pheromones of conspecific females[13] were our first glimpses of chemical signals mediating social behaviors above the level of pair-formation. These behaviors arise when densities are sufficiently high that intrasexual interactions occur regularly enough to select for specialized responses. Active aggregation of individuals and group living would comprise a higher level of social behavior, and many landmark cases exist among arthropods. As in simpler social interactions, chemical signals, often derived from sex pheromones, play a central role in the formation, cohesion, and regulation of active aggregation in many species.

Active aggregations may form if the participants attain a higher fitness when clustered than when solitary. Reasons for this increased fitness include an enhanced ability to acquire food resources that would be unavailable to lone individuals, vigilance or other means of avoiding or deterring natural enemies afforded members of groups, construction of shelters and nests, and thermoregulation. Additionally, males may congregate in leks that enhance their encounter rates with females (Höglund and Alatalo 1995). In all cases, an optimal group size or density exists, and in groups above that size the per capita fitness begins to decline with each new arrival (see Brown 1982). Thus, signals regulating aggregation are not only responsible for initiating mutual attraction of individuals, but they must also prevent additional ones from arriving once the optimal size is attained. Such control has been examined intensively—but not fully ascertained—in bark beetles (Curculionidae: Scolytinae) of North American and Eurasian forests. These insects are important forest pests, and economic concerns, as in the advertisement pheromones of Lepidoptera, have clearly motivated much of the research.

3.3.1 Bark Beetles

Holarctic bark beetles feed and oviposit within the cambium layer of declining, yet living, coniferous trees, and they are only able to conduct these activities if a great number of adult beetles are recruited to a given tree (Alcock 1982, Byers 1989a). Aggregations of adult beetles are initiated after a "pioneer" individual locates a suitable tree, bores through the bark, and emits an advertisement pheromone. These pioneers are males in some genera (e.g., *Ips*) and females in others (e.g., *Dendroctonus*). Advertisement pheromone recruits both male and female conspecifics, who also emit the advertisement. After a sufficient number have arrived, they are able to overwhelm the host tree's defenses, initiate gallery construction, mate, oviposit, and rear offspring in the galleries. If too few adults are drawn to the pioneer and incipient aggregation, the beetles succumb to the physical defenses of the host: they are incapacitated by the tree's resin flow, and the symbiotic fungi which they transport to stem that resin flow are encapsulated by necrotic tissue (Raffa and Berryman 1983, 1987, Berryman *et al.*1985).

Aggregation Mechanisms. Advertisement pheromone serves as the signal regulating the mass attacks. Components of this pheromone, in conjunction with mechanical stridulations, also serve as a mating signal facilitating pair-formation within an aggregation.[48] Pioneers appear to select particular trees on the basis of volatile compounds that signal a tree's susceptibility: trees in excellent health are generally ignored. But the variation in volatile emissions among trees would be insufficient to elicit a mass attack at the tree selected by a pioneer. Instead, the pioneer emits a bouquet of compounds (Figure 3.18), some of which are synthesized de novo by the beetle whereas others are modified from precursors that it ingested in the course of feeding on the tree's phloem. These precursors are various terpenes (e.g., myrcene, α-pinene) that may serve as chemical deterrents to a wide variety of insects (other than bark beetles).[49] The precise site at which the beetle synthesizes advertisement pheromone compounds remains unknown, but the hindgut is suspected because pheromone is disseminated from a beetle's fecal pellets. Some of these syntheses may be aided by microbial symbionts (Brand et al. 1975), as beetles treated with antibiotics may not produce the entire suite of compounds forming an attractive advertisement (Byers and Wood 1981). In essence, advertisement pheromone serves to magnify any small intrinsic differences between trees by adding modified compounds to the tree's own emissions. Therefore, a tree attacked by one or a few beetles is readily distinguished from its neighbors. Once an incipient aggregation forms, a positive-feedback process takes over: as more and more adults are drawn to the attacked tree, the flux of the emitted advertisement pheromone increases, and even more beetles arrive. Thus, large aggregations become larger, whereas small ones may die out because they do not recruit additional individuals and the few remaining ones become imprisoned in the host's resin. Resin may also interfere with the emission of advertisement pheromone from the tree.

Anti-aggregation. An optimal group size for mass attacks by bark beetles exists because competition for the limited resource, host cambium, begins to counteract the advantages of overwhelming the host with thousands of recruits. As expected, large aggregations of bark beetles cease to attract additional recruits, and they may accomplish this reversal with an "anti-aggregation pheromone," a substance that inhibits attraction toward the advertisement pheromone (Byers 1989a,b). The terpene-ketone verbenone, which is emitted by decaying hosts and also found in the hindguts of male bark beetles, may be the anti-aggregation pheromone in certain species (Schlyter et al. 1989). Some studies indicate that

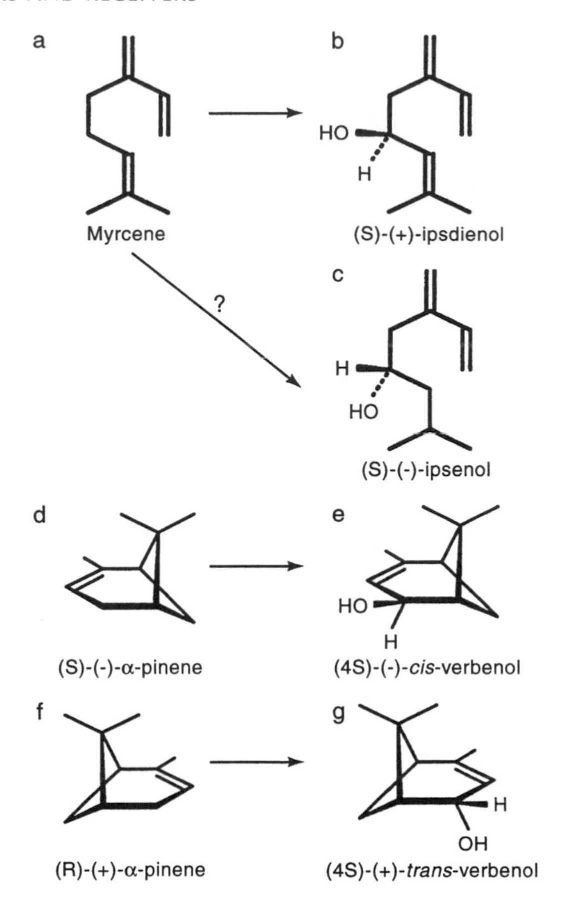

Figure 3.18 Host tree precursors and aggregation pheromones in scolytine bark beetles. *Ips paraconfusus* may convert the host tree compound myrcene (a) to the aggregation pheromone component (S)-(+)-ipsdienol (b), and possibly to an additional component, (S)-(−)-ipsenol (c). *I. paraconfusus* stereoselectively derive another pheromone component, (4S)-(−)-*cis*-verbenol (e), from the host tree compound (S)-(−)-α-pinene (d). Similarly, *Dendroctonus brevicomis* derive a (+) pheromone component, (4S)-(+)-*trans*-verbenol (g), from the host tree compound (R)-(+)-α-pinene (f) (Byers 1995).

verbenone is synthesized by conversion of terpene precursors in the host and that various bacteria and fungi are responsible for the conversion. When bark beetles colonize a host, they may introduce these microorganisms and thereby augment the host's verbenone flux. Because various bark beetle species exhibit a general avoidance of decaying hosts and rely on verbenone as a primary indication of decay, we might speculate that microbial introduction evolved as an exploitation of a pre-existing sensory perception (cf. Section 6.1): by enhancing the verbenone flux from their aggregation, beetles inform potential recruits (and heterospecific competitors) that they should seek an incipient aggregation on another host or even act as a pioneer. It is also possible that qualitative and quantitative changes to the advertisement pheromone itself occur at high population densities and cause the aggregation to level off.

Origins. Because active aggregations are beneficial to both the pioneers and the recruits, selection is expected to favor both responses to odors indicating infested resources and enhancement or modification of those odors. As argued earlier for male courtship pheromones (Section 3.2), the more parsimonious explanation is that the responses originated first: in the absence of responses to odors of infested resources, what factor would have readily selected for specialized odor enhancement? Under this proposed evolutionary pathway (see Section 6.1), the original receiver responses to infestation odors would represent conspecific cuing: the localization of inadvertent cues emanating from the activities of conspecific individuals as a shortcut to finding resources. Such conspecific cuing is a form of copying behavior (see Gibson and Höglund 1992; cf. Section 4.4.1), and its primary advantage is the avoidance of a tedious trial-and-error process.[50]

An equivalent set of receiver responses and signaler modifications may limit aggregation size. Limitation is beneficial to the signaler and negative responses to large aggregations are beneficial to receivers, unless no alternative sites remain. For both parties, the benefit is avoiding excessive intraspecific competition.

3.3.2 Oviposition Pheromones

Chemical signals mediate a variety of potentially competitive interactions among arthropods in the manner outlined above for bark beetles. For example, the oviposition, or marking, pheromones of tephritid fruit flies prevent two or more females from each depositing an egg in the same fruit: a female drags her ovipositor on the fruit surface during and after ovipositing and thereby leaves a chemical signal indicating that the fruit already has an egg. Another female subsequently visiting that fruit, or the same female revisiting it, responds to that signal by departing and seeking another fruit in which to oviposit (Prokopy 1972, Nufio and Papaj 2001). Because the second female's offspring would be younger and smaller at any given time, it would likely lose in competition for the food resource. And, whereas the first female's offspring would probably win, it would avoid competition altogether if its mother ensures that any oviposition indicators are conspicuous and unmistakable; that is, they have a high signal : noise ratio. Conspicuous signals would also be advantageous to the marker female because of those rare occasions in which the second female's offspring compete successfully for the resource. Thus, specialized marking pheromones in tephritids and other insects with similar egg-laying habits and ecologies may have evolved from chemicals left inadvertently at the time of oviposition (cf. Sections 3.1.5 and 6.1).

We might expect parasitoids also to apply marking pheromones to their hosts in order to prevent competition resulting from superparasitism (Roitberg and Mangel 1988, Godfray 1994). Such competition-reducing signaling systems have been found in both solitary and gregarious parasitoids that are egg-limited as opposed to time-limited. These parasitoids do not have eggs to spare in superparasitic ovipositions of dubious value, but they do have the time necessary for locating another host that had not been previously parasitized. Some parasitoids, however, avoid superparasitism by ovicide: a female who discovers a host that already contains a (conspecific) parasitoid egg finds that egg by probing with her ovipositor, kills it by puncturing, and then deposits her own. Are these practitioners of ovicide limited by both eggs and time (or unparasitized hosts), or is the practice especially easy for some parasitoids because a relatively high egg : host size ratio makes finding eggs deposited by the first female easy?

3.3.3 Locusts

The locust phenomenon that affects various acridid grasshoppers of Africa and Western Asia is perhaps the most striking case of active aggregation among arthropods and is also mediated, in part, by chemical signals. Here, various signals act as both releasers of aggregation behaviors and primers initiating physiological and morphological changes in aggregated individuals. These effects are collectively known as gregarization, which distinguishes aggregated individuals, locusts, from those in the solitary phase, hoppers. Several of these acridid species were responsible for the locust plagues recorded in antiquity, including the Old Testament and the Koran, but the connection between locusts and the solitary hoppers seen during non-plague years was not recognized until Sir Boris Uvarov's pioneering work in the 1920s (Uvarov 1966).

Phase Change. Nymphs and adults of both sexes undergo the gregarization process when population densities are high enough that individuals exchange visual, tactile, and chemical information (Pener and Yerushalmi 1998). The chemical stimuli include both cuticular hydrocarbons that are perceived during contact and more volatile aliphatic acids and aldehydes and aromatic compounds. Some of the latter are deposited with the feces and may reflect the insect's feeding history.[51] Reception of these varied stimuli induces a darker color (melanization) and structural changes to the head, legs, and thorax of nymphs. The head and thoracic modifications later facilitate an elevated feeding rate and long-range flight, respectively. Certain of the volatile chemicals also elicit mutual attraction and deter the members of a group that has already formed from moving apart from each other. Thus, gregarized nymphs tend to form bands that march en masse across the ground from one patch of food plants to another, and adults aggregate in flying swarms. The volatile compounds responsible for group cohesion are designated the aggregation pheromone.

Group Cohesion. Research at the International Center for Insect Physiology and Ecology (ICIPE) in Nairobi, Kenya has succeeded in isolating and identifying the major components of aggregation pheromone in the two most devastating locust pests, the desert locust (*Schistocerca gregaria*) and the migratory locust (*Locusta migratoria*). These investigations found separate aggregation pheromone systems in nymphs and adults, although adult *Locusta* will respond to nymphs as well as to adults. Both females and males respond to the adult pheromones, which are produced by both sexes in *Locusta* but only by males in *Schistocerca*. Interspecific responses too are seen, and these lead to mixed-species bands of marching nymphs (Torto et al. 1994, 1996, Njagi et al. 1996, Niassy et al. 1999).

The aromatic compound phenylacetonitrile is a major component of the aggregation pheromone system of adult male *Schistocerca*. This compound is only produced by gregarized males, but solitary and gregarized adults alike respond to it by arresting movement. Thus, solitary adults are recruited into gregarizing groups (Njagi et al. 1996).

As with the mass attacks of bark beetles, locust bands and swarms are formed by a positive feedback process (Applebaum and Heifetz 1999). The aggregations are initiated when the distribution of food resources is highly clumped and concentrates the insects into pockets of high density (Collett et al. 1998). There, visual, tactile, and contact chemical stimuli initiate gregarization (Roessingh et al. 1998, Simpson et al. 1999). These gregarized individuals then synthesize and emit the volatile components of the aggregation pheromone that draws more insects to the group. The elevated titer of aggregation pheromone in and surrounding the group accelerates maturation of young insects, who soon add to that cloud of volatiles. Chemical controls on expanding group size are unknown but may be unnecessary, as the locust aggregation is a mobile unit that stays in place only as long as the local food supply holds out. When acceptable host plants are depleted, the insects march or fly on. In the latter case, swarms rely on prevailing winds and only land after arriving over an area previously untapped for food.

The chemical ecology of locusts is clearly of interest because of the hope that such information can help to predict where and when swarms will appear, feed, and oviposit and, ultimately, to suppress their population outbreaks. These swarms are also one of the great spectacles of the natural world, and study of their evolution and adaptations would be most intriguing. Here, though, their infrequent and irregular recurrence (typically less than once every 10 years in a given locale) and migration distances (hundreds of kilometers) are unfortunate, as they would preclude most experimental studies or even systematic observations. Thus, we can only suppose that the aggregations afford their members (1) "safety in numbers" from vertebrate predators, who would be quickly satiated when visiting a swarm, and (2) the ability to avoid settling on locations that had just been depleted of food by previous insects (see Cody 1971). These proposals are consistent with the observation of mixed-species bands of nymphs, as members of the aggregations would benefit whether or not their neighbors are conspecifics.[52] Competition as well as cooperation must occur within bands and swarms, and members would be expected to vie intensely for the limited food within a resource patch. The accelerated maturation and elevated feeding rates induced in gregarized nymphs are probably adaptations for such conflict.

Because the adult aggregation pheromone in *Schistocerca* is emitted only by males (Niassy et al. 1999), we may suspect that it originated as a sexual advertisement or courtship signal. This possibility is consistent with the apparent absence of mixed-species swarms of adults. Did females then interpret certain volatile components of this signal as a reliable indication of food in the vicinity of a male or of the male's foraging ability? Did males also arrest their movement in response to the volatiles of other males because this signal normally indicated that valuable food resources were nearby and that females might be encountered as well? Answers to these questions might be forthcoming from investigations of those acridids that undergo density-dependent behavioral and physiological phases but do not exhibit the migratory phases and massive swarms of "extreme locusts" such as *Schistocerca gregaria* and *Locusta migratoria* (see Heifetz and Applebaum 1995).

3.4 Social Behavior: Royalty and Altruists

Whereas moths may be unsurpassed in olfactory navigation and sensitivity among arthropods—and all metazoans—the social insects are the undisputed masters of chemical signal diversity. Honeybees (*Apis mellifera*), for example, include no fewer than 36 different pheromone components in their lexicon, and these signals, secreted by 15 different exocrine glands (Figure 3.19), pervade every aspect of colony life (Winston and Slessor 1992, Katzav-Gozansky et al. 1997). Among the eusocial Hymenoptera (ants, bees, and wasps) and Isoptera (termites) in general, pheromones may effect or influence the recruitment of foraging workers to a new resource, the ability of the worker who initially found that resource to return to the colony, the summoning of colony members to answer the threat from a natural enemy, the distinguishing of colony members from outsiders, the fission of colonies and localization of a new nest site by members of a colony fragment (swarm), and, of course, mating. But, it is in

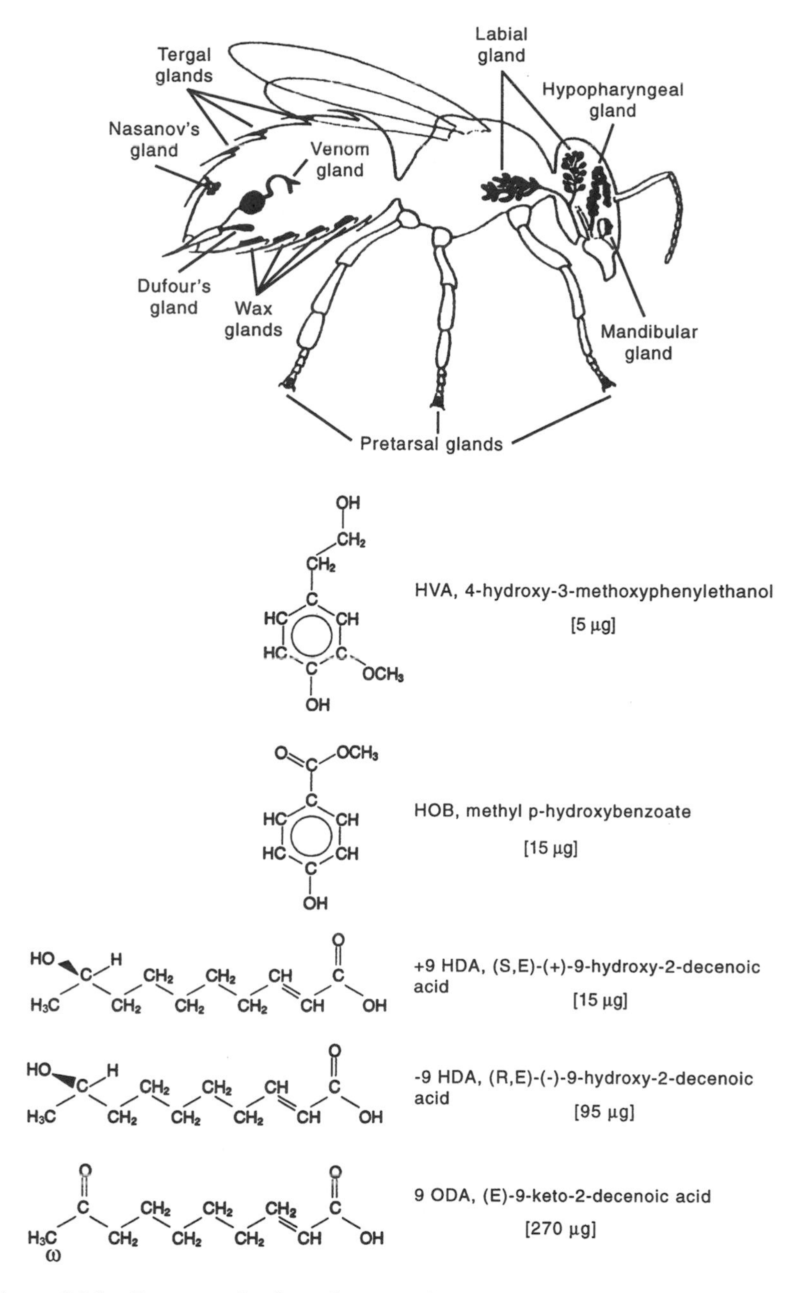

Figure 3.19 Exocrine glands and queen pheromone in the honeybee *Apis mellifera* (Apidae). Secretions from mandibular, pretarsal, venom, tergal, and Nasanov's gland all have known pheromonal functions (redrawn from Billen and Morgan 1998). Queen pheromone includes five known components, all emitted from the mandibular glands. Values in brackets are average quantities of the component normally found in a queen's glands at a given time (Winston and Slessor 1992). Mandibular gland pheromones emitted by worker honeybees have functional moieties attached at the ω position of the carbon chain rather than the ω-1 position, as in queens (Gadagkar 1996).

establishing the colony's reproductive hierarchy that social insect pheromones are most special. This function has been considered unique to the advanced eusocial insects and bears little resemblance to any of the pheromone-mediated behaviors presented thus far.

In honeybees, the olfactory message that elicits royal treatment of the colony's α-female, the queen, by the workers is known as queen substance or, alternatively, queen pheromone. Like the female advertisement pheromones of moths, queen pheromone is a multi-component substance (Figure 3.19). Three components are 10-carbon straight-chain aliphatic acids, two of which are different chiral forms (enantiomers) of the same compound; two additional components are smaller, aromatic compounds. None of these components, not even the major one (*trans*-9-keto-2-decenoic acid, 9ODA), are active alone, and removal of any one of the five components drastically reduces the behaviors elicited in worker bees. These behaviors may include (1) forming a retinue surrounding the queen and directly acquiring some of her queen pheromone on their bodies, (2) serving as messengers and distributing the acquired queen pheromone to other bees in the colony, (3) refraining from constructing enlarged brood cells for the production of future queens, and (4) deferring colony fission via swarming until a later date. All four behaviors ensure that the current queen retains her hegemony over the colony.

Analytical work similar to that which characterized moth sex pheromones identified the constituents of honeybee queen pheromone and demonstrated that it is produced by a pair of (mandibular) glands filling both sides of the queen's head (Winston and Slessor 1992, 1998; Figure 3.19).[53] The pheromone exits the glands via ducts through the bases of her mandibles and then spreads over her body. On average, a queen produces 400 μg of this mandibular gland secretion daily. Owing to retinue activity among the worker population, this daily production does not build up on a queen in an active colony. Rather, retinue bees continuously remove pheromone from the queen by licking and antennating her, and the queen normally has only 5 μg on her body at any one time. Messenger bees similarly acquire queen pheromone from the retinue bees, and secondary messenger bees then acquire it from the primary messengers. Thus, a chain-reaction effect distributes queen pheromone widely among the colony until nearly every worker is exposed. This distribution method greatly dilutes the amounts of pheromone that most workers would receive, but honeybees are extremely sensitive to queen pheromone. The estimated threshold for worker behavioral responses is only 40 pg, and the activities of retinue and messenger bees would be expected to keep most workers supplied with this quantity throughout the day and night.

In addition to eliciting the four behaviors noted above, queen pheromone also inhibits the normal ethological ontogeny in workers that progresses from hive activities to outside foraging over the initial 10 days of adult life. This inhibition is most likely primed via a pheromone/hormone interaction: queen pheromone suppresses juvenile hormone secretion, which delays the behavioral maturation of worker bees.

The most salient feature of the reproductive hierarchy in honeybees and other advanced eusocial insects is the inhibition of ovarian development in the worker caste. For many years queen pheromone, possibly synergized by secretions from other queen exocrine glands, was also suspected to prime ovarian suppression, but no experimental evidence has supported this expectation. Rather, the most likely candidate for the primer is a pheromone produced by the brood.

As with the chemical signals mediating other aggregation and grouping behaviors among arthropods (Section 3.3), queen pheromone also serves as a sexual advertisement and may have originated in this context. During her mating flight(s), a virgin honeybee queen attracts drones (male honeybees) up to 60 m distant. Behavioral and electroantennogram analyses indicate that attracted drones are responding specifically to 9ODA in the queen pheromone and that they are even more sensitive to it than workers are (Brockmann et al. 1998). Unlike workers, drones appear to possess a moth-like labeled line for 9ODA, which may explain their heightened sensitivity.

Do queen pheromone's various functions and effects on different castes provide any insight to its evolution? This question is difficult to address for reasons similar to those encountered earlier regarding locust aggregation pheromone: the amount of research on chemical communication in bees is highly skewed toward the most highly derived eusocial species, such as *Apis mellifera*, and it is nearly impossible to discern what intermediate stages in queen pheromone formulation and function may have existed between *Apis mellifera* and its solitary ancestors. Among the genus *Apis*, queens in all of the Asian species that have been examined each have unique mandibular gland secretions (Plettner et al. 1997). But, only *Apis cerana* queens include aromatic compounds and show a marked differentiation from worker secretions, as *Apis mellifera* do. Both *A. cerana* and *A. mellifera* nest in cavities as opposed to open locations, and their specialized aromatic compounds may afford more ready dispersal in the still air within the colony. Queen pheromones are also known in other eusocial Hymenoptera (Vargo 1998), but we have too little information on their chemistry, biosynthesis, and specific functions to rely on them for an evolutionary reconstruction.

Can we obtain more useful clues to queen pheromone evolution by examining its counterparts among other castes within the colony? Workers in all species of *Apis* also produce mandibular gland secretions, and these compounds appear to be secreted into brood food as a possible preservative or nutrient (Plettner et al. 1996, Robinson 1996, Winston and Slessor 1998). In *A. mellifera* the worker secretions are aliphatic acids that are distinguished from those in queen pheromone by the simplest of modifications: worker acids have keto moieties at the ω rather than the ω-1 position of the carbon chain (Plettner et al. 1998; Figure

3.19). One interpretation of these observations is that female mandibular gland secretions originated as a sexual advertisement and later acquired their social functions as the reproductive hierarchy structuring colony life evolved. The modification of the keto group's position would have accompanied these functional changes, but it is uncertain whether the worker or the queen signal we observe today represents the ancestral state (Gadagkar 1996).

The most vital and controversial issue concerning queen pheromone is its reputation as the consummate manipulator. By exuding sufficient scent to permeate the entire colony, are queens holding subject workers in a trance so that their labors benefit the queen's fitness while limiting their own? This possibility begs the question of why the workers have not rebelled against the opiate effects of queen pheromone to the extent that they are physically capable. Thus, a more plausible view is that queen pheromone is an honest signal and that workers behave in seemingly altruistic ways when they perceive it because it is in their best interest to do so (Keller and Nonacs 1993). Granted, a worker may conceivably attain higher fitness by ceasing to assist the queen and instead mating and producing offspring on her own, but such outright rebellion would undoubtedly be constrained by the extreme morphological specialization of workers in the advanced eusocial insects (Gadagkar 1997a,b). In lieu of a reversal in social evolution, workers may favor their own aims most by cooperating with the queen in various ways and interpreting her mandibular gland secretion as a reliable indication of her presence. Should that scent disappear, as when the queen dies, the workers begin to construct enlarged cells for the raising of future queens among the fertilized (diploid female) eggs laid by their queen before her death.[54] By facilitating the production of these future queens, whose genetic relationship to the workers would be either full- or half-sister, the workers can still enhance their inclusive fitness in the emergency situation. When both the scent and eggs or young brood are absent, some of the workers then begin to lay unfertilized (haploid male) eggs. Under this more dire situation, egg-laying would be a worker's only means of enhancing any aspect of her fitness.

3.5 Social Behavior: Discriminating Insiders from Outsiders

In order for a worker's altruism to benefit its inclusive fitness, the altruism must be normally donated to a genetic relative. Thus, we may expect workers in social insect colonies to treat colony members (i.e., nestmates) differently from outsiders, as the former are likely to be genetic relatives whereas the latter rarely are. Observations of various eusocial Hymenoptera and Isoptera have long supported this expectation (Fletcher and Michener 1987, Smith and Breed 1995).

In general, nestmates are allowed to enter the colony and may receive nourishment therein, whereas conspecific outsiders may be attacked and expelled. Additionally, species in which colonies maintain exclusive foraging territories, such as palaeotropical weaver ants (*Oecophylla* spp.), may extend such behavior to a wide area surrounding the colony proper. All evidence indicates that nestmates are recognized by chemical signals, as opposed to mechanical or visual ones, but further details of the recognition process are fraught with controversy.

3.5.1 Nestmate Recognition via Colony Signatures

Kin recognition theory identifies four potential mechanisms by which animals may discriminate kin from non-kin: (1) recognition alleles, the so-called "green beard effect" in which a given gene influences its bearer both to produce a signal and also to recognize it when in the receiver role; (2) location near the natal site; (3) familiarity with associates acquired during development; (4) phenotype matching, the similarity between an individual's own signals and those of another (Sherman and Holmes 1985). Of these possibilities, familiarity with associates and phenotype matching are the most likely mechanisms in the eusocial Hymenoptera and Isoptera. No findings support the occurrence of recognition alleles, and mere location near the natal site could not be effective owing to the spatial clustering of nests in many species.

Familiarity and Phenotype Matching. Both familiarity with associates and phenotype matching rely on the learning of signals that nestmates are likely to emit. In the first case, an individual learns the signals of its associates in the nest and uses this information to establish a template with which to compare the signals of individuals it encounters later. In the second case, an individual learns its own signal and uses it to establish the template. These two mechanisms are not mutually exclusive, and both may rely on signals that are environmentally and/or genetically determined. For example, a social insect may learn signals of its nestmates or itself that are influenced by environmental odors of the colony. If these environmental odors are distinct from those in all other local colonies, uniform within colonies, and evaluated with sufficient precision, only nestmates will carry it and be accorded insider status. Alternatively, a social insect may learn signals of its nestmates or itself that reflect the genotypes found in the colony. Again, if the above criteria of signal diversity among colonies, signal uniformity within colonies, and signal evaluation are met, only nestmates will be recognized as matching the template.

Environmental and Genetic Determinants of Signatures. While nestmate recognition has been observed in a variety of eusocial Hymenoptera and Isoptera, detailed analyses of the recognition process have focused on a limited selection of taxa: *Polistes* paper wasps (Vespidae) (Gamboa et al. 1986), ground-nesting *Lasioglossum* bees (Halictidae) (Greenberg 1979, Buckle and Greenberg 1981), the honeybee *A. mellifera* (Breed 1983, 1998), and several species of ants (Vander Meer and Morel 1998). These analyses found recognition based on both environmentally and genetically determined signal features. Evidence for the latter consisted of preferential assistance or recognition directed toward siblings who had been reared in different nests and with whom the recognizer had no prior contact. Here, individuals that were made to share a nest with unrelated nestmates were still recognized by their siblings when subsequently tested. Thus, it appeared as if the genetically determined signal features were not readily transferred among nestmates by physical contact. In all cases save one, the desert ant *Cataglyphis cursor* (Isingrini et al. 1985), recognizers were found to be recently molted adults at the time they learned the signal features used in nestmate recognition.

Given that chemical recognition signals must vary between colonies while retaining uniformity at any one time within a colony, from what source might they be derived and what sort of compounds are they? Several studies considered that environmentally determined colony odors might be signatures derived from floral oils and other plant materials collected by foragers. Workers would presumably absorb these colony signatures due to continual exposure within the nest. Limited evidence for this hypothesis has been found in the honeybee, but important questions remain: foragers from adjacent colonies must often collect materials from the same locales, and it is not apparent how colony-specific signatures would be maintained over time. Additionally, a forager returning to the nest with a novel material—which must represent a frequent event—should be rejected according to this hypothesis. Probably, odors from foraged plant materials only play an auxiliary role, whose importance may increase in the absence of other stimuli (Bowden et al. 1998). In the honeybee and other eusocial Hymenoptera, these other stimuli may be cuticular hydrocarbons,[55] emissions from various exocrine glands (Dufour's gland at the end of the abdomen is a likely source), and materials used in nest construction (e.g., paper or wax) that include substantial amounts of worker-produced secretions. Because workers contact the nest material for many hours each day, and many also chew on it to fashion cells, the material would serve as a vehicle transferring the secretions from worker to worker. These three candidates consist of complex mixtures of organic compounds that could provide the requisite diversity between colonies and uniformity within colonies to serve as reliable signatures. In the Isoptera, however, colony-recognition odors might be derived from symbiotic gut bacteria (Matsuura 2001).

Uniformity of the colony signature is maintained, in part, by effective transfer of its constituent compounds among workers. Effective transfer implies that all colony members acquire the signature and can be appropriately recognized, but at the same time the signature ought not to be passed on too readily lest illegitimate recipients—social parasites seeking to gain entrance and acceptance in the colony—easily acquire it. Relatively non-volatile compounds in worker cuticle or that workers incorporate into the nest structure would meet these conflicting demands. More volatile substances could be quickly absorbed by conspecific or heterospecific intruders, who would then circulate freely within the colony.

Some evidence indicates that cuticular hydrocarbons, exocrine gland secretions, and compounds in nest material do form colony signatures in various eusocial Hymenoptera, but the complete bouquet of compounds recognized and the precision with which they are differentiated from other compounds are not known in any species. Presumably, this precision would be set at a level such that very, very few chemical outliers among the colony's members are rejected but most outsiders are. That is, recognition may focus on a common feature, a set of compounds and specific ratios of those compounds, that all workers in a colony possess, that few outsiders do, and that can be discerned by the workers.

Despite the above measures, many eusocial insect colonies do harbor heterospecific infiltrators, Wasmannian mimics who masquerade as colony members by virtue of wearing the colony signature. In many cases, these parasites somehow acquire their host's signature by the same means that colony members do: absorbing nest chemicals through intimate exposure and feeding. But, some termites harbor parasites who apparently synthesize the colony signature chemicals de novo. Such syntheses may have evolved fortuitously by exploiting receiver biases of the host species (cf. Section 6.1), an opportunistic process that, when applied to receiver biases of prey species, might also lead to the evolution of aggressive mimicry (e.g., bolas spiders; see Section 3.1.5).

Does the elusive nature of colony signals reside in their complexity and the precision with which they are evaluated by receivers? Possibly, as colony signals may include many more minor components than sexual advertisement pheromones do, and the sensitivity of olfaction in some Hymenoptera (e.g., the honeybee) may approach that of male moths. Moreover, colony signal is often perceived via direct antennal contact between workers rather than in the drifting filaments of an aerial plume, and receivers may thereby detect components that comprise only a minute percentage of the entire signal. Another difficulty may arise because colony signals are learned, and the signals that are learned change as the colony's worker force and its resources change. Thus, any attempt to characterize fully the compounds comprising a colony signal and their evaluation would have to update the bioassay continuously.

Are Genetic Subgroups Recognized by Intracolony Signatures? Once it became clear that colony-specific olfactory signals exist and are the means by which workers discriminate nestmates from outsiders, not a few investigators began to explore the possibility that genetic subgroups within eusocial Hymenoptera colonies are discriminated in like manner. Genetic subgroups arise because queens may mate with multiple males (polyandry, in the parlance of eusocial insect research), and multiple related or unrelated queens and their progeny may share a single nest (polygyny) in some species. These subgroups also arise in species in which new colonies are formed by fission (swarming). When polyandry or polygyny occur, not all workers are full-sisters. Rather, workers may only be half-sisters of their nestmates, or even unrelated. Colony fission creates an analogous situation, because workers in a new colony may be daughters of either the old queen or the new queen, who is the old queen's daughter; daughters of the new queen would be either full- or half-nieces of daughters of the old queen. By the same lines of argument used in the beginning of this section, workers should, if possible, preferentially donate their altruism to full-siblings. That is, a worker should first assist her own mother to raise offspring who had been sired by her father.

Initially, a flurry of experimental findings claimed that workers in various eusocial hymenopteran species were indeed adhering to the above prediction (e.g., Getz and Smith 1983, Getz et al. 1986). These findings were believable because the same genetically determined odors allowing workers to discriminate nestmates from outsiders should be usable, albeit with more precise evaluation, for discriminating full- from half-sisters, etc. Theoretical models were even devised to predict the expected number of loci and alleles influencing genetically determined colony signatures (Getz 1981, Getz and Chapman 1987, Getz and Page 1991). Nevertheless, reconsideration of many of these findings and more rigorous experiments have failed to confirm the contention that workers recognize and discriminate among genetic subgroups within colonies and do so by olfactory signatures.

Current opinion on the functioning of altruism within eusocial insect colonies does not find the lack of subgroup recognition surprising (Keller 1997).[56] Were workers to implement the finely tuned recognition necessary for discriminating full- from half-sisters, the reduction or absence of nepotistic favors donated to distant relatives might not be offset, in the arithmetic of inclusive fitness, by the increase in favors extended to closer relatives. Additionally, recognition errors might lead to some close relatives failing to receive favors due them. On a different level, a colony comprised of non-cooperating or only partially cooperating genetic subgroups may not operate as an efficient unit, negating many of the benefits of sociality. Thus, colony members might be expected to thwart the formation of renegade subgroups, each with its own agenda, by "scrambling" recognition stimuli among all workers (Keller 1997). Scrambling does occur in colonies by transfer of recognition stimuli through physical contact between workers, by exposure to the materials with which the nest is constructed, and by the exchange of food. These factors imply that chemical signals serving for subgroup recognition, should they exist, would have to be resistant to casual transfer among workers. They would also, of course, have to be determined primarily by genotype. While some behavioral evidence supports the existence of genetically determined recognition signals in eusocial Hymenoptera and Isoptera (e.g., Adams 1991), the genetics of such signals are currently unknown, and we cannot evaluate the possibility that genetic differences between full- and half-sisters would be sufficient to generate distinguishable signatures.

Origins. Hypotheses explaining the evolution of colony olfactory signatures include origins in nest recognition, parent–offspring recognition, and inbreeding avoidance. Solitary Hymenoptera, including the ancestors of eusocial species, would confront each of these issues. Hymenopterans returning to their nests generally recognize that they are at the correct location by means of visual cues that they learned when they left. But the spatial clustering of nests in many species would lead to frequent errors, and it is likely that nest-specific chemical cues, perhaps from nest material, are learned and recognized as well.[57] This problem of nest recognition might be particularly acute in species where females maintain separate, individual nests that branch off a single, common burrow.

That parents should feed and care for their own offspring rather than those of others is obvious. Thus, females who can discriminate their offspring from others would hold a considerable advantage in nesting situations where mistakes in nest recognition might occur. Such selection pressure could lead to the evolution of recognition signals, either genetically or environmentally determined, by which parents ensure that they are provisioning their own offspring (see Gamboa et al. 1986).[58]

Extreme inbreeding is detrimental in general, and in some Hymenoptera it may present a specific problem owing to their mechanism of sex determination: females are heterozygous at the sex-determining locus, and males are either homozygous or hemizygous (haploid), but in many species the homozygous (male) state is lethal or sterile. Consequently, special selection pressure to discriminate genotypes and avoid similar ones during mating may exist in order to reduce the chance of producing homozygous offspring. Could this hypothetical avoidance have evolved to a reversed application in the context of nestmate provisioning within eusocial species?

3.5.2 Individual Recognition and Categorical Perception

The treatment of recognition signals cannot be considered complete without raising the issue of individual-level recognition. Do the same sorts of signature systems used to discriminate colony membership in the eusocial insects also afford recognition of specific individuals in arthropods? Here, we might initially exclude recognition of the queen in eusocial insects, as this behavior is more properly designated caste recognition; the confusing point is that the recognized caste normally has a membership of one. But, other arthropod behaviors do suggest that mates and neighbors are discriminated on a level that may be truly individual. Investigators working with the desert isopod *Hemilepistus reaumuri* have reported that male–female recognition in its long-term pair-bonds is maintained by chemical signatures. These signatures are claimed to be individual-specific and genetically determined (Linsenmair 1984, 1985, 1987). Aggression in various marine crustaceans (Caldwell 1985, 1992, Karavanich and Atema 1998, Berkey and Atema 1999) and some species of cockroaches (Clark et al. 1995) may be regulated by chemically mediated recognition of neighbors. Animals who have lost fights with particular neighbors behave subordinately when encountering these particular individuals at a later time. That they do not behave in a subordinate manner toward individuals with whom they had no previous experience indicates that their responses toward familiar individuals were not simply affected by their previous record of losing fights.

The scientific literature now lists hundreds of confirmed and suspected cases of individual-level recognition among animals. Such recognition may be accomplished by the discrimination of chemical, visual, or acoustic stimuli. While the number of arthropods on the list has grown markedly in the past 10 years, all cases seem based on the discrimination of chemical signatures.[59] Does this feature of individual recognition among arthropods reflect relative inadequacies in their visual and acoustic perception? We shall consider this point in the next two chapters.

I close this section by noting that from a certain perspective the issue of individual recognition is a philosophical one. Is individual recognition in animals, including humans, just another form of categorical perception in which events or objects, conspecific individuals in this case, are assigned to discrete categories based on the neural representation evoked when they are perceived (Green and Marler 1979)? Thus, when a male desert isopod recognizes a particular female as his long-term mate, is he merely identifying her because of perceived features that he has learned from past experience and that no other local individuals are likely to match? If so, then individual recognition really does not differ from other forms of discrimination, such as nestmate and caste recognition in the eusocial insects. Ruses that human identical twins play to interfere in each other's conjugal relationships, as occasionally reported by the press, suggest that this perspective on individual recognition applies to our species as well.

3.6 Social Behavior: A Call to Arms

The stored food and nest material, brood, and workers of a eusocial insect colony represent a rare concentration of resources in most habitats. Thus, most eusocial insect species must contend with a host of generalist and specialist natural enemies seeking to plunder the colony's organic wealth. The venomous stings, scythe-like mandibles, noxious chemical sprays, and glues found among eusocial insects represent, for the most part, evolutionary responses to these enemies. Such defenses may not be particularly effective when wielded by a single individual, but when hundreds or thousands of workers mount a synchronized defense, even large vertebrates can be held at bay and repelled. Synchronized defenses are possible, of course, because a large number of workers may be available. The synchrony with which worker forces are summoned and choreographed to defend the colony effectively is accomplished by specialized chemical signals normally termed alarm pheromones.

3.6.1 Collective Defense: Hymenoptera

In the honeybee, *A. mellifera*, individual workers emit alarm pheromones from exocrine glands associated with the stinger at the end of the abdomen (Schmidt 1998). The processes of alarm and group defense begin when workers performing sentry duty near the colony perceive a threat, expose their sting chambers, and return to the colony while buzzing their wings. These behaviors disseminate alarm pheromone rapidly among many workers, who, upon detecting the pheromone, exit the colony. Once outside, recruited workers are drawn to the threat by various stimuli including its movement, odors, and temperature. Autotomized stingers left in the skin of an intruder by the initial defenders continue to release alarm pheromone and thereby recruit more workers outside and also draw their attacks. Thus, a positive feedback process leads to a rapid buildup of the defense force and its strategic deployment.

Collective defense similar to that described above is found throughout the eusocial insects (Vander Meer et al. 1998). The alarm pheromones coordinating these phenomena are emitted from various exocrine glands, including the mandibular, Dufour's, and venom glands associated with the stinger. Behavioral assays have identified a wide variety of terpenoid, aliphatic, and aromatic compounds as components of these alarm pheromones. The only features common to alarm pheromone compounds are that most contain only 5–10 carbon atoms and have molecular weights ranging between 100 and 200 (Figure 3.20). Consequently, alarm pheromones are relatively volatile and have short "fade-out" times in comparison with sexual advertisement pheromones. Volatility would be critical for summoning a rapid defensive response, and a short fade-out time would reduce the possibility of false alarms, inappropriate responses to amounts of pheromone lingering near the colony long after any danger has passed.

Alarm pheromones are also more complex bouquets of compounds than sex pheromones, and may include more than 30 components. The full complement of compounds is not necessary for an alarm response, however, and even single compounds may elicit some reaction. Thus, a colony's response to a threat can

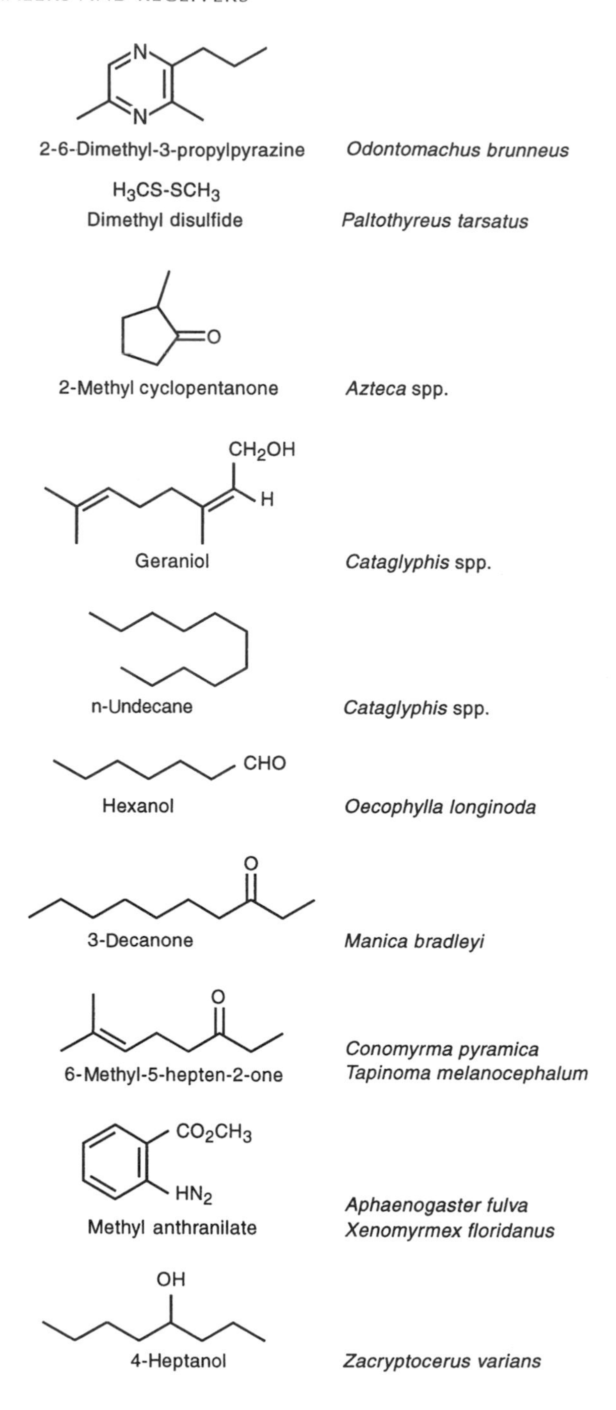

Figure 3.20 Alarm pheromones in ants. Chemical structures of alarm substances in 11 different ant species (right column) are shown. 6-Methyl-5-hepten-2-one and methyl anthranilate have each been found in two different species; two different alarm substances are shown for *Cataglyphis* spp. (Vander Meer and Alonso 1998).

begin immediately, because receivers may not defer their reaction until they have detected all or even a majority of the alarm pheromone's components. Reactions to alarm pheromones may also be graded responses in which an individual worker is more likely to react, or exhibits a more pronounced reaction level, as more components are present and detected and as the overall pheromone titer increases. For example, an elevated response level may be elicited by detection of the less volatile components of alarm pheromone. Graded responses may allow a colony's overall reaction to be commensurate with the level of perceived threat: A higher alarm pheromone titer, which is more likely to be detected by a greater percentage of the colony's workers, would be generated when the sentries making initial contact actually sting an intruder rather than merely inspecting or repelling one.

Because only a small proportion of all the compounds comprising an alarm pheromone need be present or detected, it should come as no surprise that alarm pheromones generally bear little species specificity. Different species sharing the same alarm pheromone components, however would not face the problems that they would if they were to share sexual advertisement pheromones. The volatility of alarm pheromones ensures that they would not form coherent plumes extending far downwind from a nest. Additionally, any other species nesting close enough for its workers to detect the heterospecific alarm pheromone would probably be threatened by the same danger triggering that alarm.

The interspecific differences in alarm pheromones that do occur may represent adaptations to different habitats or levels of predation. For example, among *Apis* honeybees, species that nest in the open include larger proportions of one relatively non-volatile component, 2-decenyl acetate, than is found in cavity nesters (i.e., *A. mellifera* and *A. cerana*). Do the open nesters require a more coherent plume because of their exposure to wind, and do they also suffer greater predation pressure? Among different varieties of *A. mellifera*, Africanized honeybees produce a greater titer of alarm pheromone than do European honeybees. Does this difference reflect the pressure from vertebrate predators in Africa that specialize on honeybee colonies?

The association of alarm pheromone with the sting apparatus and venom glands in various eusocial Hymenoptera offers a clue to the origin of these signals. In the manner argued for other chemical signals, worker responses to odors inadvertently associated with attacks on intruders may have been selected for strongly: such responses would facilitate the group attacks that form the very core of colony defense. Once established, receiver perception of, and response to, cues from fellow workers attacking intruders may have coevolved with certain modifications of those cues (cf. Section 6.1). Thus, increasingly conspicuous odors that represented bona fide signals rather than inadvertent cues would have been selected for. Easily recognized alarm signals would accelerate colony defense and benefit the inclusive fitness of a worker in either the receiver or signaler role.

3.6.2 *Collective Defense: Aphidoidea*

Elsewhere among social arthropods, alarm pheromones are well developed in various aphid species, where the individuals of a clonal colony may release an alarm substance if threatened or attacked (Stern and Foster 1996, 1997). The substance is released from the cornicles, a pair of tubular structures positioned dorsally on the abdomen, which also produce a noxious spray or exudate that can function in defense against some natural enemies. This dual function of the aphid cornicle in alarm and defense strengthens the argument presented above for the evolution of arthropod alarm signals by coevolution of receiver perception with defensive cues. In most cases, aphid colony members who perceive alarm substances simply disperse from the potential danger. Presumably, emission of alarm pheromone benefits the signaler's inclusive fitness because of its clonal relationships to other colony members, who may save themselves by fleeing. These benefits would outweigh any risk, which might be minimal or non-existent, that the signaler itself incurs by emitting alarm pheromone.

Some species of hormaphidid and pemphagid aphids produce specialized soldier castes and, by many accounts, are considered eusocial. Aphid soldiers do not disperse in response to the alarm pheromone, but instead may orient toward an approaching natural enemy and pierce it with their stylets (mouthparts). This collective defense response parallels more closely the function of alarm pheromones in the eusocial Hymenoptera and Isoptera than other aphid alarm behaviors do.

3.7 Social Behavior: Recruitment Trails

One of the major payoffs of the social way of life is access to kinds and amounts of food resources that would otherwise remain unavailable. Many eusocial Hymenoptera and Isoptera obtain such access by recruiting sizable worker forces to forage at food-rich patches discovered by lone foragers or specialist scout individuals. Here, the colony relies on information provided by a very small number of individuals who happen to come upon a valuable resource patch by chance. Information concerning the value and location of the patch is communicated by its discoverers to the colony's workers en masse, many of whom are stimulated by the communication and orient to the patch. Thus, local resources can be efficiently harvested and stored, thereby freeing other workers to specialize on tasks such as defense, brood care, or nest construction and maintenance. In the terrestrial eusocial insects, the ants (Formicidae) and termites (Isoptera), trail pheromones serve as the signal by which discoverers communicate to potential recruits information on resource patches.

3.7.1 Formicidae and Isoptera

In various species of ants and termites, recruitment trail communication begins with an outbound worker on an exploratory mission repeatedly touching the end of her gaster (abdomen) to the substrate as she walks (Traniello and Robson 1995, Vander Meer and Alonso 1998, Reinhard and Kaib 2001). This behavior leaves a series of faint chemical dots along her path, which form an autocommunication signal that she can later follow to return to the nest. If this scout individual has found a valuable resource on her trip, she may reinforce the dots by depositing additional chemicals on the inbound journey. These chemicals added to her path initiate the formation of a recruitment trail. Other foraging workers may detect these odors and be recruited to follow the trail to the resource and retrieve it morsel by morsel. If they too assess the resource value as high, they may add their own chemical deposits to the trail. Thus, a system of semi-permanent trunk trails leading away from a nest toward resource patches can emerge from the simple behaviors of individual foragers (Deneubourg and Goss 1989, Deneubourg et al. 1990, Franks et al. 1991): trails ending at valuable patches are chemically intensified, whereas those ending at undesirable or depleted resources are not and gradually fade away. Such recruitment trail systems may be useful for any social arthropod foraging terrestrially on patchily distributed resources that remain valuable for an extended time interval. Trails would not be useful for species foraging on dispersed and ephemeral resources. As an example, *Cataglyphis* ants of the Sahara Desert, which feed on small arthropods that have succumbed to heat, are individual foragers and do not make any use of recruitment trails.

Species and Colony Specificity?. Operationally, trail pheromone signals may be expected to combine elements of both nestmate recognition and alarm substances. A forager who follows a heterospecific trail may not locate suitable resources, and following a trail laid by a neighboring conspecific colony may result in an unfortunate aggressive encounter. Moreover, difficulties in returning to the correct (home) nest may arise. Thus, some degree of species and colony specificity in trail pheromone would be predicted. Bioassays have demonstrated such specificity in some species of ants, particularly when a choice between trail pheromone from the home nest and another is offered. In many ants, however, and in general among termites, a surprising lack of specificity has been noted: different species may include identical compounds in their trail pheromones and may readily follow each other's trails in laboratory bioassays. Nonetheless, there is little evidence that foragers experience confusion while orienting along odor trails in the field, and other mechanisms must serve to keep a colony's individuals marching on the appropriate paths. In some cases, chemical and visual landmarks, or a solar or geomagnetic compass, may augment odor trail following during outbound and inbound foraging journeys. Here, a forager who encounters an ambiguous junction between the odor trails of its colony and a neighboring one might orient with respect to a learned chemical landmark and thereby follow the correct trail. Additionally, behavioral mechanisms may serve to prevent such trail junctions from ever arising (Adams 1994). For example, different colonies or species of Nearctic harvester ants (*Messor*) may follow each other's trails in bioassays, but their trails do not intersect in the field. These ants tend to be fiercely territorial, and workers from one colony would rapidly detect and deter a neighboring colony's trail construction project extending into its domain. Colony-specific odor signatures on the ants would enable recognition of the encroaching construction workers as foreign elements.

Glandular Sources, Chemistry, and Odor Dispersion. Because workers are first alerted before they are summoned to orient along a pathway by trail pheromones, these substances might also share features with alarm pheromones. Chemical analyses and behavioral assays in various ant species support this prediction. In the dolichoderine ant *Tapinoma simrothi*, the same pygidial gland substances elicit alarm behavior when in high concentration and emitted from a point source and trail-following behavior when in low concentration and emitted with an elongate active space (Simon and Hefetz 1991). In many ant species components of the trail pheromone are secreted by the venom glands, which may also excite alarm behavior.

In the eusocial Hymenoptera and Isoptera, trail pheromones are normally secreted by various exocrine glands at the end of the abdomen: Dufour's, venom, and pygidial glands are known sources. Hindgut fluids passed by the rectum also serve as trail pheromone components in many species, and in the ponerine ant *Prionopelta amabilis* specialized glands on the tarsi of the hindlegs secrete trail pheromone (Hölldobler et al. 1992). Often, several of these sources contribute to the overall trail pheromone.

As with alarm pheromones, most trail pheromones include multiple compounds that may elicit different elements of recruitment behavior. A volatile compound(s) may stimulate the initial excitement or state of alertness that draws workers out of the nest or toward a trunk trail. Less evaporative compounds may then serve to stimulate actual walking along the trail. The compounds from glandular sources include a range of aliphatic, aromatic, and terpenoid substances of various molecular weights and diffusion coefficients (Figure 3.21). Deposited on a terrestrial trail, some of these organic compounds may be fairly resistant to rain and other forms of weathering. This property contributes to the longevity of trunk trails, and it may be particularly crucial for species in the wet tropics (Beugnon and Dejean 1992).

The actual chemical signal of the trail has both surface and aerial aspects. Depending on the porosity of the substrate and the volatility of the compound, some amount of chemical signal will adhere to the surface on which it was deposited. Owing to evaporation and degradation by physical factors such as

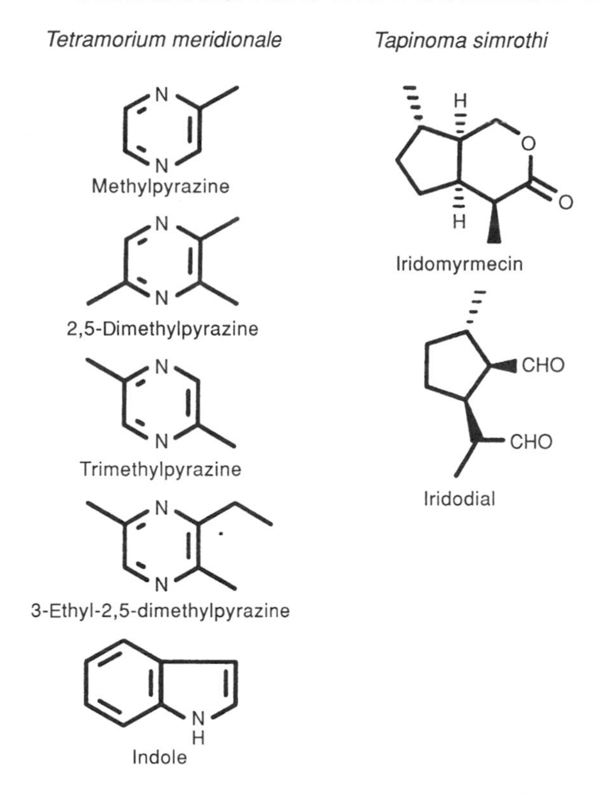

Figure 3.21 Trail pheromones in ants. Chemical structures of trail substances found in two ant species, *Tetramorium meridionale* (Myrmicinae) and *Tapinoma simrothi* (Dolichoderinae) (Vander Meer and Alonso 1998).

UV radiation, this aspect of the signal will slowly diminish over time, but until the amount dips below a threshold value, an arthropod walking along the trail may detect it via contact chemoreception. The aerial aspect of the chemical signal arises from evaporation and diffusion away from the trail, and its active space comprises a slender hemicylindrical volume centered over the trail axis. This dispersion pattern arises for several reasons. Some trail pheromone compounds are relatively volatile, and pheromone diffusing or carried by convection far from the trail will be unlikely to remain in concentrations above the receiver threshold. Additionally, an individual's pheromone titer is distributed along and emitted from a linear source, rather than one point, and the pheromone flux at any given instant from a point along that linear source (trail) would represent only a tiny fraction of an already small amount that had been deposited there. Thus, a coherent and detectable plume extending away from the trail will not form, but an active space will exist within a slender hemicylindrical volume over the trail, and an arthropod walking within its confines may detect the pheromone via olfaction.

Orientation. Unlike flying or walking arthropods tracking an odor plume established by convection (Section 3.1.4), arthropods walking along a terrestrial odor trail may orient easily by simultaneous comparison of olfactory inputs to their two antennae. Both surface and aerial aspects of the trail have steep lateral gradients, and sharp boundaries delimit the extent of pheromone concentration that exceeds a receiver's threshold. Thus, an arthropod may walk with one antenna within those boundaries and the other outside. Should the individual step astray, it can immediately return to the trail by simply turning toward the antenna that had been within the active space. Alternatively, it may walk with both antennae within the active space and correct any lateral transgression by turning away from the antenna that momentarily lost detection of the odor. And sequential comparison of olfactory input may also be used in certain situations, as when an arthropod reaches a bifurcation point in the trail. At such junctures, the two antennae may be used as a single unit to compare successively the odor concentrations of each fork.

While the above algorithms may serve to keep an arthropod on track along an odor trail, they cannot ensure that it will proceed in the correct direction. Because convection does not influence either the surface or aerial aspect of a terrestrial odor trail, the trail's polarity should not be apparent to the tracker. As expected, an ant experimentally removed from and shortly replaced on its trail exhibits no ability to select its previous heading and continue walking in that direction. However, under natural circumstances walking arthropods are not troubled by this ambiguous property of odor trails, and they somehow find resource patches and return to their nests with minimal difficulty. Given that prior trail layers left no clue of the direction in which they were walking while depositing pheromone,[60] and the tracking individuals lack inertial devices that could retain a memory of their own previous direction, how is their orientation accomplished? A likely means is by relying on additional external cues, which are usually present. As noted earlier, learned chemical and visual landmarks or compass mechanisms can reliably indicate the direction toward a resource patch or the home nest. Additionally, the forager will encounter a stream of fellow workers along the trail and may exchange tactile and chemical information with them. Thus, an outbound forager who encounters and antennates a returning worker carrying a food morsel may simply respond by continuing to move in the opposite direction of that worker. By itself, though, this mechanism would rely on the trail's initial workers walking in the correct direction and all subsequent ones following suit. A single error would be copied and would generate a string of wayward foragers.

Collective Responses and Emergence of Semi-permanent Trail Systems. Similar to alarm pheromones, graded responses to trail pheromones may exist and collectively allow a colony to exploit resource patches efficiently. The amount of pheromone that a worker deposits on a trail while returning to her nest may be commensurate with the perceived value of the resource (Beckers et al. 1993). Thus, a later recruit reaching a bifurcation in the colony's trail system may be led to the more valuable resource patch, because the fork leading to it bears a more intense signal. Positive feedback will then reinforce this bias, and most recruits will forage at the more valuable patches. This process should not be expected to match a colony's foraging trails and effort perfectly with the quality of local patches, though. Stochastic events would establish some initial biases toward average or even poor-quality patches, and foragers may continue to be recruited toward and to visit these patches for some time (see Stickland et al. 1993, 1995, Edelstein-Keshet 1994, Bonabeau et al. 1997).

3.7.2 Odor Trails for Migration and Swarming

Trail pheromones also occur in those terrestrial eusocial insects with a nomadic lifestyle and in some of the aerial eusocial insects, the vespid wasps and the apid bees, but they do not function in recruitment to valuable resource patches. Ecitonine army ants move their colony to a new bivouac every few days, and the multitude of workers are recruited to the new site by odor trails left by scout individuals (Gotwald 1995). In species of vespid wasps where new colonies are founded by fission or swarming, the segment of the old colony that emigrates to a new nest location may be recruited to that site by scent that (scout?) workers deposited at stations along the emigration route (Landolt et al. 1998). The glandular sources and chemistry of these scents are poorly known. The scent stations are typically situated on locally conspicuous vegetation and form an intermittent trail, which would not be tracked in the same manner as a terrestrial odor trail. Rather, the scent at each station may represent a signpost, and each signpost along a trail is probably found by local search, although visual observations of wasps in the vanguard must assist following individuals. In the honeybee, *A. mellifera*, swarms are led to the new nest location by pheromone from the Nasanov's gland on the dorsal side of the abdomen. Scouts who have found a suitable cavity for the new nest advertise that location by emitting Nasanov's gland pheromone, a mixture of volatile terpenoid compounds. This chemical signal attracts the attention of additional scouts, who may return to the old nest and communicate information on the value and location of the potential new site to other individuals who will participate in the swarm. Thus, Nasanov's gland pheromone is a recruitment signal, but it does not form a trail. The information on the new site is communicated by signals transmitted along various mechanical channels (Section 4.4.3), precluding the need for a continuous chemical link between the two sites.

3.7.3 Social Lepidoptera

Many social lepidopteran larvae also exhibit complex systems of chemical recruitment trails for exploiting resources or moving a nomadic colony to a new site (Fitzgerald 1995, Costa and Pierce 1997). Such communication may bolster further the growing claims that separation between the so-called eusocial insects and other social arthropods is an artificial construct (Costa and Pierce 1997, Wcislo 1997). Lepidopteran recruitment is most elaborate in the lasiocampid tent caterpillars, where individuals leaving a home base, the silken tent, lay an autocommunicative exploratory trail. As with eusocial hymenopterans and isopterans, the caterpillars supplement the trail odor on their return trip if they located desirable foliage. Unlike eusocial hymenopterans and isopterans, the caterpillars do not sustain a division of labor between specialist scouts and foragers, and individuals who locate food both forage on that food and lay information trails that recruit their siblings from the tent.[61]

In *Malacosoma* tent caterpillars, the trail pheromone is apparently secreted by epidermal cells at the end of the abdomen, but in other social caterpillars the active compounds may be incorporated in silken strands laid along the trail. Several steroid compounds have been identified as caterpillar trail pheromones, and they may be active in concentrations as low as $1\ \mathrm{pg{\cdot}mm^{-1}}$ trail. The bi-level communication system of exploratory and recruitment trails is maintained in most cases by the higher pheromone concentration in recruitment trails, but in the Eastern tent caterpillar, *Malacosoma americanum*, a different steroid compound may be used for each type of trail. These compounds are quite low in volatility and would be detected by contact chemoreception, via receptors on the maxillary palps, only. Thus, caterpillar trail pheromones only elicit orientation along trails and do not alert individuals that a food patch has just been located. This latter function may be accomplished by hostplant volatiles that waft off caterpillars returning from a successful feeding trip.

3.7.4 Origins

Given the general sophistication of recruitment trail communication in social arthropods outlined above, can the evolution of these signaling systems be inferred? Fortunately, the occurrence of recruitment signals exhibiting intermediate levels of development allows for some informed speculation. In the myrmicine ant *Myrmicaria eumenoides*, a predatory species that hunts various arthropods, scouts who locate a prey too large to be subdued individually summon nestmates by emitting a volatile compound from the venom gland (Kaib and Dittebrand 1990). Is this act not one step removed from emitting an alarm pheromone in the presence of an intruder? Emitting a substance that elicits alarm behavior may effectively recruit workers to resource patches near the nest or other large congregations of nestmates, however it would not suffice for more distant patches. Various species of ants have solved the problem of distance recruitment by "tandem running": an ant who has found a patch returns to the nest, alerts a nestmate with volatile venom or pygidial gland compounds, and then leads the individual recruit back to the patch by maintaining contact with her en route (Moglich et al. 1974; Figure 3.22). But tandem running is an awkward affair, and it could not represent an efficient means of recruiting thousands of workers on short notice. Viewed from this perspective, worker ability to detect and respond to non-contact signals emitted by returning scouts would offer a colony a distinct advantage in resource acquisition. Two levels of such detection and response occur. In the first, as exemplified by the carpenter ant *Camponotus socius*, the returning scout stimulates multiple recruits at the nest, who then follow her back to the patch en masse (Hölldobler 1971).

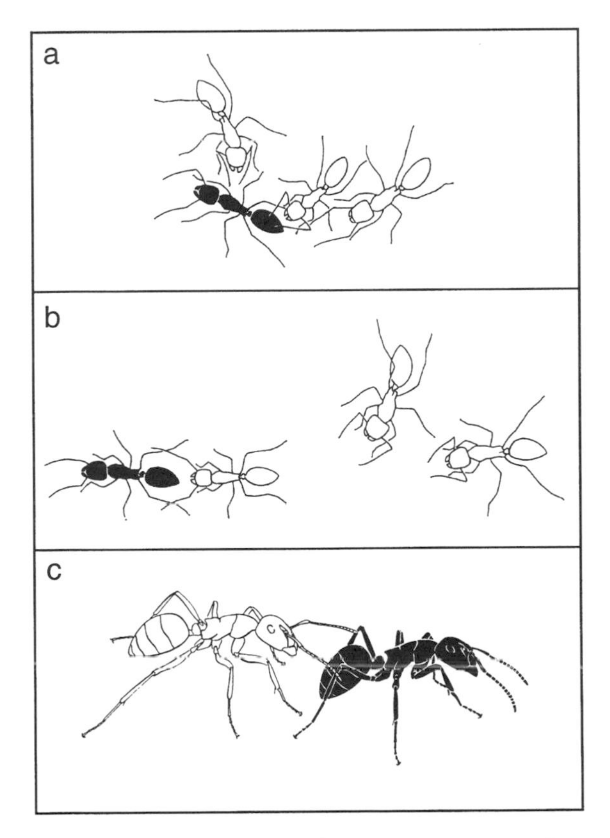

Figure 3.22 Tandem running in the carpenter ant *Camponotus sericeus* (Formicinae). (a) Forager (black) being antennated by potential recruits (unshaded). (b) Forager being followed by a single recruit back to the food source. (c) Detail of antennation during tandem running. (Reprinted from "Communication by tandem running in the ant *Camponotus sericeus*," *Journal of Comparative Physiology A*, vol. 90, p. 108, fig. 2a, B. Hölldobler, M. Moglich, and U. Maschwitz, copyright 1974 with permission by Springer-Verlag.)

In the second, of which there are many variations (Section 3.7.1), the recruited workers do not follow the scout directly. As described earlier in this section, workers stimulated by volatile compounds emitted by the returning scout are then recruited to orient along the scout's odor trail back to the patch.

The different forms of recruitment signaling seen among ants are spread across subfamilies and do not reflect phylogeny. Rather, they may reflect selection pressures imposed by ecological and social factors such as resource dispersion and colony population size. Where resources are distant from the nest or require a mass of foragers to be efficiently exploited, worker ability to detect and respond to compounds emitted by food discoverers would offer a colony a distinct advantage in resource acquisition. Here, the perceptual abilities and odors from nestmate recognition, alarm, and autocommunication signaling might have been co-opted: Could workers have evolved an orientation response to the autocommunicative trails of other individuals or to accidental deposits of Dufour's gland secretion, a common source of recognition signals and also an ingredient in many trail pheromones, left by returning foragers whose gasters dragged on the ground? And, with perception of and orientation to the autocommunicative or incidental odors of other individuals established, could specialized and less ambiguous recruitment trail signals that enhanced resource acquisition further have evolved subsequently? Earlier, we considered whether various sexual and social signals could have originated as other signals or cues that coevolved with receiver perceptual abilities (Section 3.1.5). This fundamental evolutionary mechanism (Section 6.1) is also plausible for recruitment trails, as the elicited responses, and signals, overlap those found in nestmate recognition, alarm, and autocommunication. On the other hand, origins in resource odors or other environmental cues do not appear likely. There exists no evidence that trail pheromones are simple modifications of compounds in food or other resources or that they communicate the identity of a resource beyond its location and value.

3.8 Synopsis

When Murray Blum stated in an early review (Blum 1974) that "the road to insect sociality was paved with pheromones," he foretold the array of findings summarized in the previous four sections. I began this chapter by emphasizing the relative importance of chemical communication in arthropods, and this importance is nowhere greater than among social species. While the mechanical and visual channels may supplement information exchange in some species, we are left with an indelible impression that the overwhelming majority of signals are transmitted and received along the olfactory channel.

Several factors may have led social arthropods to favor the olfactory channel. Unlike sexual advertisement by female arthropods, though, avoidance of natural enemies was probably not one of them: social Hymenoptera and Isoptera are besieged by a host of social parasites, Wasmannian and aggressive mimics, who are attracted by and exploit recognition, alarm, and trail pheromones (Howard and Akre 1995). Instead, basic features of social life, constraints on signal transmission, and arthropod sensory biology may have been deciding factors. First, life within burrows and dark cavities may preclude the use of visual signals based on reflected light. Second, short of architectural construction, chemical signals represent the only means available to animals for creating a message whose transmission continues in the absence of the sender. This feature may be critical for informing multiple receivers and reiterating that message over an extended time interval. But the most crucial advantage of the olfactory channel may be its effective capacity for transmitting information, the extensive vocabulary needed to communicate with multiple categories of individuals and in diverse circumstances. In the following chapters (Sections 4.5 and 5.4) I address the question of why the effective information capacities of other channels may be limited in arthropods.

4

Sound and Vibration and the Mechanical Channel

Communication along the mechanical channel is highly developed in only two animal phyla, the arthropods and the chordates. In arthropods as a whole, sound and vibration are not the pre-eminent modalities for signaling, but they nonetheless play vital functions in diverse taxa distributed among seven major insect orders,[1] several minor ones,[2] and some groups of chelicerates and crustaceans (Table 4.1). In these arthropod species, sound and vibration serve as sexual advertisement, courtship, aggression, defense, and social recruitment signals. Moreover, certain of these communication systems have been studied most intensively, and various focused efforts have come to represent major contributions to the physiology and evolution of sensory perception and signaling. Studies on perception of and orientation to advertisement calling songs, and production of those songs, in the European field cricket *Gryllus bimaculatus* form one of our most comprehensive analyses of peripheral and central filtering and motor control (Huber and Thorson 1985, Huber 1990, 1992). Recruitment signals, the so-called "dance language," in the honeybee *Apis mellifera* are transmitted acoustically (Kirchner 1993, Kirchner and Towne 1994), as well as via tactile and vibratory channels, and their analysis has contributed much to our understanding of signal coding (Gould et al. 1985, Gould and Towne 1987). Other acoustic arthropods have served as model subjects in landmark studies on sexual selection (e.g., Cade 1979, 1981) and speciation (e.g., Walker 1964, Hoy et al. 1988).

Advancements in general technology that could be readily adapted to research on acoustic arthropods no doubt fueled some of the success in this area. Several times during the past century, experimental breakthroughs in acoustic communication research directly followed the introduction of electronic devices. The first such instance was Johan Regen's use of the telephone in pre-World War I Vienna to demonstrate unequivocally that female crickets respond and orient to the malc calling song (Regen 1913). Unlike advertisement pheromones, there had existed little uncertainty that crickets and other acoustic insects used their songs to communicate, as the signal was conspicuous to human observers and the channel was a familiar one, but this simple, yet

Table 4.1 Mechanical Communication in the Arthropoda

Class / Order / Suborder, superfamily, or family	Signaling mode(s)	Function
Arachnida		
Araneae	vb-ss*, vb-web*, sd-nf?	sx_{ad}, sx_{ag}, sx_{co}, sx-du, soc
Acari	vb-ss	sx_{co}
Malacostraca		
Decapoda	vb-ss, sd-ff?	sx_{ad}, sx_{ag}, sx_{co}, ag
Insecta		
Orthoptera		
Caelifera	sd-ff	sx_{ad}, sx_{ag}, sx_{co}, sx-du
Ensifera	sd-ff*, sd-nf, vb-ss, wd	sx_{ad}, sx_{ag}, sx_{co}, sx-du
Blattodea	sd-nf	sx_{ag}
Isoptera	vb-ss	soc
Plecoptera	vb-ss*	sx_{ad}, sx_{co}, sx-du
Psocoptera	vb-ss?	
Hemiptera		
Belostomatidae	vb-ws	sx_{co}
Gerridae	vb-ws*	sx_{ad}, sx_{ag}, sx_{co}, sx-du, ag
Corixidae	sd-uw*	sx_{ad}, sx-du
Cydnidae	vb-ss	sx_{co}
Pentatomidae	vb-ss	sx_{co}
Reduviidae	vb-ss	sx_{co}
Homoptera		
Cicadoidea	sd-ff*, vb-ss*	sx_{ad}, sx_{co}, p-o, sx-du
Fulgoroidea	vb-ss	sx_{co}, sx-du
Neuroptera		
Chrysopidae	vb-ss*	sx_{co}, sx-du
Lepidoptera		
Pyraloidea	sd-ff	sx_{ad}, sx_{ag}
Noctuoidea	sd-ff	sx_{ad}, sx_{co}, sx-du, ap
Nymphalidae	sd-ff	sx_{ag}, sx_{co}
Lycaenidae	vb-ss	het-r
Coleoptera		
Gyrinidae	vb-ws*	
Hydrophilidae	vb-ss	sx_{ad}, sx_{co}, sx-du
Anobiidae	vb-ss*	sx_{ad}, sx_{co}, sx-du
Passalidae	vb-ss*	soc, p-o
Tenebrionidae	vb-ss	sx_{ad}, sx_{co}, sx-du
Scarabaeidae	vb-ss	sx_{co}
Chrysomelidae	vb-ss	soc
Curculionidae	vb-ss	sx_{co}
Diptera		
Culicidae	sd-nf*	sx_{ad}
Chloropidae	vb-ss?	
Drosophilidae	sd-nf*	sx_{co}
Hymenoptera		
Apoidea	sd-nf, vb-ss	soc
Formicoidea	vb-ss	soc
Vespoidea	vb-ss	soc

Signaling mode: so-ff: far-field airborne sound; so-nf: near-field airborne sound; so-uw: underwater sound; vb-ss: vibration of substrate (ground, nest material, or vegetation); vb-w: vibration of water surface; wd: wind (air current). *Widespread occurrence throughout taxon. Function: ag: general aggression; ap: aposematic warning; het-r: heterospecific recruitment; p-o: parent–offspring; soc: social; sx_{ad}: male sexual advertisement; sx_{ag}: intrasexual aggression; sx_{co}: male courtship; sx-du: male–female duetting; ?: communication suggested but not confirmed.

novel, electric device allowed an investigator for the first time to separate the signal from the signaler. Later, when magnetic recording media became available after World War II, an investigator could easily store signals recorded with the aid of a transducer and then present them as test stimuli in a playback experiment. Stored signals could also be analyzed in the time and frequency domains—a marked improvement over earlier means of characterizing the songs of insects and other animals. Prior to this time, observer impressions of insect songs were, by necessity, commonly portrayed in the scientific literature by series of notes on musical scales. The digital revolution of the 1980s and 1990s expanded the above possibilities still further. Now the investigator was afforded a seemingly unlimited ability to copy and edit the stored signals and to synthesize them de novo as well. Complex experimental designs such as interactive playback (Dabelsteen 1992) also became possible. As opposed to the instrumentation necessary for arthropod pheromone research, much of the technology used in acoustic and vibration work was developed for a wide range of scientific and commercial applications. Thus, many of the advances in this area were accompanied by reduction, not increase, in cost to the investigator. Moreover, most acoustic and vibration instrumentation is somewhat portable and may be deployed in the field.

The relative ease with which many arthropod acoustic and vibratory signals can be studied, as well as their familiarity to observers of natural history, have led to fairly thorough accounts, particularly of those signals that the unaided human can perceive. As examples, nearly complete surveys of the calling, courtship, and aggressive songs of the acoustic Orthoptera and Cicadidae (Homoptera) of eastern North America (Alexander 1956) and the Tettigonioidea of Europe (Heller 1988) have been compiled. These and other surveys reveal a striking difference between mechanical communication in arthropods and their use of the olfactory channel: mechanical communication is distinguished by a diversity of structures and mechanisms for the production and reception of signals. There is no universal structure for sound production comparable to exocrine glands or secretory epidermal cells, and hearing is not associated with a particular body segment or appendage (e.g., antennae), or organ). Vibratory signaling and reception entails yet additional structures and mechanisms. In this chapter, I introduce these diverse means of signaling and reception along the mechanical channel and then treat their various functions and evolution.

4.1 What Is the Mechanical Channel?

Communication along the mechanical channel occurs whenever an individual moves in a fashion that creates a disturbance generating an inertial force detectable by another individual. Thus, specialized tactile signals between contacting individuals, sound, and substrate vibrations would be included. Because sound and vibration may be rapidly transmitted over relatively long distances via wave motion, they are particularly useful for various sorts of advertisement signaling. Arthropod tactile signals, on the other hand, may contain a greater quantity of information bits than sound and vibration (Section 2.5), and a receiver may thereby respond to them in diverse ways. This property may render tactile signals useful for courtship and social interactions, circumstances in which long-distance transmission may not be an asset.

4.1.1 Sound: Longitudinal Compression Waves

When a mechanical disturbance is created in a compressible medium, the disturbance may spread from its point of origin via longitudinal waves known as sound. While any medium may possess some compressibility and transmit waves that oscillate in the same direction as they travel away from the originating disturbance, most biologically relevant sound waves are transmitted through the fluids in which organisms live, air and water. Because the physical properties of air and water differ greatly, sound has very different characteristics in each. It will travel roughly 4.3 times as fast in water as in air, and a sound of given intensity may be transmitted over a much longer distance in water. In general, a biological receiver can function more efficiently in water than in air, but localizing the source of a sound is easier in air than in water.

Two physical features are associated with the longitudinal oscillations of sound waves (Figure 4.1). First, fluid pressure is caused to rise and fall periodically (sinusoidally in the ideal case) at a given point in the medium; this feature is known as the pressure aspect of sound. Second, as a wave passes a given point, fluid molecules are displaced infinitesimally away from, and then toward, the propagating source; this feature is known as the particle velocity aspect of sound. The particle velocity aspect is more significant than pressure close to the propagating source but declines in relative importance with increased distance. In air, the distance at which both aspects become comparable in strength[3] is approximately equal to one wavelength. Both sound frequency and the geometry and size of the sound source will influence this distance. The region in which the particle velocity aspect of sound is relatively strong is referred to as the near field, distinguished from the far field beyond. Sound waves within the near field behave in a complex manner, and many basic principles of physical acoustics discussed below do not apply in this region. A substantial proportion of acoustic communication in arthropods occurs within the near field, though, and it behooves an investigator to measure such signals and interpret their reception with care.

Frequency (f), wavelength (λ), and velocity (c) of sound are related by the elementary wave equation,

$$f = c/\lambda \tag{4.1}$$

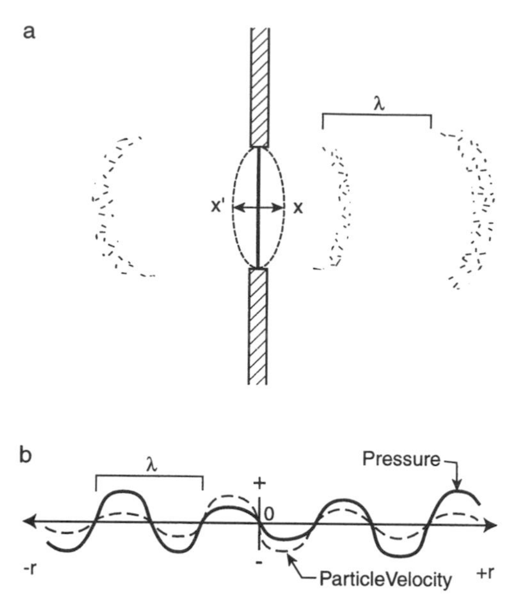

Figure 4.1 (a) Vibration of a membrane (vertical line) creates sound waves propagating away from it in both directions. Stippled arcs represent compressions (pressure above ambient), which are separated by rarefactions (pressure below ambient). View shows waves as they occur when the membrane is moving from position x to x′. (b) Pressure (solid curve) and particle velocity (dashed curve) of sound waves in (a) as a function of distance to the right ($+r$) or left ($-r$) from the vibrating membrane (0). Graph shows relative values as they occur when the membrane is moving from position x′ to x; particle velocity values below the baseline signify movement to the left.

Frequency is the rate at which waves pass a given point and is measured in hertz (Hz), where 1 Hz = one 360° cycle·s^{-1}. Wavelength is the distance, measured along an axis extending radially outward from the sound source, from the peak of one wave to the peak of the next at a given instant in time. Here, peak refers to the maximum pressure, or compression phase, of the wave. Alternatively, we could measure the wavelength between consecutive pressure minima, or rarefaction phases. Velocity of sound is the speed at which a given compression or rarefaction travels outwardly from the source.

The velocity of sound in liquids is determined by $c = \sqrt{}$(elasticity/density), where elasticity is represented by the bulk modulus, the change in pressure (Δp) needed to achieve a given proportional change in volume ($\Delta V/V_0$) divided by that proportional change (Denny 1993). Thus, sound travels slightly faster in seawater (1521.5 m·s^{-1}, at 20°C) than in fresh water (1482.3 m·s^{-1}), because the bulk modulus of seawater is slightly higher. In gases, the determination of sound velocity is somewhat different, because the gas in the compression phase of the sound wave is not in thermal equilibrium with its surroundings. Sound velocity in a gas is independent of density and pressure and depends only on temperature. In air at 20°C, sound travels at $343.4\,m{\cdot}s^{-1}$, and its velocity increases by $5.8\,m{\cdot}s^{-1}$ for every 10°C rise in temperature. Within a given medium and set of conditions, sound velocity is generally fixed,[4] and all wavelengths travel at that value.

Sound frequencies used in and relevant to arthropod communication range from several hundred hertz to ultrasound at 125 000 Hz (125 kHz), or possibly even higher. The designation of sound as ultrasound is generally made for frequencies in excess of 20 kHz, and it is entirely based on human perception: for most humans, sounds above 20 kHz in frequency are inaudible, and this cutoff value decreases with age. Wavelengths of sounds relevant to arthropods would then range from 1 m or more to as little as 2–3 mm in air; values in water would be 4.3 times longer.

The pressure and particle velocity aspects of sound waves remain in-phase in the far field (Figure 4.1), and visualizing this relationship will help us to understand the distinctive properties of the near field. At a given point in the path of oncoming sound waves, when fluid pressure decreases to full rarefaction, fluid particles move at maximum speed toward the propagating source and a particle's displacement from its resting position is expected to be zero. Similarly, when fluid pressure increases to full compression, fluid particles move at maximum speed away from the propagating source and a particle's expected displacement is again zero. But, when fluid pressure is either increasing or decreasing and equals the value that would be found in the absence of sound waves, fluid particle velocity is zero and expected particle displacement is maximum, away from or toward the propagating source, respectively. Thus, particle displacement lags pressure and particle velocity by a 90° phase angle. At points very close to the propagating source, though, fluid particles simply move back and forth in phase with the vibration of the source and match the source velocity, but concomitant increases and decreases in fluid pressure do not arise. The high particle velocity and absence of corresponding pressure deviations is the fundamental characteristic of the near field.

The amplitude of sound waves at a given point is either measured as sound pressure level (SPL) or intensity level (IL). Here, pressure always refers to the pressure deviation, Δp, the difference between fluid pressure at compression or rarefaction and the ambient pressure that would exist in the absence of sound waves. SPL at the given point may be measured as a peak value, Δp_{max}, the deviation found precisely at full compression or rarefaction, or as a value averaged over time. A root-mean-square (rms) measure of pressure deviation is typically used for an averaged value; for purely sinusoidal waveforms, $\Delta p_{rms} = (1/\sqrt{2}) \times \Delta p_{max}$.[5] The oscillating pressure deviations and particle velocities of sound waves also generate a force that acts along an axis extending radially outward from the propagating source and can thereby do work or transmit energy. For example, the membrane in a biological receiver (the tympanum of an ear) or an electronic transducer (the diaphragm of a microphone) will be caused to oscillate by energy transmitted by sound waves impinging on it. Sound intensity (I) is the rate at which this energy is transmitted per unit area at the given point of interest and is determined, in the far field, as

$$I = (\Delta p_{max})^2/(2\rho_0 \cdot c) \tag{4.2}$$

where ρ_0 is the fluid density in the absence of sound waves. Alternatively, sound intensity may be determined as a function of the amount of particle displacement, in which case

$$I = 2\rho_0 \cdot c \cdot \pi^2 \cdot f^2 \cdot A^2 \tag{4.3}$$

where A is the maximum longitudinal particle displacement[6] away from the position that would occur in the absence of sound waves; particle velocity, u, is related to A by $u_{max} = 2\pi \cdot f \cdot A$. In SI units, I is measured in watts per square meter ($W \cdot m^{-2}$).

Although both measures of sound amplitude can be provided in absolute SI units, they are usually given as relative measures that are referenced to a standard value. SPL in air is normally referenced to 20 micropascals ($1\ \mu Pa = 10^{-6}\ N \cdot m^{-2}$) rms, the average threshold for human hearing. In air at 20°C, this SPL reference value is equivalent to an intensity of $10^{-12}\ W \cdot m^{-2}$, the value normally used as an IL reference. Because the perceived loudness of sound in humans and other animals does not increase linearly with amplitude, the relative measures of SPL and IL are ordered on a logarithmic decibel (dB) scale. For SPL,

$$\text{dB} = 20 \log_{10}(\Delta p \div \Delta p_{ref}) \tag{4.4}$$

where $\Delta p_{ref} = 20\ \mu Pa$. Thus, 0 dB SPL = 20 μPa, and each 10-fold increase in Δp is equivalent to an increase of 20 dB. A 94-dB SPL sound, which is fairly loud to human ears and may be characteristic of a heavy moving vehicle or the calling song of a male field cricket at 5 cm, is equivalent to 1 Pa, but even this Δp value represents only a 0.01% deviation from standard atmospheric pressure ($\cong 10^5$ Pa): ears are extremely sensitive receiving devices! For IL,

$$\text{dB} = 10 \log_{10}(I \div I_{ref}) \tag{4.5}$$

where $I_{ref} = 10^{-12}\ W \cdot m^{-2}$. Thus, 0 dB IL $= 10^{-12}\ W \cdot m^{-2}$, and each 10-fold increase in I is equivalent to an increase of 10 dB; a 94-dB IL sound is equivalent to $2.5 \times 10^{-3}\ W \cdot m^{-2}$.

In the absence of interference, sound radiates spherically from its source. Consequently, sound energy is spread over an area that increases proportionately with the square of the distance from the propagating source, and intensity values (in SI units) decrease proportionately with the inverse square of this distance. Because SPL (Δp) is proportional to $\sqrt{I}$ (cf. equation 4.2), SPL values (in SI units) decrease proportionately with the inverse of the distance from the source. Spherical spreading implies that both IL and SPL decrease by 6 dB for each doubling of the distance from the propagating source. These relationships indicate that we may also use decibels as a relative measure to compare acoustic signals or receiver sensitivity thresholds. Thus, we may refer to a signal or threshold that is twice another in Δp as being 6 dB higher in SPL, or twice another in I as being 3 dB higher in IL.

Most microphones used to measure sound yield a voltage that is linearly proportional, within a given range, to acoustic pressure deviation (Pa) rather than to intensity ($W{\cdot}m^{-2}$). Consequently, the biological literature generally reports SPL values for signals and receiver sensitivity, and this convention is followed here.

Excess Attenuation and Distortion. Before ending this introduction to sound, it must be emphasized that the above principles assume ideal, free-field conditions for the transmission of acoustic signals. But organisms never enjoy such conditions outside laboratory anechoic chambers and instead must deal with various sorts of interference and impediments to the spherical spreading of sound waves (see Forrest 1994). In any fluid, the fluid particles themselves absorb energy from sound and also scatter the waves, thereby attenuating sound intensity. Both absorption and scattering will cause SPL to decrease by more than 6 dB as distance from the source doubles. This excess attenuation, beyond that due to spherical spreading, is more severe at higher sound frequencies (Griffin 1971, Lawrence and Simmons 1982), and is therefore a significant factor in the communication of acoustic arthropods: Nearly all species are constrained by their small size to use high-frequency signals (Section 4.2.5). At these frequencies ($\gg$1 kHz), excess attenuation is greater in drier than more humid air (Kinsler and Frey 1962),[7] a factor that may constrain the communication of diurnal species active in sunlight. Diffraction by solid objects, such as vegetation, will attenuate sound waves further, and this process too is greater at higher frequencies: in general, sound is severely diffracted by objects whose circumference exceeds the wavelength. In addition to attenuating sound, diffraction distorts spectral and temporal features. Because higher frequencies are disproportionately attenuated, the frequency spectrum of a sound close to its source will be broader than at a greater distance, where mostly lower frequencies will remain. When solid obstacles in the path of sound waves cause diffraction, the waves are split and the various fragments are directed along different paths. Because the paths that these diffracted fragments travel between the source and a distant receiver may not be identical in length, reverberation is likely to arise: what had been the sudden onset or termination of a sound pulse near its source may be extended for many milliseconds at a receiver by the asynchronous arrival of fragments of a given sound wave. Thus, reverberation can "smear" temporal features such as pulses, yielding a more continuous sound that lacks such features.

A special attenuating effect occurs where sound waves abruptly pass from one type of fluid to another and those fluids differ in their specific acoustic impedance (Z) to the transmission of sound. Here, specific acoustic impedance may be considered as analogous to resistance in an electrical circuit, and we determine it as

$$Z = \rho_0 \cdot c \tag{4.6}$$

where Z is measured in units of $kg{\cdot}s^{-1}{\cdot}m^{-2}$ (= 1 rayl) or $Pa{\cdot}s{\cdot}m^{-1}$. The latter unit indicates that Z measures the pressure : velocity ratio, the amount of pressure that must be applied at a specific point in a fluid to effect a given particle velocity. In general, only a small portion of the sound energy (intensity) in one fluid is transmitted across the boundary into another. Most of the energy is reflected by the boundary, and the amount transmitted into the second fluid (I_t) is proportional to the amount in the first fluid incident at the boundary (I_i) by

$$I_t = I_i \cdot ((4Z_1 \times Z_2)/(Z_1 + Z_2)^2 \tag{4.7}$$

where the subscripts refer to the first and second fluids. Because Z_{water} is approximately $3600 \times Z_{air}$, only 0.1% of the incident sound energy is transmitted from air into water, or vice versa. Of the sound energy that is reflected, much is dispersed into different wavelengths (see Aiken 1985). Consequently, terrestrial signalers cannot effectively transmit their calls into the water for the benefit of submerged receivers, should that need ever arise. More relevant to normal communication, however, are the smaller local differences in specific acoustic impedance that often arise within air or water due to thermal layering. Where such conditions occur, sound waves will be refracted in the direction of the layer in which they travel slower (see equation 5.2 and endnote 5 in Section 5.1.4). If air temperatures are higher adjacent to the ground than several meters above, refraction will then create a sound shadow limiting the transmission of acoustic signals at ground level. Conversely, thermal layers may reverse at night and create a sound channel at ground level in which attenuation is less than that predicted by spherical spreading. Wind, which is normally reduced at ground level, may create similar effects in certain locations: sound is generally refracted away from the ground in the upwind direction and toward the ground in the downwind direction (Dusenbery 1992).

The geometry and configuration of a sound source, such as a vibrating appendage of an organism, will also prevent waves from propagating equally in all directions and establishing the idealized property of spherical spreading. In terms of basic physical models, sound sources may be represented by vibrating pistons or pulsating, vibrating, or deforming spheres (Figure 4.2). Only a pulsating sphere, often termed a monopole source, will generate a sound field that spreads equally in all directions. The other models yield more complex patterns in which sound is propagated strongly in some directions but little or not at all in others. Arthropod sound radiators typically resemble the vibrating sphere model

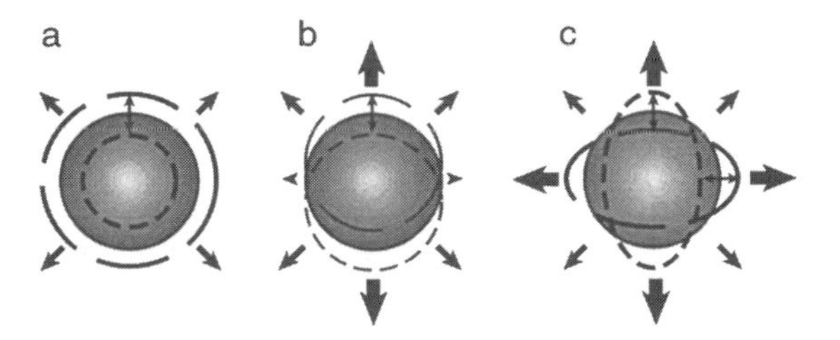

Figure 4.2 Fundamental physical models of sound sources (see Aiken 1985). (a) A pulsating sphere alternating (double-ended arrow) between large (long dashes) and small (short dashes) volumes is considered a monopole source and creates a sound field of equivalent intensity in all directions. (b) A vibrating sphere alternating between upper (long dashes) and lower (short dashes) positions is considered a dipole source and creates a cardioid sound field with intensity maxima (thick arrows) along the axis of vibration and intensity minima (short arrows) in a plane perpendicular to that axis. (c) A sphere that is alternately deformed along two perpendicular axes creates a complex sound field with maxima (thick arrows) along those axes.

(dipole sources, which produce a cardioid sound field) more so than the pulsating sphere one, and acoustic measurements of song production reveal that SPL is seldom identical in all directions.[8]

Departures from idealized conditions aside, sound nonetheless remains a most effective means for organisms to transmit and receive signals. By exploiting the compressible nature of fluids, which allows a mechanical disturbance to be transmitted by longitudinal wave motion rather than actual conveyance of the originally disturbed particles—as is necessary in pheromonal communication, for example—an organism can communicate nearly instantaneously over relatively long distances and across a wide range of directions. In Sections 4.2 and 4.3 we examine how arthropods transmit acoustic signals and encode and extract information in them. We also examine the special problems, largely associated with their small size (see Bennet-Clark 1998a), that constrain the ability of arthropods to perform these tasks and limit the energetic efficiency with which they do so.

4.1.2 Vibration: Surface, Bending, and Transverse Waves

Several kinds of substrate vibration transmitted via transverse wave motion are relevant to arthropod communication. These vibrations include surface waves on water and certain solid substrates, bending waves in plant leaves and stems and other structures of comparable dimensions and stiffness, and waves in taut strings such as the strands in silken webs. While surface and bending waves and web vibrations are not capable of the nearly instantaneous transmission over long distances that sound can provide, a small arthropod can rely on them to send messages over distances up to 1–2 m within several seconds or less (e.g., Michelsen et al. 1982).

When a point disturbance is made on the surface of water, circular ripples spread radially outward from that point (Denny 1993). If the wavelength is relatively large ($\gg$1.7 cm), most of the wave motion is maintained by gravity, which restores wave crests and troughs toward the mean water level. On the other hand, if the disturbance causes the water to oscillate at a higher frequency ($f > 15$ Hz) such that the wavelength is small, there exists substantial surface tension, and hence curvature, in the waves. Under these conditions, wave motion is largely maintained by capillary forces. A small aquatic beetle or bug on the water surface may vibrate and generate concentrically spreading waves at a frequency of 25 Hz and with wavelengths of 10 mm. These capillary waves will travel outward at approximately $0.25\,\text{m}\cdot\text{s}^{-1}$ (Figure 4.3) and may be used to transmit signals to conspecifics. Because the waves will be reflected by objects in their path, the insect may also use them as a form of echo-location for navigation or prey capture.

Concentrically spreading water surface waves, whether maintained by gravity or capillary forces, suffer an exponential decay of energy (E) over distance. In water deep enough (>2–3 cm) to avoid bottom drag, the damping coefficient $(\alpha_{10}) \cong 0.11\,\text{dB}\cdot\text{cm}^{-1}$ $[\alpha_{10} = 10 \cdot \log(E_{\text{x cm}}/E_{\text{x+1 cm}})]$ (see Lighthill 1978). Because energy per unit area of surface waves is proportional to the square of their (transverse) acceleration and, consequently, to the square of wave height, the vertical distance from wave trough to crest, wave acceleration and height decline with increasing distance from the source and eventually fall below receiver threshold. In capillary waves, energy per unit area is also proportional to the square of frequency, but high-frequency surface waves, like sound, are attenuated more rapidly than low-frequency waves. Because arthropods dwell-

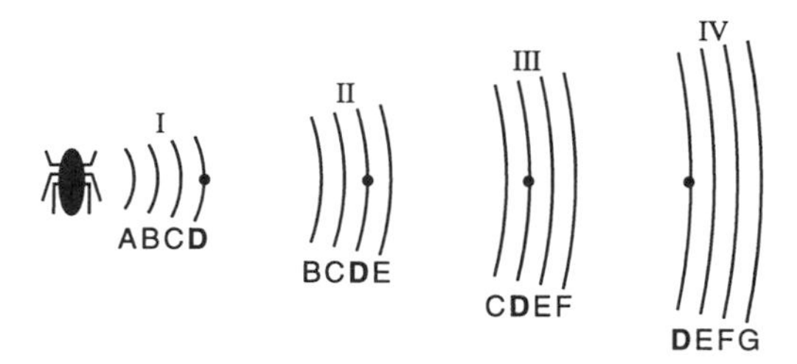

Figure 4.3 Concentrically spreading surface waves maintained by capillary forces. Views I–IV show the position of a specific group of waves at four successive instants. Arcs depict the crests of individual waves in a particular direction from the source, an aquatic beetle disturbing the water surface. At each successive instant, the trailing wave in the group disappears while a new wave forms at the leading edge. Thus, group velocity, the velocity of the leading (or trailing) edge, exceeds wave velocity. Letters designate specific waves; the dot on wave D highlights retreat of a given wave within the group over time.

ing on the water surface mostly generate waves at relatively high frequencies ($f > 20$ Hz), their communcation is restricted to relatively short distances (<1–2 m; Wilcox 1995).

Wave velocity as discussed above refers to the celerity with which a particular phase of a given wave moves radially outward. This value, generally termed c (cf. equation 4.1), is not the velocity with which a vibratory signal would be transmitted across the water surface, however. Vibratory signals would be transmitted either slower or faster than c; the specific value depends on whether gravity or capillary forces dominate. When an impulse disturbs the water surface, a group of waves is generated. Signals would be transmitted at the velocity at which the modulation envelope of this group travels, and this group velocity, c_g, does not equal the celerity of a given wave, c (Figure 4.3). This non-intuitive result arises in gravity waves because the leading wave of a group gradually declines in height and disappears, at which time a new wave appears at the trailing edge of the group and gradually increases in height. Thus, the modulation envelope of the wave group travels slower than any of its component waves. For pure gravity waves, $c_g = 0.5\ c$. In capillary waves a reversed process occurs (Figure 4.3). Here, it is the trailing wave of a group that gradually declines in height and disappears, while the new wave appears at the leading edge of the group and grows in height. Thus, the modulation envelope travels faster than any of its components, and for pure capillary waves, $c_g = 1.5\ c$.

The dispersive property of water complicates the transmission of surface waves further. A sinusoidal wave initiated at a given wavelength is dispersed by water into a range of wavelengths. Unlike most sound, surface waves of various wavelengths travel at different celerities. Thus, when an impulse generates a wave group, the radial distance from the trailing to the leading wave of the group lengthens as distance from the source increases. In very shallow water (depth $< \lambda/6$), however, the waves are not dispersed into different wavelengths, and they spread elliptically rather than concentrically.

Several types of surface waves may also spread concentrically from impulses imparted to solid substrates (Figure 4.4; Fletcher 1992). Rayleigh waves include both longitudinal and transverse components of wave motion, with particle motion occurring in a plane perpendicular to the surface and aligned with the direction of wave propagation. The transverse component arises from compression of the substrate near its surface and is a function of elasticity as measured by the substrate's Poisson ratio.[9] Love waves are similar to Rayleigh waves except that particle motion occurs in a plane parallel to the surface. Both Rayleigh and Love waves are transmitted at high group velocities, $\gg 100\ \mathrm{m \cdot s^{-1}}$ within some solid media, but on many surfaces damping is severe and an impulse of considerable force would be needed to transmit these waves. Thus, arthropod communication via surface waves is probably restricted to a narrow range of terrain. On substrates with reduced damping ($\alpha_{10} < 0.3\ \mathrm{dB \cdot cm^{-1}}$) (e.g., compact wet sand) some larger arthropods can strike the ground with sufficient force to

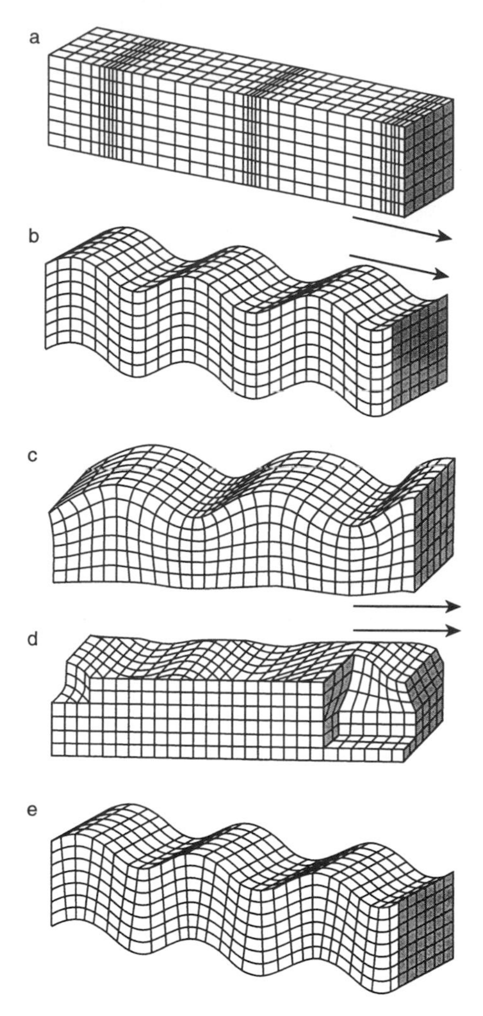

Figure 4.4 Vibrational waves in solid media. (a) Compression (P) waves, or sound: longitudinal waves occurring throughout the interior of a solid medium. The vector of wave motion lacks a component transverse to the direction of wave propagation (arrow). (b) Shear (S) waves: transverse waves occurring throughout the interior of a solid medium. The wave vector may be vertical (as depicted) or horizontal, and it lacks a component longitudinal to the direction of wave propagation. (c) Rayleigh (R) waves: surface waves in which the vector of wave motion has both longitudinal and vertically transverse components (as in vertical shear waves described in b). (d) Love (L) waves: surface waves in which the vector of wave motion has both longitudinal and horizontally transverse components (as in horizontal shear waves described in b). (Reprinted from "Vibrational communication in the fiddler crab, *Uca pugilator*. 1. signal transmission through the substratum," *Journal of Comparative Physiology A*, vol. 166, p. 346, fig. 1, B. Aicher and J. Tautz, copyright 1990 with permission by Springer-Verlag.) (e) Bending waves: waves occurring in stiff rod- or plate-like structures and including both longitudinal and transverse components (as in Rayleigh and Love waves described in c and d; unlike Rayleigh and Love waves, bending waves affect the entire structure). Illustration depicts bending wave with vertical transverse component.

generate signals which conspecifics within 1–2 m may perceive and respond to (e.g., Horch 1975, Aicher and Tautz 1990).

Bending waves arise from impulses imparted to thin rod-, bar-, or plate-like structures that are clamped at one end and thereby kept stiff (Fletcher 1992). Thus, branches, stems, and leaves of the vegetation that many arthropods inhabit readily serve as a medium through which these waves can be transmitted and carry signals (Barth 1997, 1998). Nesting materials of social insects, such as the wax or paper comb of bees and wasps, may also transmit bending waves. When a point disturbance occurs in the structure, wave motion propagates away from that point in the form of sinusoidal bending or rippling of the entire structure, not only its surface. Like Rayleigh and Love waves, bending waves include both longitudinal and transverse components (Figure 4.4e). Bending wave celerity increases proportionately with the square roots of both frequency and structure thickness. Again, high-frequency waves are attenuated more rapidly, but many arthropod bending wave signals are transmitted with relatively low frequencies ($\ll$1 kHz) and may travel effectively. For example, a wandering spider (*Cupiennius*; Araneae: Ctenidae) on an elongate, monocotyledonous leaf may vibrate and generate bending waves at a frequency of 30 Hz that travel for the length of the leaf with celerities (c) ranging from 4 to 40 m·s^{-1}; higher values of c would occur at the base of the leaf, which is thicker and stiffer.

As with surface waves, groups of bending waves are generated by an impulse imparted to the structure, and signals would be transmitted at the velocity (c_g) that the modulation envelope of a wave group travels. In general, $c_g > c$ for bending waves, and for slender rod-like structures, $c_g = 2\ c$. Because bending waves are also dispersed by the medium through which they travel, the modulation envelope expands as distance from the disturbance increases. Unlike waves radiating on the surface of water, bending wave amplitudes may be attenuated very little in various kinds of stiffened vegetation and at the frequencies typical of arthropod signals; losses of 0.3–0.4 dB·cm^{-1} have been reported for wandering spider vibratory signals on the vegetation they typically inhabit. Thus, detectable signals may be transmitted 1–2 m along stems and leaves via bending waves in a fraction of a second (Keuper and Kuhne 1983).

Echoes and distortion, however, can be major problems for signal transmission through some types of vegetation and for receivers who must extract information from the reflected and distorted signals (Čokl et al. 2000). Bending waves can be reflected at certain locations, which may confuse a receiver's perception of the direction to the source, and the dispersion of both outbound and reflected waves would lengthen modulation envelopes of the signals that more distant receivers are exposed to (Miklos et al. 2001). In vegetation structures that are thicker in the center than at the edges, waves would travel faster along the midline, and a receiver straddling the structure sideways would also perceive lengthened modulation envelopes. Complex branching structures, as in dicotyledonous vegetation, possess multiple nodes at which sharp impedance differences in the substrate could modify bending waves in yet other ways (McVean and Field 1996). The perception of and responses to bending waves subject to these various alterations remain unexplored.

The vibratory waves most familiar to us are probably those occurring along taut strings: elongate elastic structures that are under tension and clamped at each end, as the strings of musical instruments. Arthropods that build silken webs may live on structures that are effectively taut strings, and they may transmit and receive vibratory signals along the webbing (Barth 1982, 1998). These signals are best known in spiders that weave orb webs, whose vibrations may be transmitted via transverse, as well as longitudinal and torsional, waves along the strands of silk. Owing to the high elasticity[10] and low mass per unit length of spider silk, the transverse vibrations may suffer very little attenuation as they travel across a web (Denny 1976, Masters 1984a,b). Troublesome aspects arise in analyzing web-borne vibratory signals, though, because these are rather complex structures, quite unlike the simple taut strings used to depict wave motion in physics texts. Most strands intersect many other strands between their end points, and the number and nature of these intersections will influence wavelength, amplitude, and velocity of outbound, and reflected, waves in that strand. Additionally, when a given strand is plucked, the impulse generates a vibration in that strand, in other strands of the web, and in the web as a unit. These different vibrations are difficult to monitor simultaneously, and their interrelationships are not understood.

Cautionary Notes. A general point for the study of arthropod vibratory signals is that they must be measured with instruments that do not themselves affect the signal. Because substrate accelerations and displacements in these signals are normally quite low (e.g., ≈ 0.1–$0.2\ \mathrm{m \cdot s^{-2}}$ and $<1\ \mu\mathrm{m}$ in tettigoniid bending wave signals; Keuper and Kuhne 1983), signal characters can easily be altered by instruments designed for stronger vibrations. For example, an instrument attached to a plant stem or the silk strand of an orb web to measure its vibration would likely represent a load on that structure, influencing its tension and hence its vibration. Ultrasensitive acceleration transducers, accelerometers, have improved the accuracy of some measurements, but the introduction of laser-Doppler vibrometry has been most helpful. Thus, the displacement of the substrate can be monitored optically without any weighting or other mechanical constraint on that movement (Masters and Markl 1981).

A final caveat for examining communication along the mechanical channel concerns our interpretation of how a signal is transmitted and received. Confusion may arise when an arthropod stridulates or tremulates and thereby generates both substrate vibrations and an airborne sound. Often, the receiver can only detect the substrate vibration, in which case the concomitant airborne sound might be just an incidental byproduct of the signaler's activity. While it may be easy for us to monitor the temporal pattern of vibratory signals with an acoustic transducer that detects the incidental sounds or whose diaphragm is stimulated by the vibration, these recordings are not a substitute for accurate measurements of substrate displacements and motion, which constitute the actual signals. This point again underscores the indispensable role that knowledge of receiver function plays in an accurate interpretation of communication (cf. Section 3.1.2).

4.2 Signal Transmission: Friction, Tymbals, Percussion, Vibration, and Tremulation

Arthropods exhibit a diverse array of sound- and vibration-transmitting mechanisms, including friction between body parts, percussion on the substrate or between body parts, vibration of appendages, expulsion of tracheal air, and drumming or tremulation on the substrate. The majority of these mechanisms have exploited the hardened exoskeletons of arthropods, but even species that are mostly soft-bodied may generate mechanical signals that are transmitted over relatively long distances (e.g., Patek 2001). Specialized sound-transmitting structures can apparently evolve when the circumstances permit, and they are well developed in such unexpected groups as noctuoid and pyraloid moths (Conner 1999). Any sclerotized body part, regardless of its size or location, may be modified into a signal-transmitting structure. Alternatively, some soft-bodied forms, such as lacewings (Henry 1980, 1986) and the larvae of lycaenid butterflies (DeVries 1990, 1992), compensate for their paucity of sclerotized structures by vibrating the substrate and relying on the mechanical characteristics of that structure to transmit a bending-wave signal.

Like pheromones, mechanical signals are characterized by their quality and intensity (amplitude). Whereas the various chemical compounds—components—and their proportions represent pheromone quality, the frequency of acoustic or vibratory wave motion represents the quality of a mechanical signal. This character is termed the carrier frequency, and it is the primary rate of modulation of pressure or displacement of the fluid medium or substrate. In rare cases, such as the calling songs of crickets, a signal may have a nearly singular carrier frequency (Figure 2.2b) and is heard as a pure tone, but a broadband frequency spectrum is the norm (Figure 2.2e). For example, a broadband spectrum would characterize frequency in a group of surface or bending waves dispersed by the water or structure, respectively (Section 4.1.2). The characters of arthropod mechanical signals are normally depicted by oscillograms (graphs of amplitude, usually SPL, versus time) and frequency spectrograms (graphs of amplitude versus carrier frequency, for a given time interval; Figure 2.2).

Because the frequency spectra of arthropod mechanical signals seldom change substantially over the course of time while produced (but see Figure 4.5),[11] these two graphs can together reveal the fundamental signal characters. This simplicity contrasts with the frequency sweeps and other frequency modulations found in vertebrate signals, which are normally depicted by sonograms (graphs of frequency versus time, with amplitude coded by color or shading).

Our treatment of chemical signal transmission (Section 3.1.2) briefly considered the possibility that temporal modulation of pheromone emission might impart an additional character, tempo, or regular pulsing, to the signal. I rejected this possibility because diffusion, wind, and turbulence would probably prevent a pheromone plume from retaining the pulsed pattern everywhere except close to the source. Temporally modulated mechanical signals, however, would retain much of their pulsed pattern except in environments where reverberation or dispersion is severe, and receivers may rely heavily on the information in these patterns (Section 4.3.5). Thus, mechanical signals are usually defined by three fundamental characters, and in many species the temporal pattern is as critical for receivers as carrier frequency and amplitude are. Several temporal patterns may be ordered hierarchically (Figure 2.2d), and the various levels of temporal patterning are under neuromuscular control. This control is as integral a part of mechanical signal transmission as the structural mechanisms yielding the basic sound or vibration waves.

In this section, we shall examine the major devices used by arthropods to generate acoustic and vibratory signals. We also discuss the various mechanisms that regulate carrier frequencies, intensity, and temporal patterning of

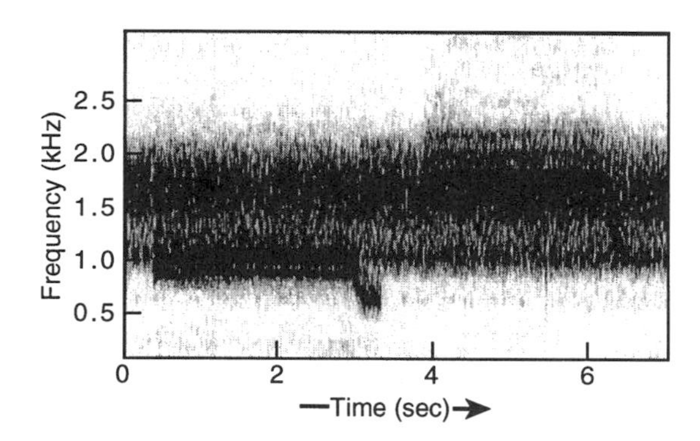

Figure 4.5 Sonograms of male calling songs of the periodical cicadas *Magicicada tredecim* (0.3–3.3 s; dominant frequency ≅1.1 kHz) and *M. neotredecim* (3.9–6.5 s; dominant frequency ≅1.8 kHz). Graphs depict frequency versus time, with relative acoustic energy represented by shading darkness. For both species the sonogram depicts the call of an individual male against the background mixed-species chorus. Note the downward frequency sweep at the end of the individual calls. (From Marshall and Cooley 2000; sonogram courtesy of David C. Marshall; reprinted from *Evolution*, vol. 54, with permission of the Society for the Study of Evolution.)

these signals. These devices and mechanisms have been shaped and constrained by arthropod body sizes and physiologies, pre-existing receiver mechanisms (see Section 6.1), the physical and biotic environments through which the signals are transmitted, and the nature of the interactions served by signaling.

4.2.1 Signal Apparatus: Airborne Sound in the Far Field

How does an insect whose mass is less than 1 g transmit a sound detectable by a receiver tens of meters away? There is no single answer, and we begin by examining stridulation in the European field cricket *Gryllus bimaculatus*, one of the most thoroughly studied acoustic insects. Male *G. bimaculatus* attract females with a calling song generated by elevating the two forewings (tegmen; tegmina pl.) 30–90° above their horizontal resting position and transversely rubbing one against the other approximately 30 times per second (Huber and Thorson 1985). This tegminal/tegminal stridulation is a frictional mechanism, and it generates sound because a transverse "file" of "teeth," the pars stridens, on the lower surface of one tegmen moves across a raised longitudinal vein, the "scraper" or plectrum, on the upper surface of the other tegmen (Figure 4.6). Both tegmina are thereby caused to vibrate at high frequency, and sound is radiated from a triangular membranous region, the harp, at the base of each tegmen.[12] Although the left tegmen normally lies over the right one while at rest and during calling, both tegmina are fully equipped with a file and scraper and the insects may occasionally switch forewing positions and continue to call (Kavanagh and Young 1989). Given this morphology, the inclination to use the left file and the right scraper implies that a form of directional asymmetry—wingedness—may exist in cricket populations.

The above structural device and stridulatory mechanism are found throughout the Gryllidae (crickets) and the closely related orthopteran families Gryllotalpidae (mole crickets) and Haglidae (hump-winged crickets). Tettigoniidae (katydids, or bushcrickets as they are known outside North America) use a similar device save that the left and right tegmina are dimorphic and cannot be switched. The left tegmen bears the transverse file, and the right one bears the longitudinal scraper, which is associated with a sclerotized U-shaped "frame" surrounding a region of thin cuticle, the "mirror." Stridulation causes the frame to vibrate as a tuning fork while the associated mirror radiates sound. A more divergent variation is found in some acridid grasshoppers, where the scraper is a raised vein on the tegmen as above, but the file is a longitudinal row of "pegs" along the inner surface of the hind femur (Elsner 1974, 1983, von Helversen and von Helversen 1983). Both left and right tegmina and hind femora are equipped with these structures, and sound is produced when one or

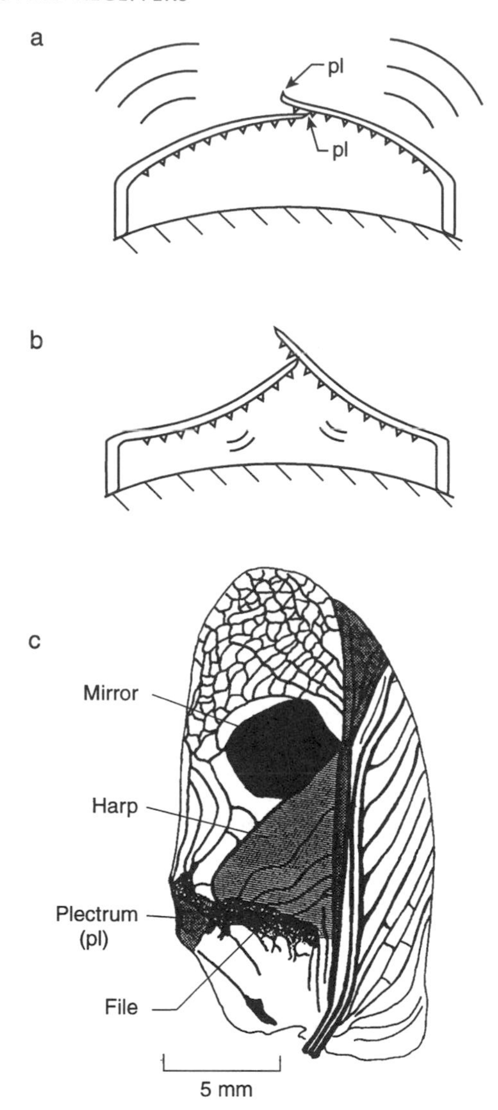

Figure 4.6 File-plectrum stridulatory mechanism in crickets. (a) Sound is produced during wing closure as the plectrum (pl) on the lower wing and a file tooth on the lower surface of the upper wing move toward one another, causing the sound-radiating structures on both wings to buckle convexly outward and generate a sound wave. (b) As the plectrum on the lower wing jumps past the file tooth on the upper wing, both wings buckle inward. Thus, sound waves transmitted by the two wings remain in phase, avoiding destructive interference. Arcs depict compression phase of each wing's sound wave at a given instant (see Bennet-Clark 1989). (c) Lower view of front wing of male *Gryllus campestris*, showing major structures responsible for sound production (file, plectrum) and radiation (harp, mirror). (Reprinted from Bennet-Clark 1989, in Franz Huber, Thomas E. Moore, and Werner Loher (eds.), *Cricket Behavior and Neurobiology*. Copyright © 1989 by Cornell University. Used by permission of the publisher, Cornell University Press.)

both hindlegs are moved up and down rapidly, either alternately or simultaneously. Hindleg movement causes the adjacent tegmen to vibrate and radiate sound.

The above examples of file/scraper stridulation among acoustic Orthoptera generate relatively intense sounds, are widespread in several major groups of the order (Section 4.5), and have been analyzed intensively. More than a dozen additional forms of file/scraper stridulation occur in other Orthoptera, and elsewhere among the insects yet more variations are found. These forms involve nearly every pair of body segments and appendages that may conceivably be rubbed together. For example, males of the odontiine pyralid moth *Syntonarcha iriastis* stridulate with considerable amplitude by rubbing a file on the genitalia across a scraper on the 8th abdominal sternite (Gwynne and Edwards 1986). Readers interested in the full range of these variations and their phylogenetic distributions should consult the compendia provided by Dumortier (1963) and Ewing (1989).

Tymbals are another major category of specialized structures with which insects can generate high-intensity airborne sounds. These devices produce sound in a manner similar to metal click toys or an empty aluminum can that is buckled in and out. They are best known in cicadas (Homoptera: Cicadidae), where males bear a pair on the sides of the anterior segment of the abdomen. Cicada tymbals are sclerotized dome-like structures to which a powerful, fast muscle is attached (Young and Bennet-Clark 1995, Bennet-Clark 1997, 1998a). Contraction of this muscle causes the tymbal to buckle concavely and emit an intense, brief click composed of multiple sound waves; its relaxation allows the tymbal to resume its convex shape. A tensor muscle maintains the normal convexity of the tymbal, which thereby retains stored elastic energy. Buckling releases this energy suddenly, and some is converted to the sound waves. In many cicada species the tymbal consists of parallel, vertical ribs, each connected to a posterior plate where the tymbal muscle attaches. Here, muscle contraction causes the individual ribs to buckle concavely in sequence from posterior to anterior, and a separate click is produced as each rib buckles. In general, sound is radiated both outwardly and inwardly from the cicada tymbal, but the latter is amplified greatly by a large air sac filling most of the abdomen. It may be broadcast to the outside through the abdominal wall or specifically through the tympana, the eardrums on the ventral side of the anterior abdominal segment (see Young 1990). In Section 4.3.6 we discuss how cicadas avoid the problems of permanent damage to the receiver apparatus and hearing loss that could potentially arise from using the tympana as both ears and sound radiators.

Similar tymbal devices are found in other Homoptera (Auchenorhyncha; Claridge 1985) and in the burrower bugs (Hemiptera: Cydnidae; Gogala 1970). These devices, however, are used by relatively small insects to generate bending waves in the substrate (Section 4.1.2). Various Lepidoptera also bear minute tymbals, which are used to generate airborne ultrasonic signals. In some noctuoid moths the tymbals are scale-less convex areas, "blisters," of thin cuticle on the sides of the posterior thoracic segment (Blest et al. 1963, Dunning and Roeder 1965). They may consist of up to 60 striae, which function as the ribs in cicada tymbals. Brief pulses of ultrasound are produced as the striae buckle, but here pulses occur during both inward and outward buckling.

Somewhat different tymbal mechanisms are found among galleriine pyralid moths (Spangler et al. 1984, Spangler 1988a). In males of the lesser wax moth, *Achroia grisella*, a tymbal is located on the anterior side of each tegula, a small sclerite at the base of the forewing. These tymbals are also minute regions of thin cuticle devoid of scales and backed by an air sac, but they generate ultrasound via wing movement. The males remain stationary while fanning their wings through a 20° arc approximately 45 times per second. A cuticular knob on the underside of the tegula contacts the forewing base during each upstroke and downstroke of the wings. Each contact causes the tymbal to buckle, inward on the downstroke and outward on the upstroke, and emit a 100-μs pulse of highly damped ultrasound. Males of the rice moth, *Corcyra cephalonica*, employ a similar apparatus except that each tymbal includes nine striae that buckle individually and in sequence (Spangler 1987), as in cicadas and noctuoid moths. Therefore, contact between the tegula and the forewing base during a wing upstroke or downstroke yields up to nine pulses of ultrasound.

Other insect mechanisms for producing airborne acoustic signals in the far field include the expulsion of tracheal air, as in the Madagascan hissing cockroach (*Gromphadorina portentosa*; Nelson and Fraser 1980), and various forms of percussion between body parts. In certain cases, the percussive mechanisms do not appear to entail specialized structures but simply rely on modifications of the movements of appendages. For example, some cicada species may clap their forewings together or bang them on the substrate (Sanborn and Phillips 1999), and the acridid grasshopper *Paratylotropidia brunneri* snaps its mandibles together (Alexander 1960). But, in Australian whistling moths (*Hecatesia* spp.; Noctuidae: Agaristinae) males produce a high-frequency sound via castenet-like knobs on the costal margins of the forewings, which are rapidly clapped together during flight in some species and at rest in others (Bailey 1978, Alcock and Bailey 1995). This clapping action sets the distal ends of the wings into a high-frequency vibratory mode that generates the sound.[13]

4.2.2 Signal Apparatus: Airborne Sound in the Near Field

Near-field sound plays a major communicative role in the courtship and social interactions of various insects, particularly Diptera and Hymenoptera.

However, these signals may be inaudible to observers, and they must be recorded with specialized microphones that are sensitive to the particle velocity aspect of sound rather than its pressure aspect (Bennet-Clark 1984). Thus, we have much less knowledge of near-field acoustic signals than those occurring in the far field.

Mechanisms that generate far-field sound invariably generate some sound in the near field as well. But in cases where signals bearing enhanced near-field characteristics would be advantageous, as where pr-existing receiver mechanisms detect particle velocity rather than pressure (see Section 6.1), where receivers are extremely close, or where transmitting signals over a long distance would be undesirable, a signal apparatus different from those described in the previous section may be expected. Here, arthropods might rely on mechanisms that generate relatively low-frequency vibrations in a broad, flat surface: near the radiator surface, sound intensity calculated from particle velocity (cf. equation 4.3) would greatly exceed that calculated from pressure (cf. equation 4.2; Figure 4.1), and this domination by particle velocity would hold over a longer distance than it would in higher frequency sound broadcast from a minute radiator (Section 4.1.1).

Wings vibrated at several hundred hertz, a frequency below that used by insects transmitting far-field acoustic signals, would be an efficient mechanism with which to generate high-intensity, near-field sound. A small organism could not use this mechanism to generate far-field sound efficiently (Section 4.2.5), but it could establish relatively large particle displacements and high particle velocities within several millimeters of its wings. These particle displacements and velocities would attenuate very rapidly over distance, though; from a dipole source (Figure 4.2), displacement would decrease proportionately with the cube of distance (Fletcher 1992, Römer 1998). Accordingly, various Diptera (Culicidae, Drosophilidae) and Hymenoptera (Apidae) use wing vibration at 100–600 Hz to broadcast sexual and social messages over short distances (<1 cm) in situations where the receivers are usually located within hearing range (Bennet-Clark 1971, Bennet-Clark et al. 1980). At a range of 2.5 mm, the particle velocity of *Drosophila* courtship signals may be as high as $3.8 \times 10^{-3}\,\mathrm{m{\cdot}s^{-1}}$, which represents a sound of 95 dB IL (for particle velocity, 0 dB IL $= 7 \times 10^{-8}\,\mathrm{m{\cdot}s^{-1}}$). In contrast, the SPL (or IL, as determined from Δp) measured at this range may be 44 dB lower. For recruitment signals in the honeybee, *Apis mellifera*, the contrast may be even greater (140 dB IL for particle velocity versus 92 dB IL for pressure). Both Diptera and Hymenoptera rely on indirect flight mechanisms and myogenic control of wing movement to beat their wings efficiently at the frequencies used in vibration, and they may be preadapted for this form of signaling (see Section 6.1 on the role of motor constraints in the evolution of signal mechanisms).

4.2.3 Signal Apparatus: Underwater Sound

Four orders of insects (Odonata, Hemiptera, Trichoptera, Coleoptera; Aiken 1985) and two families of marine crustaceans (Alpheidae, pistol shrimps; Ritzmann 1973; Palinuridae, spiny lobsters; Meyer-Rochow and Penrose 1976) are known to produce specialized sounds underwater. Several other families of crustaceans occasionally generate underwater sounds (Hawkins and Myrberg 1983), but no evidence indicates that they are anything other than incidental byproducts of other activities. Like near-field airborne sounds, underwater arthropod sounds have not been intensively studied: these signals too are inaudible to observers—unless the observers are underwater themselves so as to avoid the great difference in specific acoustic impedance between air and water—and they must be recorded with microphones (preferably, specialized hydrophones) designed to withstand submersion. The reduced attenuation and longer wavelengths of sound in water present additional recording problems: Sounds are strongly reflected by solid surfaces at the boundary of the water body, and much communication may take place in the near field.

Among aquatic insects, most sound-generating mechanisms are stridulatory. The head, thorax, legs, forewings (elytra of beetles), and abdomen may be used, but in all cases the sound radiator is a heavily sclerotized body part bearing substantial mass (Aiken 1985). These features may reflect that more kinetic energy is needed to overcome the inertia and viscosity of water than of air and to initiate vibration in this medium. That is, a membranous wing region, the harp of a cricket, for example, would probably not be effective underwater.

Among marine crustaceans, acoustic mechanisms are frictional (Palinuridae) or percussive (Alpheidae). Spiny lobsters (*Palinurus* spp.) rely on a frictional "stick-and-slip" device that is effective in generating sound even when the animals are soft, as during their molt cycle (Patek 2001). The plectrum (antenna) moves across the file (antennular plate on head, below the eyes) as a bow does across the string of a violin: it sticks in place until the force of sliding friction exceeds that of static friction. The pistol shrimp *Alphaeus heterochaelis* generates very intense clicks by snapping the two segments of an enlarged chela together. A "plunger" on the distal segment is propelled into a "socket" on the basal segment during the snapping action (Ritzmann 1974), which generates a high-speed jet of water that momentarily reduces pressure sufficiently to cavitate the fluid. Sound results as the bubbles created by cavitation collapse under returning pressure (Versluis et al. 2000).

4.2.4 Signal Apparatus: Substrate Vibration

In most cases, arthropods transmit substrate vibrations without relying on specialized morphological devices for generating and radiating the signals. Bending wave signals that are transmitted through vegetation or other similar structures are usually generated by grasping the structure and tremulating the body at frequencies ranging from 10 to 100 Hz. The vibration is transferred from the body to the substrate via the legs (e.g., Morris 1980, Keuper and Kuhne 1983, Dierkes and Barth 1995). This mechanism takes advantage of the body's mass to generate sufficient energy for overcoming the substrate's inertia. Specialized devices may be unnecessary because the signaler relies on the mechanical properties of the substrate to transmit a signal with sufficient intensity and appropriate carrier frequencies.

Various spiders, however, use specialized stridulatory devices to generate substrate vibration, and some of these mechanisms concomitantly yield airborne sound (Barth 1982, 1997). The devices may be coupled to the substrate via stiff spines such that stridulatory activity is converted to vibrations transmitted to a distant receiver (e.g., Rovner 1975). But, it is also possible that in some species the airborne sound generated by a signaler's stridulation sets the substrate of a distant receiver into vibration.

Among insects, many small-sized Homoptera (Cicadellidae, Membracidae, Fulgoroidea) and some lycaenid butterfly larvae also generate substrate vibrations via specialized devices. Here, the devices, tymbals (Homoptera) or stridulatory structures on the dorsal surface of the head and thorax (Lycaenidae), are modified such that they generate little or no airborne sound but instead transfer much of their mechanical energy to the substrate.[14] The homopteran devices, which resemble the sound-producing tymbals of cicadas, minimize sound generation because they are not backed by an abdominal air sac (Ewing 1989). Can these small insects, <8 mm in body length, communicate more effectively or over a greater distance by vibrating the substrate than would be possible via sound production? Probably, given the difficulty that a small organism has in broadcasting all but the highest ultrasound frequencies and the relatively large amount of excess attenuation that high frequencies would suffer. Additionally, both signalers and receivers in these species dwell on plant stems and twigs, structures that generally are very conducive to the transmission of bending waves (Michelsen et al. 1982). Thus, a miniature signaler can use vibration to channel most of its signal energy linearly to the potential locations of receivers, whereas the same energy broadcast as sound may spread more or less spherically, with much escaping to areas where receivers never occur. Directing the message along specific pathways rather than broadcasting it indiscriminately may also reduce vulnerability to natural enemies (see Belwood and Morris 1987, Henry 1994) or other illegitimate receivers, such as males attempting to interlope in the courtship activities of conspecifics (Section 4.4.2). These advantages parallel those indicated above for the specialized generation of near-field sound signals (Section 4.2.2).

Surface waves on water, ripple signals, are usually generated by simple vibration. Either a small object may be grasped and then rapidly tremulated, as in some water striders (Hemiptera: Gerridae; Wilcox 1972), or the water surface may be directly agitated by rapid vertical movement of the body (Hemiptera: Belostomatidae; Kraus 1989) or legs (Coleoptera: Gyrinidae; Hemiptera: Gerridae; Wilcox 1995). On solid surfaces, Rayleigh and Love waves are usually generated by percussion: drumming, rapping, or other forms of striking the ground with a body segment or appendage of substantial mass (e.g., fiddler crabs, *Uca*; Crane 1975). Some insects and wolf spiders that generate percussive substrate vibrations may protect their point of contact with a pad- or plate-like cuticular device that acts as a shock absorber and enhances the impact to the substrate (e.g., stoneflies, Stewart et al. 1991; tenebrionid beetles, Lighton 1987).

Spiders can efficiently generate a great variety of vibratory signals by plucking the strands of their webs (Barth 1982), but this mode of signaling is potentially fatal unless special measures are taken. Webs are generally constructed and tended by females, who may remain in the hub of the web and interpret certain vibratory modes in the strands as twitching by trapped prey. Thus, a male initiating courtship by transmitting vibratory signals to the female in the hub must present himself stealthily or risk being eaten. Males may achieve relative safety during courtship by adding tactile signals to their repertoire, which distinguishes them from prey, or by attaching a specialized mating thread to the web and transmitting their vibratory signals along this strand (Robinson and Robinson 1978). Presumably, the male can more easily evade a ravenous female along the mating thread than on the main orb web, but it is also possible that vibrations transmitted along mating threads are distinctive in frequency, amplitude, tempo, or the direction of propagation (see Section 6.1). Social spiders face the same dilemma, and various species have also solved it via specialized web construction that magnifies differences between vibrations caused by trapped prey and by neighboring webmates (Burgess 1976, 1979).

4.2.5 Signal Characters: Carrier Frequency

Following a mechanical impulse, a fixed structure will vibrate at a natural frequency that is determined by its size and stiffness or tension. In the special case where a fixed structure is driven by an external, forced vibration that equals its natural frequency, it will continue to vibrate at that frequency and the magnitude of its vibrations will be enhanced. A structure driven as such is said to resonate, and its natural frequency may also be termed its resonant frequency. For an acoustic insect's vibrating sound radiator, the insect's metabolic energy will be more efficiently converted to vibratory motion when that structure resonates than when driven at higher or lower frequencies (cf. Ryan 1988). In the latter cases, much of the metabolic energy driving the forced vibration will be dissipated as heat rather than vibratory motion. Because a membranous structure's natural frequency increases proportionately with the inverse square root of its area, small vibrating membranes such as the harp and the frame/mirror apparatus in the wings of crickets and katydids, respectively, have high natural frequencies.[15] Shell-like structures such as tymbals behave similarly, except that their natural frequencies increase proportionately with the inverse of area and may be higher than the natural frequencies of flat structures of equivalent size.

After an acoustic insect's metabolic energy is converted to vibratory motion of its sound radiator, that vibratory motion must be converted to pressure deviations in the surrounding fluid in order to transmit sound waves into the far field (cf. Bennet-Clark and Daws 1999). This latter conversion also depends on vibration frequency and size of the radiator: in general, the efficiency with which acoustic energy is radiated from a small vibrating source, less than one wavelength in diameter, is proportional to the frequency and the source radius (Fletcher 1992). Both the conversion efficiency of metabolic energy to structural vibration and of structural vibration to radiated sound indicate that acoustic insects will usually maximize the intensity of their signals by establishing relatively high vibration frequencies in their sound radiators. As expected, actual vibration frequencies always exceed 0.8 kHz and are ultrasonic in many katydids and moths. But insect muscles cannot contract more than 1000 times per second, and most are limited to much lower rates, so acoustic insects must resort to indirect methods for driving their sound radiators at the vibration frequencies that would generate sound efficiently. This issue does not arise, however, in the specialized transmission of near-field sound, which simply exploits the flight mechanism in most cases and is actually more efficient at low frequencies (Section 4.2.2).

Frequency Multipliers. Many stridulating insects, such as Orthoptera, have solved the problem of driving their sound radiators at a high vibration frequency by employing a frequency-multiplier device. The calling songs of the European field crickets *Gryllus campestris* and *G. bimaculatus*, whose wing harps have natural frequencies ≅4 and 4.5 kHz, respectively, are thoroughly analyzed examples of this process (Nocke 1971, Huber and Thorson 1985, Bennet-Clark 1989). A *G. bimaculatus* male may open and close its wings at $30\,s^{-1}$ and produce a single 16-ms sound pulse during the closure phase of each 33-ms wing cycle; in general, one muscle contraction effects one wing closure. If the scraper on the right wing is struck by 72 of its file teeth on the left wing during a wing closure (see Figure 4.6), or sound pulse, a toothstrike rate of $4500\,s^{-1}$ ($= (1/0.016\,s) \times 72$) results. Provided that the file teeth are evenly spaced and each toothstrike imparts a single mechanical impulse to the wings, the harps will be driven at 4.5 kHz during wing closure (see Koch 1980). The carrier frequency of this forced vibration equals the harp's natural frequency, and the 4.5-kHz sound generated will be resonant: it will be broadcast with relative efficiency and intensity.

Remarkably, the frequency-multiplier device may continue to generate nearly resonant sound even as temperature changes and rates of muscle contraction follow suit and yield faster or slower pulse rates. In *Gryllus* spp. an escapement mechanism has been identified in which the wing's mechanical properties control the scraper's advance from one file tooth the next (Elliot and Koch 1985). Thus, when temperatures increase and wing muscles contract faster, they do not yield substantially briefer wing closures and higher toothstrike rates.[16] Similar mechanisms may be responsible for the relative insensitivity of carrier frequency to temperature among resonant singers throughout stridulating Orthoptera (see Walker 1962, Prestwich and Walker 1981, Doherty 1985a, Pires and Hoy 1992a).

Undamped versus Damped Vibrations. Crickets are notorious for producing calling songs whose carrier frequencies are extremely narrowband (see Figure 2.2b) and perceived as nearly pure tones; it is the purity of these tones that, owing to human psychoacoustics, makes their songs pleasing to our ears. Most other stridulating orthopterans do not produce such melodious songs, and their harsher, often cacophanous, sounds may be attributed to broader frequency spectra. In katydids and acridid grasshoppers, these different spectra reflect the way in which vibration of the wing is damped following a toothstrike (Ewing 1989). Cricket harps are damped very little following a mechanical impulse, and the toothstrike rate may match the structure's natural frequency. Because the particle velocity and pressure deviations of sound waves correspond with the motion of the vibrating structure, these two features of cricket stridulation yield nearly pure tones (Figure 2.2b) and allow the harp to act efficiently as a resonator rather than as a simple sound radiator (Bennet-Clark 1999).[17] The katydid frame, however, usually exhibits highly damped vibrations following an impulse. Consequently, the spectra of katydid song frequencies are generally created by both the toothstrike rate and the damped vibrations, and a broad frequency spectrum that includes harmonic and other tones (sidebands) can be found even if the toothstrike rate matches the frame's natural frequency. These broad frequency spectra may represent more than simple artifacts of wing biomechanics, though, as field recordings indicate that such spectra can preserve a calling song's specific tempo as it is transmitted over a long distance more effectively than a pure tone can (Römer and Lewald 1992; Section 4.1.1). This effect may arise because a given frequency can suffer greatly from reverberation or other forms of distortion and degradation in a specific environment and at a particular moment, but severe distortion and degradation are unlikely to affect all of the frequencies in a broadband signal.

Despite features that should ensure the transmission of pure-tone calls, some crickets also produce calls that include a range of frequencies. In *Oecanthus* spp.

the frequency range results because the left and right wings or their harps have different natural frequencies and the transmitted call combines both (Sismondo 1993). And, crickets that do produce pure-tone calling songs may also transmit other songs at different, usually higher, carrier frequencies. These frequencies may be generated by specific regions of the harp (Figure 4.6) or vibratory modes, which might be induced by body posture or the angle at which the wings are held. Courtship songs in *Gryllus* spp. are typically transmitted at carrier frequencies several times higher than the resonant frequency used in the calling song. The higher carrier frequency is not simply a harmonic of the 4.5-kHz resonant frequency, which is not present, and the precise manner in which it is attained is puzzling.

Not all acoustic insects employ frequency-multiplier devices to convert a low muscle-contraction frequency to the high natural frequency of the sound radiator and a vibration frequency at which it would efficiently radiate sound energy. Species with tymbal organs may simply cause these structures to buckle, which yields a train of vibrations at the natural frequency. These tymbal frequencies range from values as low as 0.8 kHz in the Australian bladder cicada (*Cystosoma saundersii*) to over 100 kHz in pyraloid moths (e.g., Heller and Krahe 1994). In some cicadas (e.g., *Cystosoma australasiae*), the abdomen is held in a particular posture and degree of inflation such that its air volume is tuned to the same natural frequency as the tymbal (Bennet-Clark 1998b). Thus, the air volume represents a resonant acoustic load, which tunes the carrier frequency of the sound generated by the tymbals to a very narrow bandwidth and greatly amplifies its transmitted intensity.

Carrier frequencies in substrate vibration signals are often quite low, as in waves created on the water surface ($3 \leq f \leq 100$ Hz; Wilcox 1995). Frequencies in bending wave signals may be higher (up to several kilohertz; Keuper and Kuhne 1983, Cocroft et al. 2000), and if so they may be attained by exploiting the mechanical properties and natural frequency of the substrate, just as tymbals are caused to vibrate at high frequencies. Additionally, some bending wave signals are generated by vibration of the signaler's stridulatory or tymbal devices, and the substrate vibration frequency near the source may be influenced by the device's frequency.

4.2.6 Signal Characters: Intensity

As described above, a widespread manner in which acoustic insects sustain high signal intensity is by driving vibration of the sound radiator at its natural frequency or a frequency at which it can efficiently radiate sound. The disadvantage of this approach is that such frequencies are typically high and may be attenuated rapidly by absorption, scattering, and diffraction. For example, in air at 25°C and 25% relative humidity, a 20-kHz sound, typical of many katydid calls (Heller 1988), will be subject to ≈6 dB excess, or atmospheric, attenuation over a 10-m distance (Lawrence and Simmons 1982; see also Griffin 1971). Were 1 kHz sound used, attenuation over 10 m would be <1 dB greater than that due to spherical spreading. On the other hand, the high-frequency sounds favored are more easily localized by receivers than low-frequency ones would be, and size constraints would render many arthropod receiver organs rather insensitive to low frequencies (Section 4.3.1).

In addition to using frequencies for which the conversions of metabolic energy to radiator vibration and of radiator vibration to sound waves are efficient, acoustic insects may exploit various environmental features to reduce the loss of sound energy during transmission (cf. Endler 1992). The most common method is to call from an elevated perch such that diffraction due to vegetation and sound shadows next to the ground are reduced or avoided altogether (Paul and Walker 1979, Arak and Eiríksson 1992). But, in some cases insects may take advantage of special sound channels that can exist during certain times at ground level. Males of the South African pneumorid grasshopper *Bullacris membracioides* can effectively transmit their advertisement signals (98 dB SPL at 1 m) over a 1.5-km range by calling at night, when such sound channels normally occur (van Staaden and Römer 1997). However, the advantages gained in transmission distance by these strategies would have to be weighed against vulnerability to predation at the selected calling stations (see Arak and Eiríksson 1992), signal interference from other acoustic insects calling at the selected times (Latimer and Broughton 1984, Greenfield 1988, 1993, Römer et al. 1989, Schatral 1990a), and receiver activity and preferences in order to assess their ultimate value.

Beaming. Acoustic signals cannot be directed along as narrow a path as bending waves on plant stems and leaves, but the calls of acoustic insects are not perfectly omnidirectional either. Thus, a signaler might control the position of its body and beam its maximum sound intensity toward potential or known receivers. In acoustic insects that have two sound radiators on either side of the body, such as acridid grasshoppers, a signaler might stridulate with the hindleg that is ipsilateral to a receiver and thereby transmit calls 6 dB higher in SPL than those generated by the contralateral hindleg (Bailey et al. 1993a, Michelsen and Elsner 1999). Whereas stridulating with both hindlegs simultaneously may increase SPL further, this signaling activity may not pay off energetically (Section 4.2.9). Additional cases of beaming rely on ground reflection and architectural construction (see below).

Reduction of Acoustic Interference. Most other methods for sustaining high signal intensity are morphological devices, behaviors, and architectural construction by which signalers avoid acoustic interference and impedance differences near the sound radiator(s). For example, crickets have a sound-radiating harp on each forewing, and stridulation would be expected to deflect the left and right wings in opposite directions. Were this to occur, though, the sound waves from each wing would destructively interfere, canceling each other. This problem is prevented by the unique curvatures and bucklings of the left and right forewings, which cause both harps to be deflected simultaneously in the same (upward) direction by each toothstrike (Figure 4.6; Bennet-Clark 1989). Consequently, the sound waves from each wing are constructively added, and SPL may be amplified by several decibels. Acridid grasshoppers also have a sound radiator on each wing, but the wings are held in a vertical orientation surrounding the body. This configuration ensures that sound waves from each wing primarily radiate in opposite directions away from the body, and specialized devices for eliminating destructive interference may not be critical.

Whether an acoustic insect has two sound radiators or one, another form of interference may arise when sound waves are simultaneously generated from both sides of a radiator. If the radiator is effectively a dipole source (Figure 4.2b), the waves on either side will be 180° out-of-phase, and they too may destructively cancel each other unless kept apart near the source. The wings themselves may help to abate this "acoustic short circuiting" provided that they are held roof-like over the body, the sound radiator is located near the base of the wing, and a distance of at least $\lambda/4$ lies between the radiator and the free edge of the wing (Ewing 1989).[18] Thus, some katydids which have large wings and call at high carrier frequencies may be safeguarded. Crickets, which generally call at lower carrier frequencies, may resort to other methods. Tree crickets (subf. Oecanthinae) often call from notches and holes in leaves and thereby use foliage as a baffle that effectively extends the free edge of the wing by several centimeters. Some species may even chew the holes and notches and create their own baffles (Prozesky-Schulze et al. 1975, Forrest 1982).

Species calling near the ground may suffer from interference between waves propagated upward from the wings and waves propagated downward, which will be reflected upward by the substrate. But, the short-tailed cricket *Anurogryllus muticus* contends with, and actually exploits, this phenomenon by digging and then straddling a shallow depression in the soil while calling (Forrest 1982, Bennet-Clark 1989). Its forewings are then approximately $\lambda/4$ (= 13 mm) above ground, and a sound wave propagated downward and reflected upward interferes destructively with the subsequent downward wave before it can cancel a wave propagated upward and away from the insect (Figure 4.7). Any of this acoustic energy not canceled below the insect will then add constructively to waves propagated upward from the wings. The specific $\lambda/4$ elevation therefore maximizes the amount of acoustic energy propagated vertically upward, but it reduces that propagated horizontally (Fletcher 1992). Presumably, *A. muticus* beams its signals to potential receivers flying overhead. For insects calling with low, narrow-band carrier frequencies, reflection from the ground could generally be used as an advantage in both horizontal and vertical transmission of signals, provided that sound shadows and foliage or other

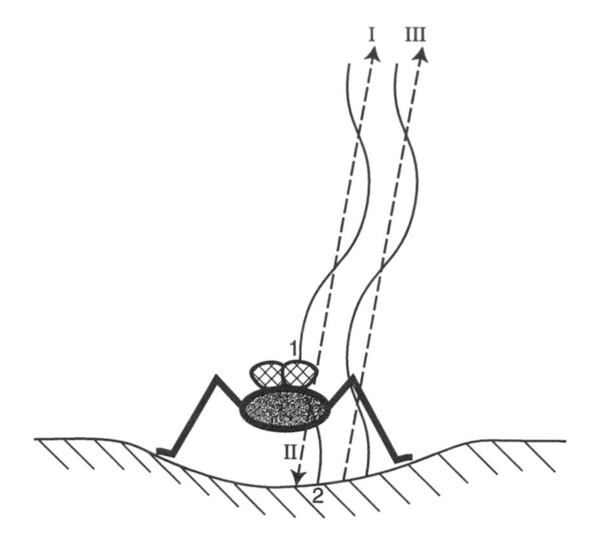

Figure 4.7 Avoidance of destructive interference in the short-tailed cricket *Anurogryllus muticus*. Rear view of singing cricket positioned over a depression in the ground. Distance from the wings (cross-hatched areas, (1) to the ground (2) is approximately $\lambda/4$. Thus, sound waves propagated downward (II) and reflected upward (III) remain in-phase with sound waves propagated directly upward from the wings (I). Sine curve along path of sound wave indicates phase (see Forrest 1982).

sources of diffraction are lacking and elevation is kept as low as possible. At distances within $\lambda/6$ of the ground, the total acoustic energy radiated may be increased as much as twofold over the free-field value (Fletcher 1992). On the other hand, signalers calling from slightly higher elevations ($\lambda/4$–λ) normally have various intensity nulls in their sound fields that are caused by destructive interference. These null spots are one reason why observers often have trouble finding certain acoustic insects calling from open and seemingly locatable sites in the field.

Conversion of Mechanical to Acoustic Energy. Perhaps the most general difficulty preventing acoustic insects from transmitting intense calls is that their sound radiators are too small for efficient conversion of vibration energy to acoustic energy in the far field. To transmit sound efficiently, a radiator should be $\lambda/3$ or larger in radius (Olson 1967), but such dimensions would be prohibitive for most acoustic insects. For example, a cricket would need a harp 4.5 cm in diameter to radiate a 5-kHz call efficiently. But even if such a device were morphologically feasible, its natural frequency would be far lower than 5 kHz unless under extraordinary tension. Thus, our cricket appears to be constrained from both directions: Its 7-mm diameter harp resonates at 5 kHz, yet it would be most efficient radiating sounds at 32 kHz or higher, which would be severely affected by excess attenuation. Is its only recourse simply to continue the inefficient transmission of acoustic energy (Section 4.2.9)?

Possibly, but various morphological and architectural devices do exist that can assist acoustic insects to overcome the problem of inefficient conversion of the radiator's vibration energy to sound energy in the far field. We have already seen how cicadas may use air volumes in their abdomens as resonant acoustic loads and then radiate the resonant sound from the tympana rather than the much smaller tymbals. Some aquatic hemipterans (water boatmen; Corixidae) use a respiratory air bubble carried beneath the water surface and held next to the stridulating structures in a similar fashion. The bubble, which is much larger than the stridulated structure, is caused to vibrate at its natural frequency, which may be close to the vibratory frequency of stridulation (Theiss 1982, Theiss et al. 1983). Thus, the bubble may resonate, and it can generate underwater sound detectable by corixid receivers as far as 10 m away (Theiss and Prager 1984). However, the volume of the bubble decreases as it loses oxygen due to the insect's respiration, and the bubble's natural frequency, and that of the radiated sound, increase. Conceivably, though, corixid bugs might receive mechanical feedback reflecting the growing discrepancy between their stridulatory and bubble frequencies and then increase their stridulatory rate accordingly, maintaining bubble resonance (cf. Elliot and Koch 1985; Section 4.2.5). Other amplification devices seem to rely more on specialized design than overall size. Morris and Mason (1995) describe a neotropical agraeciine katydid in which the cap-like pronotum covers the forewings during stridulation. The volume enclosed between the pronotum and wings acts as a Helmholtz resonator (cf. Bailey and Stephen 1978; Section 4.3.5), and the calling song is thereby amplified 10 dB.

The most effective device used by small animals for increasing radiation efficiency is one well known to musicians and public speakers: an exponentially flared horn. Flared horns can efficiently couple a small mechanical sound radiator in the horn's throat with the free-field air at its mouth. Some vertebrate vocalizations are intensified in this fashion, and among acoustic insects the mole crickets have evolved this solution via burrow architecture (Bennet-Clark 1970, 1987, Ulagaraj 1976). Mole crickets call from the bottom of horn-shaped burrows that they construct in the soil, and these devices may enhance their song SPL by as much as 24 dB along the burrow axis. Here, air in the wide entrance to the burrow, which may meet the $\lambda/3$ criterion, serves as the sound radiator to the free field. As acoustic impedance[19] at a point within a horn is inversely proportional to the horn's cross-sectional area at that point, the gradual flaring of the burrow avoids sharp discontinuities in acoustic impedance between the bottom and entrance. Consequently, for a range of sound frequencies little energy is lost in transmission from the insect's radiator, which behaves like a piston at the burrow's throat, to the free field outside the burrow despite the large overall impedance difference between these two points (cf. equation 4.7). Because mole crickets normally shape their burrows such that the range of efficiently transmitted frequencies includes the carrier frequency of their calling songs, air in the burrow entrance can resonate, making the 24-dB SPL gain possible. This method for efficiently coupling a small mechanical sound radiator with free-field air also operates reciprocally,[20] and horn-shaped devices figure prominently in ear design. We encounter horns again in our treatment of arthropod hearing and receiver apparatus (Section 4.3.5).

4.2.7 Signal Characters: Tempo

Most arthropods modulate the amplitudes of their acoustic or vibratory signals such that they are transmitted as a series of discrete units, or pulses, each representing a packet of waves that corresponds with a single repetitive action (i.e., wingstroke or legstroke). Pulses may be grouped into higher-order temporal units and, more rarely, divided into lower-order ones. Such pulse tempos are critical for species recognition (e.g., Schul et al. 1998) and further mate assessment by receivers in many cases, but in others they are merely incidental byproducts of the movement and central rhythm that generates and controls signaling.[21]

Electronic devices that simultaneously record stridulatory movements and the sounds generated have shown that amplitude modulation reflects the periodic movements of the wings or legs (Walker et al. 1970, Walker and Dew 1972, von Helversen and Elsner 1977). In stridulating Orthoptera, discrete pulses are often generated because sound is not transmitted continuously throughout a cycle of wing or leg movement. Crickets normally transmit sound only during wing closings, and the gaps between pulses correspond with wing openings. Duty cycle, which may be defined as the proportion of a call during which sound is actually transmitted, is therefore approximately 50%.[22] Katydids, in contrast, may transmit some sound during wing opening, and their duty cycles are generally higher (Walker 1975a). Brief silent intervals invariably occur, though, when the wings reverse direction. Acridid grasshoppers may also transmit sound during both leg upstrokes and downstrokes (Elsner 1974). Because the two hindlegs may be moved simultaneously, but not in synchrony, sound can be broadcast continuously during a bout of leg movement, and discrete pulses may not be apparent. In general, the sounds of acoustic insects are much more continuous than those of vertebrates, whose vocalizations must be coordinated with breathing.

Higher-order temporal units of arthropod signaling may be generated because cyclic wing and leg movements are often organized into regular bouts. For example, a cricket or katydid may produce five consecutive cycles of wing movement, followed by a rest interval of similar length. This pattern of movement would yield a series of 5-pulse "chirps," which would be distinguished from a continuous "trill" composed of the identical pulses but transmitted without regular rest intervals. Lower order temporal units arise in some katydids and acridid grasshoppers that move their wings and legs in a complex fashion during each complete cycle of movement (Walker and Dew 1972, Elsner 1974, Walker 1975a). Here, a series of partial wing openings or leg upstrokes may occur during the interval of wing closing or the leg downstroke, respectively. Such movements lead to complex patterns of amplitude modulation in which the sound emitted during one complete cycle of movement appears as several pulses, which may or may not be equivalent in duration, amplitude, etc. In most species, the pulses that result from each complete cycle of movement are transmitted faster than human acoustic fusion frequencies ($\approx$15–20 Hz), and observers recognize the higher order temporal units only. Consequently, many higher order units have been assigned onomatopoetic designations in the literature, but the wealth of these terms creates much confusion. I shall avoid them where possible and remain consistent with the above terminology where necessary.

The tempo of calls and vibrations produced by tymbal organs is similar to that in stridulation. Pulses are generated by the tymbal bucklings, or, in cases where tymbals consist of multiple units, by bucklings of individual ribs or striae. Among cicadas, muscles operating the left and right tymbals may contract in alternation, thereby doubling the pulse rate.[23] In galleriine pyralid moths, these muscles do not alternate, but tymbals normally generate pulses during both inward and outward buckling (Spangler 1988a). Additionally, the left and right tymbals may buckle in slight asynchrony (Spangler et al. 1984, Jang and Greenfield 1996). Thus, pulse rates may be 4 × muscle contraction rates, or higher in species wherein individual striae buckle.

Motor control that yields the tempo of acoustic and vibratory signals is effected by the regular contraction of locomotory muscles, typically those associated with flight (Elsner 1983, Josephson 1985, Pfau and Koch 1994). During signaling, however, these muscles generally contract faster and are controlled by different nerve networks than those used during locomotion (Hennig 1990). Mutual inhibitory interactions may occur between these networks. The central pattern generators controlling signal rhythms normally reside in the thoracic ganglia but are controlled by command interneurons descending from the brain (Huber 1990, Hedwig 1992, 1996, 2000, Ocker and Hedwig 1996, Hedwig and Heinrich 1997, Heinrich and Elsner 1997). In various Orthoptera these generators may be derived from the flight rhythm generator (Elsner 1983, von Helversen and von Helversen 1994).

Pulse rates in acoustic and vibratory signals range from 10 to 1000 Hz. The lower limit is set by inefficient muscle operation at low contraction rates (Bennet-Clark 1989), and the upper limit represents the maximum contraction rate of asynchronous muscle that is myogenically stimulated, as in the wingbeat of small Diptera. In stridulating Orthoptera, an upper limit ≅ 250 Hz is set by contraction rates of synchronous, neurogenically stimulated muscle. The higher rates (≥200 Hz) that are found in some katydids are facilitated by short twitch durations in these muscles and may be sustained by endothermy 10–15°C above ambient temperature (Stevens and Josephson 1977, Heller 1986). This heating begins prior to stridulation. In cicadas, pulse rates several times higher than in Orthoptera can occur because the muscles may contract asynchronously with neural firing, tymbals may produce pulses during both inward and outward buckling, and the left and right tymbals may alternate. Cicadas too may augment muscle contraction rates by endothermy, which they achieve by basking or flight prior to calling (Sanborn et al. 1995).

Acoustic and vibratory pulse rates are generally dependent on ambient temperature (e.g., Walker 1975b,c, Skovmand and Pedersen 1983, Pires and Hoy 1992a, Souroukis et al. 1992), except in species that are strongly endothermic. Rates may vary linearly with ambient temperature within a range of moderate values (15–30°C), but in hump-winged crickets (Haglidae), which are adapted to cold temperatures, the linear relationship may extend down to 0°C (Morris and Gwynne 1978). At the other extreme, cicadas, which are diurnal and may be considered thermo- or heliophilic, can exhibit a linear relationship over a range extending up to 40°C (Josephson and Young 1979). The rate at which higher order temporal units, such as chirps, are delivered may also vary with temperature, but the slopes of these relationships do not necessarily equal those for pulses (e.g., Walker 1962).

Temperature Coupling. In various species, females prefer a specific range of male pulse or chirp rates (e.g., Perdeck 1957, Walker 1957) rather than exhibiting an "open-ended" preference for faster rates (Section 4.4.1, Chapter 7). Female preferences focusing on a specific range of a signal character impose stabilizing selection on the population, and they generally function in the context of species recognition as opposed to assessment of mate quality (Gerhardt 1991). Preferences for specific tempos may shift to faster values at higher temperatures, and the rates preferred by females can approximate those typically produced by males under these conditions (Figure 4.8). This signaler/receiver phenomenon is known as temperature coupling (e.g., Walker 1957, von Helversen and von Helversen 1981, Doherty 1985a, Pires and Hoy 1992a; see Section 7.1), and it arises in various poikilotherm signaling systems based on rhythmic activity. For example, acoustic anurans (Gerhardt 1978) as well as optically signaling arthropods (cf. Section 5.2.1) exhibit varying degrees of temperature coupling.[24] Some investigators have argued that temperature coupling reflects a genetic mechanism that can influence the evolution of signaling systems (Section 7.1), but some current findings (Ritchie et al. 2001) suggest that it is merely a coincidence.

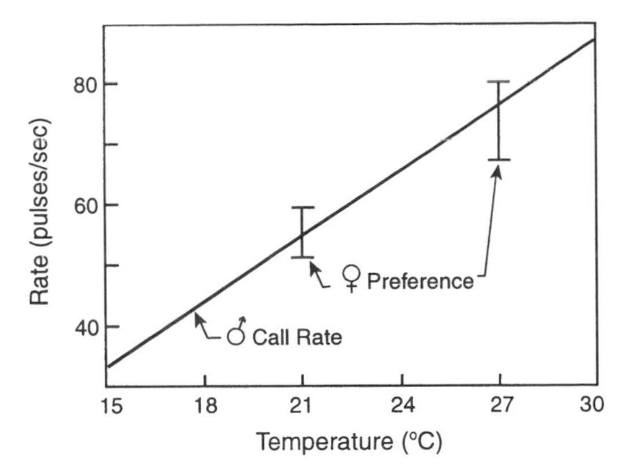

Figure 4.8 Temperature coupling in the tree cricket *Oecanthus niveus*. At higher temperature females prefer male calls that are delivered at faster pulse rates. The shift in female preference is commensurate with the linear increase in male pulse rate at higher temperature (Walker 1957).

4.2.8 Signal Production: Feedback Mechanisms?

The influences of neural reafference, self-stimulation from sensory (auditory) feedback, and learning on vertebrate signaling, particularly the calls of songbirds, are well known to animal behaviorists and neuroethologists. Do arthropods similarly modify their acoustic and vibratory signals in response to such input? On the one hand, there is negligible evidence that arthropods must perceive the calls of conspecifics, or even their own calls, in order to produce an accurate signal as soon as they are developmentally capable. Orthopterans who are reared in isolation from other individuals or who are deafened and thereby deprived of auditory feedback will produce songs that are essentially normal in carrier frequency and tempo,[25] as will individuals who are first to mature during the spring in the temperate zones, where no singing adults survive the winter. Most signal modification or variation by individuals can be attributed to changing environmental conditions (Whitesell and Walker 1978, Cade and Wyatt 1984, Souroukis et al. 1992), such as photoperiod or temperature, or to increasing sclerotization as adults age, which may yield higher carrier frequencies (e.g., Snedden et al. 1994). While arthropods may receive proprioceptive information during signaling via fields of sensory hairs, or hair plates, and campaniform sensilla that allows them to adjust body, leg, or wing posture (Elliot 1983) and maintain a normal song intensity, these adjustments do not affect signal tempo and the feedback generally does not come from their transmitted signals.[26]

There is, however, a growing body of evidence showing that certain temporal signal characters are strongly affected by the social environment. In general, acoustic insects will call at a faster chirp rate or more regularly, produce longer or more intense chirps, or even switch to an entirely different type of call when in the presence of calling conspecifics. They may also time their calls with respect to those of their neighbors and create a structured chorus. These adjustments occur within the context of sexual selection in most cases, and we might presume that comparable adjustments also occur among some arthropods producing vibratory signals. Like tempo, these adjustments are a major distinction between the mechanical and olfactory channels, and we shall return to this feature in our treatment of functions and adaptations (Section 4.4).

4.2.9 Signal Production: Energy Requirements

Because the transmission of acoustic and vibratory signals is accompanied by overt, repetitive movements of the body or its appendages, a measurable amount of energy expenditure would be expected during signaling. Biologists have sought to measure these amounts accurately in order to determine the cost of signaling relative to other activities or options for pair-formation and to learn whether the most effective signalers in a population expend the most energy. To date, accurate measurements for acoustic and vibratory signaling energy have been obtained in several stridulating Orthoptera (Prestwich and Walker 1981, Kavanagh 1987, Bailey et al. 1993b, Hoback and Wagner 1997, Hack 1998), in a cicada and a pyralid moth that use tymbal mechanisms (MacNally and Young 1981, Reinhold et al. 1998), in a tenebrionid beetle that transmits Rayleigh waves on the desert surface by striking the ground with its abdomen (Lighton 1987), and in a wolf spider (Lycosidae) that drums on dry leaves (Kotiaho et al. 1998a). Metabolic measurements were made via flow-through respirometry wherein oxygen consumption rates ($\dot{V}o_2$) during signaling were calculated and compared with resting, or basal, rates. These studies showed that factorial metabolic scopes of signaling (= signaling metabolic rate ÷ basal metabolic rate) range from 2.75 to 25. Lower values occur in species producing discontinuous calls, such as chirps, as opposed to continuous calls, such as trills. Factorial metabolic scopes of signaling are less than those of flight, but signaling may continue for many hours each day and, thus, can account for more than 50% of an individual's daily energy budget. The lowest value reported, 2.75 in the galleriine pyralid moth *Achroia grisella*, occurs in an insect that may signal for 10 hours per night and does not feed as an adult. Consequently, a highly efficient signaling mechanism, in this case a wing-activated tymbal on the tegula, may have been strongly selected.

Once the rates of energy expended by acoustic signaling (= signaling metabolic rate − basal metabolic rate) and of acoustic energy transmitted by signaling (= $4\pi r^2 \cdot I$, for a free-field measurement taken at a distance r from the signaler;

I = intensity) are calculated, the acoustic efficiency of signaling can be determined as acoustic energy rate ÷ signaling energy rate. This value would represent the efficiency with which the signaler's muscle power is converted to sound. Acoustic efficiencies have been reported for several species (Kavanagh 1987, Forrest 1991, Bailey et al. 1993b, Reinhold et al. 1998), and, as expected from the points made earlier (Section 4.2.5), they are generally low. Normally, a high proportion of energy is lost as the vibration energy of a signaler's sound radiator is converted to acoustic energy transmitted to the far field, and efficiency values range from 0.1 to 6% in most species.[27] Efficiency values reported for mole crickets, however, are as high as 35% (Kavanagh 1987), which probably reflects the large acoustic gain provided by their horn-shaped burrows.

The exceptionally high values reported for mole crickets raise the question of whether signaling energy rate is the most accurate term for the denominator in the efficiency equation. Mole crickets must expend considerable energy excavating and shaping their burrows, without which their rate of acoustic energy transmission would be only 1/250 as great. Thus, it might be appropriate to add this construction energy, pro-rated over time, to the metabolic energy expended during actual signal transmission, in which case the calculated efficiency would be much lower. Following this logic, should any energy expended prior to signaling that is necessary for or enhances transmission of the signals be included? For example, a stridulatory or tymbal mechanism invariably demands some investment of energy during immature development, energy that could have been allocated to other structures or activities. While it may be difficult to express this investment in the same currency as immediate metabolic energy expended on the motor activities that generate signals, doing so would allow energetic comparisons among communication channels. It has been customary to assume, albeit with little substantiation, that acoustic and vibratory signaling are relatively expensive compared with chemical signaling (Section 3.1.5). But, we might obtain a different view if the metabolic energy needed to obtain the precursors for a courtship pheromone and then synthesize and sequester it were added to the small amount of work done transmitting the compound.

4.3 Signal Reception: Membranes and Hairs

Organisms transduce mechanical signals to neural firings via mechanoreceptors, devices which move or are otherwise deformed by the inertial forces of incoming signals. Thus, mechanoreceptors may operate as reversed signaling devices, and we may expect them to be constrained by the same factors that limit signal transmission. For example, receptors for airborne sound in the far field must vibrate in response to a signal's pressure deviations. Just as a sound radiator is restricted by its size and other physical properties to transmitting certain frequencies, a sound receptor is restricted to responding to a particular frequency range. Outside this range, stimuli must be increasingly intense in order to stimulate the receptor. Owing to their small size, arthropod sound receptors are normally restricted to stimulation by relatively high frequencies (> 2 kHz).

In this section, the various devices with which arthropods transduce far- and near-field airborne sound, underwater sound, and substrate vibration to neural messages are described. We then examine how the peripheral and central nervous systems process input from these devices, extract information, and effect motor responses to this information.

4.3.1 Airborne Sound in the Far Field: Pressure and Pressure-Difference Receivers

Terrestrial organisms face a major impedance-difference problem in acoustic perception: the air through which sound signals are transmitted and the organism's body have very different specific acoustic impedances, and most incoming sound energy should be reflected by the body's surface before it could excite a mechanoreceptor and trigger a neural response. Mammals and some other terrestrial vertebrates overcome this potential difficulty with their specialized middle ear bones and associated devices, which act to amplify the forces from airborne sound impinging on the outer ear several thousandfold. This amplification effectively reduces the impedance of the inner ear to a value similar to that of air. Among arthropods, various acoustic insects have evolved a far simpler solution by using thin membranes backed by enclosed volumes of air. These devices, known as tympanal organs, exploit the insect's cuticle and tracheal system to create a structure whose specific acoustic impedance roughly approximates that of the surrounding atmosphere. Thus, sound waves can readily vibrate the tympanal membrane with sufficient amplitude to trigger the specialized sensory neurons that attach nearby.

Tympanal organs all share three features: (1) a thin cuticular membrane, the tympanum, which has high resonant frequencies and whose outer surface contacts the surrounding atmosphere; (2) an air-filled chamber, part of or derived from the tracheal system, behind the tympanum; and (3) specialized sensory neurons, individually housed within scolopidial units numbering from one (notodontid moths) to as many as 1000 (cicadas), which are attached to the air chamber wall or to the tympanum itself. As such, they are a specialized type of chordotonal organ, structures found throughout the insect body which respond to mechanical deformation and generally function in proprioception. Unlike vertebrate ears,[28] however, insect tympanal organs are by no means homologous, and they have repeatedly evolved in no fewer than eight orders

(Fullard and Yack 1993, Hoy and Robert 1996, Yager 1999). Moreover, phylogenetic analyses indicate multiple evolutionary origins within several of these orders. Among the insects, tympanal organs are located on various body segments and appendages, including the foreleg tibiae (Orthoptera: Ensifera), thoracic pleura (Hemiptera; Lepidoptera), thoracic sternum (Mantodea: Yager and Hoy 1986; Diptera: Lakes-Harlan and Heller 1992), abdominal pleura (Orthoptera: Caelifera; Lepidoptera), abdominal sternum (Homoptera; Lepidoptera), abdominal tergum (Coleoptera: Spangler 1988b, Yager and Spangler 1995), cervix (Coleoptera: Forrest et al. 1997), mouthparts (Lepidoptera: Sphingidae: Göpfert and Wasserthal 1999), and wing bases (Neuroptera; Lepidoptera: Hedylidae, Nymphalidae). Some organs in Coleoptera are concealed except during flight and consequently escaped detection until quite recently. It is likely that further investigations will reveal additional examples. The function of tympanal organs in hearing has been confirmed by simply occluding or otherwise loading the tympanum and then testing the insect's behavioral or neurophysiological response to acoustic stimuli. Such confirmation is critical, because structures have been mistakenly designated as ears in the past based solely on appearance and gross anatomy.

Despite bearing functional tympanal organs, not all of the above groups use the acoustic channel for communication: Ears are definitely more widespread than sound-transmitting devices, a point that I shall return to in considering the origins of acoustic and vibratory communication (Section 4.5).

Tympanal organs transduce the energy of far-field sound to neural messages by operating as pressure or pressure-difference receivers. Air lies on both surfaces of the tympanum, and this membrane therefore moves in response to differences between external and internal pressure. In insect ears where the air chambers lack direct connections to the outside atmosphere, the tympanum operates as a pressure receiver that simply responds to the deviations in external pressure caused by oncoming sound waves (Figure 4.9a). Compression phases cause the tympanum to warp inward such that pressure on both surfaces is equalized; conversely, rarefaction phases cause it to warp outward. This vibratory movement is transferred to the scolopidial units, which may be directly attached to the inner surface of the tympanum. Sensory neurons in the units then fire if sufficiently stimulated, which can occur in response to a tympanal displacement as small as 0.1 nm (Michel 1974, Iwasaki et al. 1999). Pressure receiver ears are found in Lepidoptera, Neuroptera, Coleoptera, and several other groups.

In other insect ears, the air chamber behind the tympanal membrane is directly connected to the outside, and the membrane therefore operates as a pressure-difference receiver that responds to the difference between pressure increments, and decrements, on its outer and inner surfaces (Figure 4.9b). These pressure differences arise because sound waves reaching the inner surface are normally delayed in phase with respect to their arrival at the outer surface. In a technically sophisticated experiment using miniature loudspeakers,

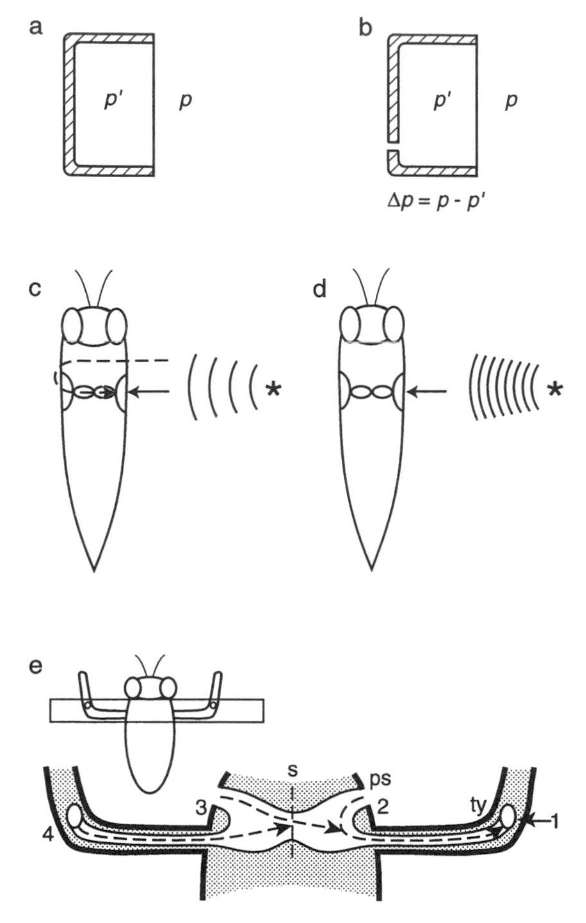

Figure 4.9 Tympanal organs serving as far-field sound detectors. (a) Pressure detector: membrane displacement and movement are primarily influenced by changes in external pressure (p) associated with the compressions and rarefactions of approaching sound waves. (b) Pressure-difference detector: membrane displacement and movement are influenced by changes in both external (p) and internal pressure (p') directly associated with approaching sound waves. Owing to the longer path over which sound waves may have to travel in order to reach the inner surface of the membrane, a phase difference can exist between pressure at the outer and inner surfaces and these two pressures will normally differ; in general, $\Delta p \neq 0$. (c) The grasshopper ear functions as a pressure-difference detector for relatively low frequencies, which are not severely attenuated on the side of the body contralateral to the stimulus (asterisk) and therefore reach the inner surface of the tympanal membrane via the dashed path, but (d) it functions as a pressure detector for higher frequencies, which the body diffracts strongly (see Figure 4.11). (e) The cricket ear functions as a pressure-difference detector in which sound waves reach the inner surface of the tympanal membrane via three paths: via the ipsilateral prothoracic spiracle (ps, 2), via the contralateral prothoracic spiracle (3), and via the contralateral tympanum (4). The central septum (dashed line, s) attenuates sound arriving via paths 3 and 4. Path 1 represents sound waves reaching the outer surface of the ipsilateral tympanum (ty).

Kleindienst et al. (1983) demonstrated that the neural messages transduced by pressure-difference ears are specifically triggered by tympanal vibration and not by pressure changes within the air chamber resulting from sound waves. By positioning loudspeakers facing both the outer and inner surfaces of the cricket tympanum and broadcasting identical signals from each, it was shown that neural messages ceased when the two loudspeakers were 0° in-phase: under this condition, the broadcasts interfered destructively at the tympanal membrane, and all of its movement was canceled (see also Larsen et al. 1984).

Various aspects of cricket ears have been analyzed thoroughly, and careful biophysical measurements indicate how these surprisingly complex devices operate as pressure-difference receivers. The paired tympanal organs of crickets are located in openings in the tibiae of the forelegs, and each includes a separate anterior and posterior membrane. Only the posterior membrane appears to play a major role in hearing (Larsen 1987), as it is larger, thinner, and backed by a tracheal branch, and it is the one that vibrates in response to sound. The tracheal branch leads up through the foreleg femur, to the prothorax, and finally to the outside via the prothoracic, or acoustic, spiracle. Tracheal branches from the left and right ears are separated within the prothorax by only a thin median septum penetrable by sound (Michelsen and Löhe 1995). Thus, each ear effectively has four acoustic inputs: (1) the outer surface of its tympanum via the foretibial opening and the inner surface via (2) the ipsilateral acoustic spiracle, (3) the contralateral tympanum, and (4) the contralateral acoustic spiracle (Larsen and Michelsen 1978; Figure 4.9e). Tympanal motion is determined by the sum of pressures from these four inputs (Michelsen et al. 1994a,b). For sound waves arriving from the same side as a given ear, the path taken via input 1 is shorter than the other three. This difference in path length means that sound waves entering via inputs 2, 3, and 4 lag waves entering via input 1 by a phase angle that depends on the carrier frequency and the position of the sound source. When the phase angle between external (input 1) and internal (sum of inputs 2, 3, and 4) pressure differs significantly from 0°, tympanal motion results and is transferred via a cuticular structure to the scolopidial units on the wall of the tracheal branch just behind the membrane. The intricacies of this four-input system are critical for the localization of sound sources, and we analyze them further in Section 4.3.5.

Minor variations of the four-input cricket ear are found in other ensiferan Orthoptera. Mole crickets have very similar devices except that the scolopidia are attached to the tympanal membrane itself. In katydids, the tracheal branches of the two ears are usually separated by a substantial mass of tissue, leaving them with only two effective sound inputs per ear: (1) the outer surface of the tympanum via the foretibial opening and (2) the inner surface via the ipsilateral acoustic spiracle (Bailey 1993). In acridid grasshoppers, the air chambers behind the tympanal membranes do not have direct connections to the outside, but their arrangement nonetheless allows the ears to operate effectively as two-input pressure-difference receivers for lower sound frequencies (Figure 4.9c,d). Here, the tympanal organs are located on the pleura of the anterior abdominal segment, and a series of air sacs may intervene between the left and right organs to form an acoustic passage (Miller 1977). Thus, sound may reach a grasshopper ear via (1) the outer surface of its tympanal membrane and (2) the inner surface by entering through the contralateral ear. Cicada ears function similarly to acridid grasshoppers (see Fonseca and Popov 1997), except that much of the abdomen may be filled with a single large air sac that serves as the passage between the ears. This arrangement may be complicated in the males of some cicada species because the abdominal air sac is also used as a sound resonator during signaling (Section 4.2.1). Sound reaches the inner surface of the cicada tympanum primarily by entering through the contralateral ear, but this input is augmented by sound entering through the tymbals in males and through the metathoracic spiracles in females (Fonseca 1993).

In several cases, insects are known to hear far-field sounds with chordotonal organs that lack associated tympanal membranes and tracheal air sacs (e.g., Lakes-Harlan et al. 1991). Cockroaches (Shaw 1994) can detect some airborne far-field sounds with chordotonal vibration sensors in their legs (cf. Section 4.3.4), and the pneumorid grasshopper *Bullacris membracioides* hears a wide range of sound frequencies with a series of stretch receptors in its abdominal segments (van Staaden and Römer 1997, 1998). Morphological studies of these atympanate organs have been particularly valuable for inferring the evolution of tympanal hearing among insects (Section 4.5).

4.3.2 Airborne Sound in the Near Field: Particle Velocity Receivers

Because a tympanal organ responds to either external pressure or the difference between internal and external pressure, it would be of little value for detecting sound very close to its source, where pressure deviations are minimal (Figure 4.1). For a sound receiver to work here, it would have to operate as a velocity or displacement receiver and respond to the infinitesimal movements of fluid particles. Additionally, these receivers should be sensitive to sound frequencies <1 kHz, as it is primarily these frequencies that establish significant near-field conditions extending more than several millimeters from the source.

Near-field sound receivers of terrestrial arthropods are typically minute sensory hairs found on various parts of the body and its appendages (Figure 4.10). Their primary function is to detect incidental low-frequency sounds of natural enemies, such as those generated by wingbeats (Gnatzy and Heusslein 1986, Gnatzy and Kämper 1990). Examples of near-field receivers include the trichoid sensilla on many caterpillars (Tautz and Markl 1978) and on the cerci of crickets (Kämper and Kleindienst 1990), and possibly the specialized trichobothria on

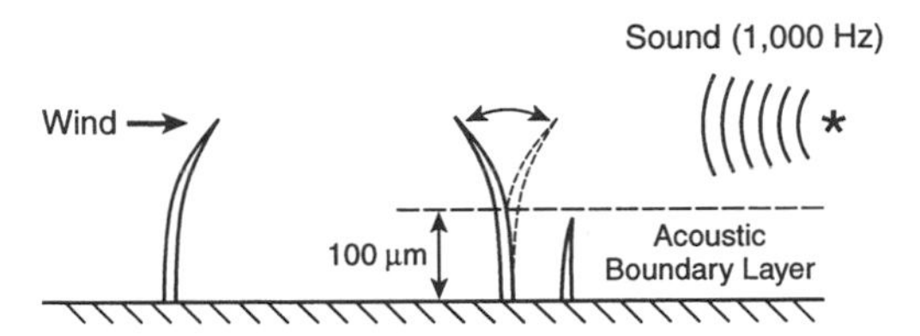

Figure 4.10 Trichoid sensilla serving as near-field sound detectors. Both wind and sound deflect these hairs, but only sound will establish vibration and trigger action potentials in the phasic neural receptors at the base of the hair. Trichoid sensilla serving as particle velocity receivers must extend beyond the acoustic boundary layer in order to detect sound.

spiders (Barth and Holler 1999, Barth 2000). These hairs can be deflected several degrees from their normal positions by particle velocities (u_{max}) as low as 1 mm·s^{-1} (approximately 84 dB IL for particle velocity in sound waves; cf. equations 4.3 and 4.5). Such deflections may stimulate sensory neurons attached to the base of the hair. But, wind as well as near-field sound waves will generate air particle velocities above threshold values, and sensory hairs must respond differentially to these sources of particle movement in order to function as ears. This discernment is afforded by the hair's natural frequency and the phasic response (e.g., Barth and Holler 1999) of the sensory neurons. Sound waves close to the natural frequency will cause the hair to resonate, which creates oscillatory deflections of sufficient amplitude to establish a continued neural response. Wind, however, would deflect a hair in a constant direction, and the neuron would fire only at the very onset of the stimulus.

Boundary layer conditions and the mechanical properties of sensory hairs influence their performance such that only low-frequency sounds are perceived. Consequently, these devices are effective at detecting specialized near-field signals but would not perceive most far-field ones. Because sound entails fluid movement, a narrow acoustic boundary layer in which sound is severely attenuated exists adjacent to a solid surface (Section 3.1.2). This layer is approximately 100 μm thick for a 1-kHz sound in air, and a sensory hair would have to extend beyond this layer to operate effectively as a near-field sound receiver (Figure 4.10). A hair's natural frequency decreases proportionately with the inverse square of its length, though, and hairs long enough to operate as sound receivers respond to low (<1 kHz) frequencies only. This relationship holds despite the narrowing of the acoustic boundary layer at higher sound frequencies.

More elaborate near-field sound receptors that function in detecting conspecific sexual and social signals are found in the antennae of certain Diptera and Hymenoptera. These receptors, named Johnston's organs after their nineteenth century medical physician discoverer (Johnston 1855), are specialized chordotonal organs located in the pedicel, the second antennal segment, and they are stimulated by vibration of the more distal flagellum of the antenna. Because the antennal flagellum will vibrate independently of the body in response to near-field sound, a flying insect's use of its Johnston's organs is not restricted by the general difficulty preventing reliance on mechanoreception for detecting wind (Section 3.1.4). Consequently, an insect may readily use these organs to detect near-field sound while in flight. In many Diptera, the antennal flagellum or its attached arista is clothed with minute stiff hairs that confer a specific natural frequency to the antenna. These plumose antennal structures will resonate in response to near-field sound transmitted at that frequency. In *Drosophila melanogaster*, the arista-bearing flagellum rotates around its axis in response to sound, thereby positioning itself such that maximum vibrational energy is transferred to the auditory receptors in the Johnston's organ (Göpfert and Robert 2001).

4.3.3 Underwater Sound: Who Perceives It and How?

Because wavelengths of sound are 4.3× longer in water than in air, a greater portion of the sounds that aquatic arthropods are exposed to will be in the near field than is the case for terrestrial species. Thus, sensory hairs on the body might be used by some arthropods to detect the particle velocity of underwater sound waves. However, in most species the threshold particle velocities that can stimulate these hairs are probably too high to allow detection of any but the most intense sounds close to the source (Hawkins and Myrberg 1983).[29] This factor invites reconsideration of the function of many underwater sounds produced by arthropods (Section 4.2.3). Are these sounds merely incidental, which seems unlikely given the energy that would be saved were the activities performed without acoustic generation, and specialized stridulatory devices, or are they intended for non-arthropod receivers?

Far-field underwater sound receivers in arthropods are best known in the semi-aquatic corixid bugs (Prager and Streng 1982, Prager and Theiss 1982). But, even these ears may function to some extent as particle-velocity receivers. Although corixids have thoracic tympanal organs, these devices bear an external, sclerotized knob that may be stimulated to vibrate by the displacement of surrounding fluid particles (see Michelsen and Larsen 1985). The insect's respiratory air bubble surrounds its tympanal organ and converts the pressure of underwater sound waves to greatly amplified fluid particle displacements at the bubble, as accomplished by similar structures, derived from the swim bladder, in the ears of many fish.[30] As in corixid signal production (Section 4.2.3), amplification provided by the air bubble may be greatest for frequencies in the range of the conspecific signal. But, the frequency of a conspecific's acoustic signals change over time, and a finely tuned ear will not always provide maximum sensitivity. The unusual asymmetry in corixid hearing—best frequencies of the left and right ears differ by 600 Hz (Prager and Larsen 1981)—may partly overcome this problem by expanding their frequency range (Ewing 1989).

It is worth noting that corixids have descended from terrestrial ancestors and, being semi-aquatic, still retain flight and aerial dispersal. Thus, the apparently dual nature of their auditory function, pressure-sensitive tympana fitted with knobs that can respond to particle velocity, may reflect the demands of life above and below the water surface as well as their evolutionary history.

4.3.4 Substrate Vibration: Subgenual and Metatarsal Organs

A great many arthropods are extremely sensitive to the vertical displacements and accelerations of vibrated substrates. In general, the sensory organs that respond to these vibratory features are strategically located in the legs, and deformation or movement of receptor units in the organs triggers neural firing. Individual receptor units may be sensitive to a restricted frequency band, but several types of units with different sensitivity bands can exist within a vibratory sensory organ. This diversity of frequency sensitivity may allow a receiver to distinguish conspecific signals from vibratory cues associated with prey and predators (Section 4.3.5).

Among insects, vibratory sensory organs are normally internal and found just below the femoral/tibial (genual) joint in all six legs (e.g., Cocroft et al. 2000). These subgenual organs are chordotonal but lack tympanal membranes and tracheal air sacs, which distinguishes them morphologically from tympanal ears. Nonetheless, subgenual organs in *Periplaneta* cockroaches may respond to lower frequency airborne sounds broadcast at high intensity (Shaw 1994). Such dual sensitivity has also been reported in decapod crustaceans (ghost crabs, *Ocypode*), which were described as having hearing ability (Horch, 1971, 1974, Popper et al. 2001).[31] However, it is important to note that the identified sensory devices, the subgenual organ in *Periplaneta* and Barth's organ in *Ocypode*, are not insulated from substrate vibration in the manner of a tympanal ear, and receivers may respond similarly to both substrate and airborne stimuli. In ensiferan Orthoptera, which bear tympanal ears in the subgenual region of the forelegs, the receptor units that respond to substrate vibration are adjacent to those that respond to sound (Kalmring et al. 1990, 1997, Bailey 1991). These two groups of receptor units form a complex "tibial organ" in the forelegs, and recordings in some katydid species show that low-frequency (<2 kHz) airborne sound will excite the vibratory receptors. In addition to chordotonal receptors, campaniform receptors in the tibiae of some insects may serve to detect vibrations of much lower (<200 Hz) frequency. Species that live and communicate on the water surface interact with a different vibrational environment than terrestrial forms generally do, and some diverge from the subgenual model of vibration perception. Whirligig beetles (Gyrinidae), for example, detect conspecific ripple signals with sensory organs in the antennae (Kolmes 1985).

Among spiders, most of which rely extensively on vibratory information, the sensory receptors are slit sensilla, elongated pits in the body wall of the metatarsi and tarsi (Barth 1997, 1998). Dendrites of sensory neurons attach to the thin cuticle at the base of these minute pits and register deformations that it experiences. While slit sensilla are found on many parts of the body and serve as proprioceptive strain gauges, those sensilla at the distal ends of the metatarsi and tarsi are specialized for responding to vertical, and horizontal, movements caused by substrate vibration. The multiple sensilla on the metatarsi are arranged perpendicular to the leg axis and form what is known as the metatarsal lyriform organ, the spider's primary vibration detector. As with certain insects and decapod crustaceans, some spiders may be capable of perceiving high-intensity, low-frequency airborne sound with these organs.

4.3.5 Information Extraction

Arthropods can evaluate four physical aspects of mechanical signals: amplitude, carrier frequency, tempo and other temporal features, and direction toward the source (Michelsen and Larsen 1985). These measures may provide information that identifies a natural enemy, comprises part of a species' mate recognition system, or functions within the context of sexual selection to indicate the quality or ability of a potential mate or the intention of a rival male.

Amplitude. The amplitude of a sound or vibratory signal influences the amount of displacement, velocity, or acceleration of a tympanic membrane, hair, or chordotonal vibration detector. Greater displacements, velocities, or accelerations are encoded as higher rates of action potentials in the associated sensory neurons.

In general, arthropod pressure and pressure-difference receivers for far-field sound are relatively insensitive. Most thresholds are in the 30–40 dB SPL range at the best frequency, the frequency for which sensitivity is highest.[32] These values are roughly 30 dB higher than typical avian and mammalian thresholds. However, some arthropod near-field sound receptors are much more sensitive, and the Johnston's organs in the antennae of male mosquitoes, which fill the entire antennal pedicel, may have thresholds as low as 0 dB IL at the frequency ($\approx$380 Hz in *Aedes aegypti*) of the conspecific female signal (Göpfert et al. 1999; Göpfert and Robert 2000).[33] Arthropod vibration receptors may also be extremely sensitive and compare favorably with those of vertebrates. Displacement thresholds of the metatarsal lyriform organs of spiders and the subgenual organs of cockroaches and ensiferan orthopterans may be as low as 10^{-7} and 10^{-8} cm, respectively, at their best frequencies (Barth 1997, 1998; cf.

Wilcox 1995 for sensitivity to ripple waves on the water surface). And, at the low frequencies (≈100 Hz) used in advertisement and courtship signals, the acceleration thresholds of wandering spider lyriform organs are approximately $0.8\,\text{cm·s}^{-2}$, considerably lower than the male and female signal amplitudes (≈$1.0\,\text{m·s}^{-2}$ near the signaler). Spider vibration receptors usually operate as displacement receivers at low frequencies and as acceleration receivers at higher (>200 Hz) frequencies.

Dynamic ranges for arthropod mechanoreception are also usually restricted. The range over which sounds of dissimilar amplitudes differentially excite the tympanum normally extends only 40–60 dB above threshold, less than half the range in humans and many other vertebrates. Moreover, individual scolopidial units may increase their rate of firing action potentials over only a 20–30 dB range above their thresholds until the neural response is saturated. The ear's full dynamic range is achieved because its receptor units differ in their absolute sensitivities, and more intense sounds excite both the sensitive and insensitive units. Neural messages from these receptor units are then processed centrally within the thoracic ganglia (cf. Section 3.1.3 for the possible olfactory analog).

Within the dynamic range, though, both Orthoptera and Lepidoptera may distinguish between sounds differing by as little as 1–2 dB (Bailey et al. 1990, Jang and Greenfield 1996). Amplitude discrimination is generally effected by a simultaneous mechanism wherein neural stimulation at each ear is compared, but sequential comparison and discrimination on an absolute basis may also occur (Figure 2.4). The last may exist in even the simplest of insect ears. For example, noctuoid moths may distinguish between weak and intense ultrasounds with a tympanal organ supplied with only two neurons: sounds low in SPL at the location of the moth only stimulate its sensitive (A_1) neuron, whereas sounds high in SPL, which may signify immediate danger from predatory bats, stimulate, and saturate, both the sensitive neuron and a second, relatively insensitive one (A_2; Roeder 1967a). Absolute intensity discrimination based on differential sensitivity of individual receptors is also found in the more complex cricket ear (Nolen and Hoy 1986; for katydids, Schulze and Schul 2001). In both moths and ensiferan Orthoptera, weak ultrasounds elicit directional steering away from the source, whereas intense ultrasounds elicit non-directional diving responses.

Size constrains the sensitivity of arthropod tympanal organs to far-field sound just as it limits the ability to produce and transmit intense sounds. Stimulation of scolopidial units and their sensory neurons is influenced by the forces imparted by sound waves to the tympanal membrane. Because force = pressure × area, a sound of given SPL will impart a greater force and be more likely to evoke a neural message if the tympanal membrane is relatively large (see Bailey 1998, Gwynne and Bailey 1999, Surlykke et al. 1999). These membranes are limited, of course, by arthropod body sizes, and enhanced sensitivity to sound may not offset biomechanical disadvantages that disproportionately large tympana would impose. Additionally, the tympanum represents a vulnerable part of an arthropod's exoskeleton, and many species retain this membrane within an enclosed chamber that may reduce its sensitivity while offering protection. Various ensiferan orthopterans, however, alleviate this constraint by relying heavily on the input to the inner surface of the tympanal membrane via the acoustic spiracle, which may be greatly enlarged. Here, the tracheal passage leading from the acoustic spiracle to the tympanum may be shaped as an exponential horn, which can act as an "ear trumpet" to amplify the sound by as much as 10–15 dB (cf. Section 4.2.6). Species relying on this amplification mechanism must then contend with effects on the tympanal organ's function as a pressure-difference receiver and its ability to localize the sound source (Section 4.3.5, Source Location).

Carrier Frequency. Contrary to popular belief, arthropods are not "tone deaf." On a crude level, mechanical filtering differentiates among carrier frequencies of both near- and far-field sounds. The antenna and the tympanum may act as band-pass filters in that sound frequencies below and above the frequencies that effectively vibrate them are not transduced to action potentials (Section 2.2).

Many acoustic insects are also capable of limited frequency analysis of transduced sounds by means of the "place principle," wherein peripheral receptor units in different locations detect or are maximally sensitive to different frequencies. The place principle is most extreme in mantises, where low-frequency sounds and ultrasound are detected by separate mesothoracic and metathoracic ears, respectively (Yager 1996a). More commonly, scolopidial units within a single ear are tonotopically arranged such that units in different locations respond maximally to different frequencies. In acridid grasshoppers the 60–80 scolopidia are divided into four groups, which are distinguished by the region of the tympanum to which they attach (Michelsen 1971). The different tympanal regions vary in thickness and tension, and hence vibratory mode and natural frequency (Stephen and Bennet-Clark 1982). Thus, incoming sound waves of different frequencies excite different sensory neurons. Two of the four scolopidial groups have best frequencies of 2–4 kHz, while the remaining two are most sensitive to 6 kHz and 10–40 kHz, respectively. Additionally, individual scolopidia within a group may vary in their best frequency, and some scolopidia may have more than one sensitivity peak (Michelsen 1971).[34] Such variance might reflect the complex vibratory modes of the grasshopper tympanal membrane.

Frequency analysis in ensiferan orthopterans (katydids and crickets) functions similarly to grasshoppers, except that it is not clear how sound waves of different frequencies stimulate different scolopidia based on their placement. Michelsen and Larsen (1985) claimed that the tympanum in ensiferans is tuned such that it vibrates equivalently in response to a broad frequency range (1–30 kHz) and does not play a role in frequency analysis. Possibly, the different best frequencies of the various scolopidia arise from differences in their mechanical attachments within the foreleg. Scolopidia in both katydids and crickets are arranged linearly in a structure known as the crista acustica, which is positioned near the tympanal membrane in the tibia (Oldfield 1982, Oldfield et al. 1986). Best frequencies of the scolopidial units decrease more or less regularly in the proximal direction along the crista acustica, and the most proximal units, closest to the femur, are sensitive to frequencies ≅2 kHz. Frequencies that have particular behavioral significance, as in conspecific calling and courtship songs or the echo-locations of insectivorous bats, are often represented by multiple units (e.g., Schul 1999), which may have more extensive arborizations with the central nervous system in the prothoracic ganglion (Pollack 1998). Thus, sensitivity to these critical frequencies can be heightened.[35]

The frequency sensitivity of vibration receptors in ensiferan subgenual organs follows a similar pattern. Various receptor types have different frequency tuning curves, with best frequencies extending below 500 Hz. Some of the vibration receptors tuned to higher frequencies (>1 kHz) may also respond to broadcasts of airborne, far-field sound (Kalmring et al. 1990).

As with olfactory perception of pheromone compounds (Section 3.1.3), a labeled line organization generally characterizes the processing of sound, and vibration, frequencies. Sensory neurons from the various scolopidia connect via interneurons to different regions of the metathoracic (grasshoppers) or prothoracic (katydids, crickets) ganglion, thereby creating a central frequency map. In some cases, frequency tuning of individual scolopidia may be sharpened by lateral inhibition at the interneuronal level wherein a given interneuron's response to frequencies below its best frequency is inhibited by stimulation of the interneuron whose best frequency is the next lower value (Pollack 1998). Nonetheless, frequency tuning of individual scolopidia is rarely very sharp (but, see Fonseca et al. 2000),[11] and analysis of across-fiber patterns may play a supplementary role in frequency analysis (cf. Section 3.1.3). Otherwise, it could be impossible for an acoustic insect to disentangle frequency from amplitude.

Tempo and Other Time Domain Features. Hearing in acoustic insects resides primarily in the time domain rather than in the frequency domain. Whereas frequency analysis is generally crude, the ability to distinguish sounds of different pulse or chirp rates is usually quite high (e.g., Perdeck 1957, Walker 1957, Popov and Shuvalov 1977, Schul et al. 1998). For example, a field cricket may not make frequency distinctions finer than between 5 and 10 kHz, but it can easily distinguish between pulse rates of 36 and 38 Hz.[36] Such discrimination of pulse rates is possible because activity in the tympanal neurons, mass rates of action potentials, can be phase-locked to individual pulses and then processed in higher neural centers. Fine differences in pulse or chirp duration and duty cycle may be distinguishable as well (Doherty 1985b, Stumpner and von Helversen 1992, Eiríksson 1993, Dobler et al. 1994a[37]). Investigations of acridid grasshoppers show that they may detect the presence of silent gaps within pulses (Kriegbaum 1989) and the rise and decay of amplitude during a pulse (von Helversen 1993). These latter features may be useful in signal localization (Krahe and Ronacher 1993; but, see von Helversen 1998).

Various neural and mechanical constraints limit the precision with which temporal acoustic characters can be evaluated. Pulse rates faster than the neural time resolution, which is the maximal rate at which action potentials can be transmitted and phase-locked to individual pulses (≈200–250 Hz; e.g. Surlykke et al. 1988),[38] may not be recognized or distinguished from one another. Mechanical time resolution, which reflects the speed at which tympanal vibration dampens following an impulse, is much faster, however (Schiolten et al. 1981). Consequently, some temporal features that are much shorter than 4 ms (reciprocal of the neural time resolution), such as brief pulses and gaps within pulses (Surlykke et al. 1988), may be evaluated based on the presence and magnitude of a tympanal response. Here, evaluation is possible because the nature of a tympanal response may influence the probability that an action potential is fired. For example, a pyralid moth might distinguish 100-μs from 200-μs pulses of ultrasound because the latter, having more total energy, stimulate greater tympanal displacements (see Jang and Greenfield 1996). Thus, a given 200-μs pulse is more likely to trigger an action potential, and a train of 200-μs pulses will thereby elicit an EPSP of greater amplitude in an ascending interneuron.

Evaluation of and responses to acoustic characters slower than the neural time resolution also rely on central neural processing. Such processing occurs primarily in the metathoracic ganglion in grasshoppers, but in crickets and katydids it occurs mostly in the brain (Pollack 1998). The processing of male calling songs by female *Gryllus bimaculatus* is a well-studied example (Huber and Thorson 1985). Both the tympana and peripheral acoustic neurons exhibit phase-locked responses to a wide range of pulse rates, but the insects may orient only toward songs containing pulse rates, within the chirps, ranging from 20 to 36 Hz, the species' range.[39] This selectivity of pulse rates results from band-pass filtering produced by the combined activity of two classes of brain interneurons (Schildberger 1994).

Substrate vibration receptors may similarly extract information on pulse rates, lengths, and related temporal features of vibratory signals (Barth 1997, 1998). As with acoustic communication, these time domain characters often indicate a vibrator's species (Henry 1994), sex (Wilcox 1979), and other attributes. Particle velocity receptors that detect near-field sound and other fluid disturbances may be generally less sensitive to temporal characters. However, some findings indicate that receptors in cricket cercal hairs can perceive and differentiate among low-frequency (<30 Hz) pulse rates and other temporal features (Kämper 1985, Heidelbach et al. 1991). This sensory ability may be critical during courtship, where the signaler and receiver are very close.

Source Location. The problems that being small pose for communication along the mechanical channel are nowhere greater than in localizing the source of a signal. In some situations, as where receivers are very close to signalers and chemical or visual information are present, localizing the source of a mechanical signal may be trivial. But, communication along the mechanical channel is often characterized by rapid transmission of information over long distances that occurs in the absence of other stimuli, and a receiver may have to localize the distant source of that information quickly in order to use it effectively. Accordingly, many arthropods, and acoustic insects in particular, have evolved specialized localization mechanisms that function moderately well despite the size handicap. Notable exceptions, which occur in non-communicative contexts, include various flying insects that respond to intense bat echo-locations by simply dropping to the ground, possibly the result of a reflex circuit (see Section 4.4.1). Under such circumstances, localization of the source may be unnecessary and actually require more time for central neural processing than would be safely available.

Animals typically localize the source of far-field sounds by simultaneously comparing the time of arrival, the intensity, or the phase of sound waves arriving at paired receptors. But, only the latter two comparisons would be realistic localization mechanisms for acoustic insects (but, see below). The distance separating the two ears is generally too short to provide an inter-aural time difference (ITD) that could be processed by the nervous system. For a field cricket whose two ears are separated by 1 cm, a sound arriving from the left or right would arrive at the contralateral ear only 29 μs after arriving at the ipsilateral one (Figure 4.11b).

Acoustic insects whose ears operate as pressure receivers may rely on the inter-aural intensity difference (IID), provided that the sound frequency is relatively high. Because the insect's body diffracts sound waves, the ear contralateral to sound arriving from a given side will be in a sound shadow where effective intensity is attenuated below that at the ipsilateral ear (Figure 4.11a).[40] However, this attenuation would only be significant where the diffracting body's diameter is greater than $\lambda/3$ (Section 4.1.1). Thus, most acoustic insects cannot use the IID as a localization mechanism for sounds below 10 kHz in frequency, and small species (diameter <1 cm) or those whose ears are on narrow appendages rather than on the body would be restricted to even higher frequencies. As many acoustic insects transmit calls that would be affected by these frequency and size restrictions, determining how they localize sound, which many do with reasonable proficiency, has been a major problem in bioacoustics.

Autrum (1940) first suggested that if sound reaches both surfaces of the tympanal membrane and allows the ear to operate as a pressure-difference receiver, localization might be afforded by the inter-aural phase difference (IPD). Later biomechanical investigations, in which tympanal displacements were measured accurately with non-intrusive laser–Doppler vibrometry, bore

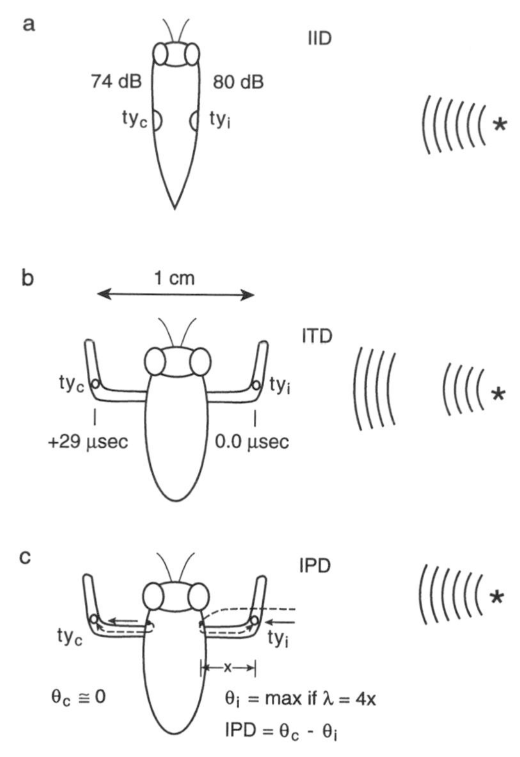

Figure 4.11 Mechanisms for localizing the azimuth of far-field sound sources. (a) Large insects with ears on the sides of the abdomen or thorax may localize a high-frequency sound source (asterisk) by means of an inter-aural intensity difference (IID). Here, the body creates a sound shadow at the contralateral tympanum (ty_c) in which sound pressure levels are 6 or more decibels lower than at the ipsilateral tympanum (ty_i). (b) Organisms with widely spaced ears may localize amplitude-modulated sounds because of an inter-aural time difference (ITD): the interval between arrival of an intensity transient at the ipsilateral and contralateral tympana. But for an insect whose tympana are separated by 1 cm, the maximum ITD is only 29 μs. (c) Insects with pressure-difference ears may localize sound by means of an inter-aural phase difference (IPD): sound waves approaching the inner and outer surfaces of the ipsilateral tympanum travel over paths of different lengths and therefore are out of phase (phase difference between external and internal sound pressure $= \theta_i$) and generate substantial tympanal displacements, whereas sound waves approaching the inner and outer surfaces of the contralateral tympanum travel over paths of comparable length and therefore are approximately in phase ($\theta_c \cong 0$) and cancel each other. Thus, a substantial IPD ($= \theta_i - \theta_c$) arises for certain sound frequencies and is perceived as a large IID.

this prediction out and demonstrated how the IPD localization mechanism functions (Michelsen et al. 1994a,b). In the field cricket *G. bimaculatus*, sound waves reach both surfaces of the tympanal membrane, and they are generally out of phase because the three paths taken to reach the inner surface are each often longer than the path to the outer surface (Section 4.3.1). Consequently, pressure differences between the outside and inside of the tympanum arise and displace the membrane correspondingly (Figure 4.11c). At each ear, the pressure difference (external SPL − internal SPL) oscillates at the same frequency as the incoming sound waves, and the maximum amplitude of this pressure difference depends on the direction from which sound waves arrive. For sound arriving ipsilaterally to a given ear, the phase angle between internal and external pressures at that ear (θ_i) may be considerably greater than at the contralateral ear (θ_c). This IPD ($= \theta_i - \theta_c$) implies that larger pressure differences, and consequently greater tympanal displacements, occur at the ipsilateral ear than the contralateral one. As the sound source is gradually moved from the ipsilateral position to a location directly in front of the insect, the IPD decreases until phase angles and pressure differences are equivalent in both ears. Thus, various small arthropods—and vertebrates—may localize the sources of low-frequency sounds by essentially assessing the inter-aural difference in maximum tympanal displacement, but this difference only arises because of the IPD.

Further analysis of the IPD localization mechanism in pressure-difference receivers indicates that it will not work for all sound frequencies (Michelsen et al. 1994a,b; cf. Oldfield 1980, Pollack et al. 1984). At low frequencies the time between successive sound waves ($= 1/f$) may be so long that phase angles are too small to generate detectable pressure differences in an ear, even when sound arrives from the ipsilateral side (Figure 4.11c). Conversely, at high frequencies the time between successive sound waves may be so brief that phase angles are large enough to generate equivalent internal and external pressures, leading to cancellation and little or no tympanal displacement. In *G. bimaculatus* the IPD localization mechanism is restricted to sound frequencies close to 4.5 kHz, the frequency of the calling and aggressive songs. Apparently, these songs are tuned such that the ear, as it is constructed, can efficiently localize them (or, vice-versa? see Section 4.5). The median septum separating tracheal branches leading to the left and right ears provides much of this tuning by delaying the transmission of sound waves approaching from the contralateral side. If the septum is perforated, phase angles between internal and external pressures at an ear are less than 1/3 as long and pressure differences sufficient to generate substantial tympanal displacements do not arise (Michelsen and Löhe 1995).

A central processing feature found in *G. bimaculatus* and other crickets may improve the accuracy of the IPD localization mechanism. Two identified interneurons in the metathoracic ganglion, the omega neurons projecting laterally from the left and right tympanal nerves, are inhibited by input from the contralateral tympanal nerve and thereby enhance the contrast between peripheral neural stimulation at the two ears (Selverston et al. 1985, Schildberger 1994). Thus, a form of lateral inhibition magnifies the inter-aural difference in maximum tympanal displacement (Figure 4.12).

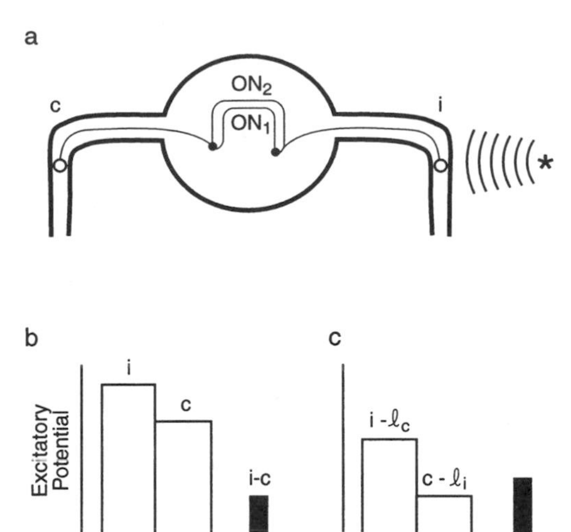

Figure 4.12 (a) Lateral inhibition provided by the omega neurons (ON_1, ON_2) in the prothoracic ganglion enhances the ability to localize a sound source (see Selverston et al. 1985). (b) Excitatory potential in the ipsilateral ear (i) exceeds that in the contralateral ear (c) by i-c. (c) Via lateral inhibition (*l*) from the omega neurons, excitatory potential in each ear is reduced by an amount proportional (0.5 in the illustrated example) to the excitatory potential in the opposite ear. Thus, the excitatory potential differential between the ipsilateral and contralateral ears (solid bar) is increased.

Based on their acoustic spiracles and tracheal acoustic passages, katydid ears superficially appear to operate also as pressure-difference receivers and rely on an IPD localization mechanism. A closer analysis, though, reveals several complicating factors and indicates yet other ways in which acoustic insects may circumvent the size problem. First, most katydid species communicate with high-frequency sounds that would be subject to diffraction by the body but not by the narrow fore tibiae, where the tympana are located. Second, in many species the acoustic tracheae are shaped like exponential horns and thereby amplify SPL at the internal surface of the tympanal membrane by 10–15 dB (Heinrich et al. 1993). Therefore, most sound energy influencing the tympanum enters through the acoustic spiracle rather than the fore tibial opening. Because the body is much wider than the fore tibiae, it can diffract and attenuate high-frequency calling songs, and katydids can rely on an IID localization mechanism. But, in some katydid species the fore tibial openings are narrow slits surrounded by elaborate pinna-like cuticular structures. Here, sound entering through these openings probably does play a critical role in localization (Bailey and Stephen 1978, Mason et al. 1991). Air in the small chamber between the slit and the tympanal membrane may vibrate as in a Helmholtz resonator and maximally stimulate the membrane when the slit faces oncoming sound waves whose frequency approximates the chamber's natural frequency.

Thus, differential orientation of the fore tibiae may establish an IID with which the sound source can be localized.

The above solutions to the size problem allow insects whose inter-aural distances are generally 8 mm or longer to localize sounds. Can much smaller insects rely on any of these mechanisms, or are they precluded from localization altogether? Small endoparasitic species in two dipteran families, Sarcophagidae and Tachinidae, rely on phonotaxis to localize and attack calling males of various acoustic insects, so effective mechanisms must exist. However, rough calculations and examination of anatomical features indicate that neither IID nor IPD mechanisms operate: inter-aural distances in these flies are only 0.5 mm, and sound approaching from the left or right would require only 1.5 μs to travel from the ipsilateral ear to the contralateral one. Moreover, the bodies of these flies are too small to diffract the 4–6 kHz calling songs of some of their hosts (crickets and cicadas), and there is no morphological indication that their ears operate as pressure-difference receivers. Instead, localization seems to be accomplished because the two tympanal organs are connected by a mechanical bridge whose center rests on a flexible pivot. This device delays tympanal displacements at the contralateral ear by a 90° phase angle (≈50 μs for a 5-kHz sound) relative to the ipsilateral one (and also reduces vibration amplitude in the contralateral ear by up to 12 dB); a peripheral neural mechanism amplifies this time difference further to yield a 320-μs ITD, which would be sufficient for central processing to act on. Ironically, the tiniest acoustic insects, with the shortest inter-aural distances, may use an ITD localization mechanism! A complete biomechanical analysis of this extraordinary apparatus may be found in Robert et al. (1996, 1998, 1999) and Robert and Hoy (1998).[41]

Regardless of whether sounds are evaluated with an IID, IPD, or an inter-aural bridge ITD mechanism, it is generally assumed that the receiver compares the inputs at each ear and turns toward the stronger or first input in order to localize the sound source (Schildberger 1994).[42] Thus, the paired nature of receptors appears to play a much more important role in hearing than in most aspects of olfaction. By using a "turn toward the stronger or first input" algorithm, an acoustic insect may generally track a sound source much more efficiently than most insects can find the source of a pheromone or other odor, terrestrial odor trails perhaps representing the major exception. Rather than undertaking the lengthy casting flights or circuitous runs necessary to track an aerial odor plume, an acoustic insect can usually approach a sound source in its horizontal plane with relatively few lateral deviations (e.g., Mason et al. 2001). Once the heading at which intensity is equivalent in both ears (IID or IPD = 0) is found, the acoustic insect need only maintain this course to locate the sound source. As with olfaction (Section 3.1.3), an optomotor response assists the tracking (Böhm et al. 1991): insects provided with visual input can localize the source more directly because the optical cues are a spatial reference with which the course heading can be finely adjusted. Deviations from a 0° course heading inevitably occur, but these departures are corrected by turning back toward the more stimulated ear. Thus, acoustic insects may zigzag toward the sound source, but their paths are much straighter than those often seen in olfactory tracking (cf. Figure 3.10). Grasshoppers can maintain a relatively straight course toward a sound source because their IIDs and IPDs change drastically as they turn away from a 0° heading. In crickets, the IPD changes less, and they therefore tend to meander more while tracking a sound.[43] Neither grasshoppers nor crickets, however, can gauge the magnitude of turn necessary to face a sound source if the IID or IPD is not zero (Rheinlaender and Blätgen 1982, Pollack 1998). In contrast, hearing in katydids is considerably more directional, even in "open-loop" tests wherein stimulus intensity remains constant over time, and these insects can turn commensurately with the IID and face a sound source within 10–15° of a 0° heading.[44] Nonetheless, their directional hearing is still an order of magnitude less accurate than that of the best vertebrates.

The IID, IPD, and ITD localization mechanisms as described above would allow an acoustic insect to determine whether a sound source is located to the left or right and, to varying degrees, by what approximate angle. But, effective tracking may also require discerning the elevation of a sound source and whether it is to the front or rear. Theoretically, full localization as such could be accomplished with four widely spaced ears, a feature not found in any species. Some vertebrates partially resolve these ambiguities via specialized pinna structures or asymmetric orientation of the ears, but among acoustic insects the problem has been investigated in only a few species. Acridid grasshoppers possibly eliminate the front/rear ambiguity by covering one tympanum with the adjacent femur (Adam 1983). In an insect oriented 0° toward a sound source, this posture would strongly attenuate high-frequency sounds reaching the occluded ear from the front. Thus, the possibility of facing 180° away from the sound and orienting opposite to the source is eliminated. Other acoustic insects may use their wings in similar fashion to eliminate either front/rear or up/down ambiguities (Roeder 1967b).

Sequential Comparison? In the previous chapter on chemical communication, I indicated the likelihood that various arthropods rely on sequential comparison of olfactory input in certain circumstances to localize the sources of odor plumes and trails (Sections 3.1.3 and 3.7.1). Given the relative efficiency with which acoustic insects can use simultaneous comparison of input to their paired ears to localize sounds, sequential comparison may not seem necessary here. However, certain experiments probing the mechanisms with which crickets track sound sources suggest otherwise. Using a locomotion compensation device that tested phonotaxis in an open-loop paradigm, several investigators found that monaural females may retain limited ability to track the male calling song (Schmitz et al. 1988, Schildberger and Kleindienst 1989). Does this ability imply that acoustic insects too can use a sequential comparison mechanism to localize sounds? While the critical experiments have yet to be performed, pressure-difference receivers and tympanal chambers functioning as Helmholtz resonators are, unlike pressure receivers, inherently directional, and many insects should be able to localize signals by moving a single organ to and fro until the intensity of a perceived stimulus is maximal. Thus, an acoustic insect might continue to localize signals in the event of injury to one ear or where front/rear and up/down ambiguities are otherwise difficult, yet necessary, to resolve.

Localization of Near-Field Sound and Vibratory Signals. Arthropods may also localize the sources of near-field sounds and various types of vibratory signals. These signals can often be localized more easily and accurately than far-field sounds because the speed of signal propagation may be relatively slow and more than two receptor organs are normally used. Near-field sounds may be essentially triangulated by evaluating which of the body's hair receptors are oscillated and the planes in which those oscillations occur. Because the particle velocity of sound waves will only oscillate a hair receptor if the waves approach laterally, the body locations of oscillated hairs can be used to discern the direction to the source. Here, localization may be augmented by several specialized features of hair receptors: Individual hairs may vary in the plane in which deflection is possible, sensory neurons attached to a hair may only respond when deflection occurs in a particular plane, and several sensory neurons may be attached to a given hair, each responding to deflection in a different plane. As with frequency analysis (Section 4.3.5), across-fiber patterns of neural activity from the various hair receptors, and neurons, would have to be evaluated centrally in order to separate source direction from amplitude. Insects using Johnston's organs may triangulate the sources of near-field sounds in an analogous manner. These chordotonal organs in the antennal pedicel include thousands of sensory neurons whose collective activity can indicate the planes of vibration of the more distal flagellum. Because the two antennae can project from the head at different angles, ambiguities in the source direction may be eliminated.

Localizing the sources of bending waves and surface waves on water may be far simpler than tracking far- or near-field sounds. These waves travel at considerably slower speeds than sound, and their potential sources are restricted to only one or two dimensions. Thus, the times of arrival of the leading edge of a wave group at the subgenual receptors in the various legs can be compared and evaluated centrally. For example, a wandering spider facing toward the base of a leaf may position its front and rear legs 2 cm apart. Bending wave signals may arrive with a group velocity of $40\,\mathrm{m\cdot s^{-1}}$ from a source at the leaf base and be received at the front legs $500\,\mu\mathrm{s}$ prior to reception at the rearmost ones. This time difference would be sufficient for central processing, and the spider would construe it as a source located basally rather than distally on the leaf.[45] Arthropods on the surface of water may similarly discern the specific direction toward the source of surface waves by comparing arrival times at the various legs.

Ranging. Localization of a distant signal would ideally include determining the distance to the source as well as the direction. Because hearing in many vertebrates includes superior frequency analysis and sound frequencies are differentially attenuated during transmission, biologists have long entertained the possibility that animals may "range" the source of acoustic signals: a given signal has a particular frequency spectrum at a close range, but this spectrum gradually loses more and more high-frequency components as distance from the source increases. Theoretically, a receiver may judge this distance by comparing the frequency spectrum of the perceived signal with a stored template of the unattenuated signal. Thus, humans may judge distance to a lightning storm by the sound spectrum of thunder, and several studies on birds suggest that some species possess limited abilities to range the songs of conspecific neighbors (e.g. Naguib et al. 2000). Do similar ranging abilities occur among acoustic insects? The broadband songs of katydids and the potential for frequency analysis in their hearing suggest that they too might be capable of ranging. However, the frequency spectra in unattenuated calling songs often vary among individuals within a population, and specific environmental features such as vegetation would further increase the variance in frequency spectra that a receiver may perceive among neighbors all calling at a given distance. It is therefore unclear how a receiver would reliably discern a signaler's distance from the frequency spectrum it perceives, just as it would encounter problems in discerning distance from the amplitude of a perceived signal. How can an intrinsically intense (or broadband) but distant neighbor be distinguished from a weak (or narrowband) but nearby one? To date, ranging abilities based on perceived frequency spectra have not been unequivocally demonstrated in any acoustic insect. If they operate at all, it may be for comparing widely different receiver–signaler distances only.

On the other hand, arthropods perceiving certain types of substrate vibrations may be able to range these signals with relative ease and accuracy. When two different types of waves traveling at different velocities are propagated by a given disturbance, that disturbance could be ranged by evaluating the time interval between arrival of the faster and slower wave groups (cf. Section 3.1.4 for discussion of the hypothetical olfactory analog). This phenomenon arises in certain solid substrates, and sand scorpions are reported to rely on it to range distances to potential prey (Brownell 1977, 1984). Concentrically spreading surface waves on water may also be ranged because curvature of the wave front, which should be detectable by receptors in three or more legs, would reliably indicate to a receiver that the source is close. Some species might apply such ranging abilities to conspecific signals in order to respond appropriately toward potential mates and rivals.

Localization versus Recognition: Parallel and Serial Processing. When receivers localize signals, we may expect them to ensure that the stimuli they respond toward are appropriate. For example, a sexually receptive female should only embark on phonotactic orientation toward a mating signal recognized as conspecific (Chapter 7), male, and, possibly, transmitted by an individual of superior "quality" (Chapter 6). Earlier (Section 3.1.4), I noted that many moth species do not initiate and sustain olfactory orientation unless they perceive the entire pheromone signal. Such behavior suggests that signal recognition and localization are processed serially and that localization acts on the neural output of the recognition process. Some moth species, however, do perform orientation maneuvers in response to incomplete pheromone signals. Are localization and recognition of pheromone processed separately and in parallel in these species?

The relationship between signal localization and recognition has been specifically investigated in acoustic orthopterans, and various organizations of these operations are indicated. Grasshoppers may initiate orientation in response to certain songs that lack critical features of conspecific mating signals and would not elicit courtship, implying parallel processing of the two operations (see von Helversen 1984). Crickets, however, will not initiate orientation unless the song includes all of the features that would elicit close-range courtship and mating behavior (see Doherty 1991). These processing schemes may reflect different evolutionary origins of acoustic communication (von Helversen and von Helversen 1995, Pollack 1998; Section 4.5) and that acoustic grasshoppers are primarily diurnal and can normally rely on visual information as well as song to recognize and localize a mate.

4.3.6 Reception Interference along the Mechanical Channel

The accurate localization of a signal and retrieval of information on its amplitude, frequency, and tempo depend on clear reception and minimal interference along the channel. Just as various forms of physical interference impede the transmission of mechanical signals, external and internal factors may prevent a receiver from recognizing, evaluating, or localizing a signal. For examples, thermal layers in the air, vegetation, and other sounds in the environment may refract, attenuate, reverberate, mask, and otherwise alter the temporal and spectral features of acoustic signals (Section 4.1.1). These factors may influence the specific locations and times where and when conspecifics transmit their signals (Römer and Lewald 1992, Römer 1993, 1998). External interference may also select for calls whose temporal and spectral features are more resistant to environmental modification, such as the broadband calling songs of katydids (Section 4.2.5).

Additional interference can occur along the mechanical channel within the animal, and a receiver may have more difficulty avoiding these problems. Because the air chamber behind the tympanum is adjacent to or actually forms a functional part of the tracheal system in many acoustic insects, respiration may interfere with hearing (Hedwig et al. 1990, Meyer and Elsner 1995, Meyer and Hedwig 1995). In acoustic insects whose tympana are located in the fore tibiae, such as crickets and katydids, walking may present another form of internal disturbance. Some observations indicate that these insects reduce the impact of walking disturbance by making repeated stops during phonotactic orientation (Schildberger 1988, 1994). Thus, signal evaluation and course corrections may be possible. Flight likely presents a similar problem for many insects, whose tympana might be deformed by wing movement, but this disturbance might be less easily circumvented by intermittent locomotion! Other forms of internal disturbance probably affect the reception of substrate vibration and surface wave signals by arthropod subgenual organs. An anatomical solution for reducing interference from certain types of bodily activity is to insulate the signal receptor within a rigid capsule. This measure is particularly evident in the tympanal organs of cicadas (see below). In arthropods whose paired receptors suffer a high level of fluctuating asymmetry (cf. Section 4.4.1), the ability to localize a sound source could be impaired (see Faure and Hoy 2000b), but it is also possible that learning might overcome this disabiity.

While it is conventional to assume that disturbance from these external and internal factors represents background noise that interferes with signal reception, some findings indicate that background noise may actually enhance a receiver's ability to detect a stimulus. This phenomenon, known as stochastic resonance, occurs because limited amounts of background noise can amplify the signal : noise ratio in certain circumstances (Douglass et al. 1993). Most biological studies of stochastic resonance have focused on the detection of stimuli generated by prey and predators (Russell et al. 1999, Greenwood et al. 2000). Its application to communication in arthropods and other animals, where it might be particularly significant along the mechanical channel, has heretofore been overlooked.

Incompatibility of Signaling and Sensory Perception. Perhaps the most entrenched and troublesome form of internal interference is that posed by an individual's own signaling. The activity of signal production may deform receptor organs, or an individual's transmitted signals may mask incoming ones. In either manner, a receiver's sensory perception may be severely compromised while it signals. Is this problem inevitable, or do various mechanisms exist with which it can be avoided?

Along the olfactory channel, signaling and sensory perception may not be incompatible in many cases because signaler and receiver functions are clearly separated by sex (e.g., female advertisement pheromones; Section 3.1) or caste (e.g., queen pheromone; Section 3.4). And, where a given individual might act as either a signaler or receiver, as in males emitting courtship pheromone or in social insects using odor trails, the possibility for interference is often minimized. An individual's olfactory receptors would normally not be exposed to its own signals because the odors are projected in a specific direction behind or away from itself, or context and other information would prevent autointerference. For example, a male emitting a volatile pheromone while courting a female may not be distracted by himself, despite his ability to perceive the substance, because he also receives both mechanical and visual stimuli from the female.

Some long-range mechanical signals, however, present more formidable autointerference problems. Far-field sound from a small organism may spread more or less spherically, which would lead to a signaler stimulating its own ears. Theoretically, self-stimulation might overload the signaler's sensory system or cause it to make inappropriate escape or attack responses toward itself. The former could result in injury to the tympanum in species producing intense calls, and the latter could occur in species wherein calling has a territorial function. Behavioral evidence from grasshoppers and katydids (Brush et al. 1985, Greenfield 1990, Greenfield and Minckley 1993) suggests that acoustic insects do experience lower hearing sensitivity during the exact time intervals when they emit sound. This insensitivity may result from sensory neural circuits that are inhibited by the motor programs effecting calling (Wolf and von Helversen 1986), but more likely hearing is simply masked by the signaler's own calls or calling activity mechanically deforms—and desensitizes—the tympanum (Hedwig 1990). These latter possibilities are well known in anurans (Narins 1992) and bats (Henson 1965, Goldberg and Henson 1998), and masking certainly plays a role in human conversation! In cicadas that broadcast their calls through the tympana (Section 4.2.1), though, desensitization in males results, in part, because the tympanic nerve is isolated in a separate capsule that protects the receptor from intense sound (Bennet-Clark 1998b).[46]

Whether a simple result of masking or a specialized protective measure, the desensitization of hearing may place a signaler at a significant disadvantage in various situations. Hearing is often necessary for assessing conspecific rivals and detecting natural enemies, and its impairment over extended time intervals could lead to disastrous consequences. Many acoustic insects pause intermittently during long calling bouts, and several observers (Greenfield 1990, Eiríksson 1992, Faure and Hoy 2000a) have proposed that these brief silent gaps represent a specific adaptation that improves, ever so slightly, the compatibility of signaling with sensory perception. On a broader scale, these selection pressures might have also led to the evolution of discontinuous, chirping songs from continuous trills.

4.4 Functions and Adaptations: Sexual Advertisement, Aggression, and Social Interaction

Both signal properties and receiver abilities indicate that communication along the mechanical channel offers several distinct advantages. Unlike communication by chemical signaling, many forms of mechanical signals are transmitted at high velocities over relatively long distances via wave action. Moreover, the signaler may exert some control over the direction in which a mechanical signal is transmitted, and it may commence and cease signaling very suddenly. Because some types of mechanical signals are not dispersed into different wavelengths and velocities by the medium through which they are transmitted, a signal's tempo may be preserved and thereby perceived by a distant receiver. Receivers may discern with relative efficiency the direction from which a mechanical signal is transmitted, and they may also extract some reliable information on the signal's amplitude and frequency as well as its tempo. These factors would render mechanical signaling eminently suited for various sorts of advertisements wherein the identity, and hence location, of a given advertiser and a reliable indication of its state or quality, in addition to its species and sex, are critical to receivers. Thus, males may be expected to use mechanical signals for their sexual, territorial, and aggressive advertisements, and a survey of the arthropods reveals that males in various groups do (see Table 4.1).

4.4.1 Pair-Formation: Reliable and Localizable Advertisements

In mating systems characterized by male signalers and female searchers, males are often under selection to match or exceed certain characters of their neighbors' mate-recognition signals (Chapter 6). Failure to do so may result in markedly reduced mating and reproductive success. Signal competition (see West-Eberhard 1984) between males would be particularly strong when they are clustered in space, as may occur in lek mating systems or when suitable habitat exists in patches, because females will then normally be exposed to the signals of several males simultaneously. Here, a male's signals should be readily localizable by female receivers, and any distinctive characters of his signals should be separable from those of his neighbors. Otherwise, a superior male signaler may not be rewarded with a high female encounter rate, and, conversely, a choosy female may not localize a superior signaler. Were males to use chemical advertisement signals solely, both of these problems would likely arise: the pheromone plumes of neighboring males might coalesce into a single cloud within which individual odor sources would be difficult to localize. And, odor sources distinguishable by a high flux rate or an unusual chemical composition would seldom be perceived as such at a distance. In reversed mating/signaling systems as found in most species of moths, these drawbacks may not matter to female signalers, who only need to attract a few males and may gain by passively filtering the superior searchers among them (Section 3.1.5). But, where males are the signalers, they are generally expected to maximize the number of attracted females and to avoid using signals that are prone to deception. In a hypothetical mating system wherein male sexual advertisements are pheromones whose flux rates are commensurate with quality, females may not discriminate against an inferior signaler positioned slightly downwind from a superior one. As explained later in this section, mechanical signals would normally be more resistant to cheating of this nature.

Female Preference for High-energy Signals. When females are attracted to conspecific male acoustic or vibratory advertisements, laboratory tests and other controlled experiments show that they often respond preferentially to signals that differ in one or more characters from the mean value(s) in the male population (see Gerhardt 1994). Such preferences, which may be open-ended, indicate the presence of directional selection, and mate choice, as opposed to stabilizing selection. While the latter sometimes occurs in mating preferences, it is more likely to be indicative of choice operating only within a less restrictive species-recognition context (Section 3.1.5; cf. Ryan and Rand 1993). In general, those signal characters that are subject to directional selection exhibit relatively high coefficients of variation between individuals (dynamic characters; cv >0.10), whereas characters subject to stabilizing selection are relatively invariant (static characters) (Gerhardt 1991; e.g., Shaw and Herlihy 2000). Among arthropods and other animals, many of the characters subjected to directional female preferences are based on signal energy or power. For example, females may orient toward males who produce longer signaling bouts, more intense signals, longer signals, or signals repeated at a higher rate (Table 4.2). Most of these data are from acoustic advertisement signaling, but several examples from vibrational signaling are also reported.

At a proximate level, high-energy signals may be more attractive to females because they impart greater stimulation to receptor organs and evoke a higher rate of action potentials. At an ultimate level, the female attraction may represent various processes of sexual selection, including, but not limited to, Fisherian (arbitrary) and good-genes (viability indicator) selection for particular males (Section 6.2). Under the Fisherian sexual selection model, females would select males solely because they, and the majority of females in the population, deem the signals of the chosen males to be attractive. Presumably, the attractive signal features are heritable, and a choosy female will likely produce sons who enjoy above average attractiveness and mating success. This outcome would be ensured if the female preference too is a heritable trait and the non-random process of female choice establishes linkage disequilibrium, a genetic covariance, between genes regulating the signaling and preference traits.

Under the good-genes model, the same assumptions hold, but females selecting and mating with males distinguished by high signal energy or power produce male, and female, offspring of higher viability[47] as well as attractiveness. Females may focus on energy-based signal characters because these traits are reliable indicators of a sire's viability that is heritable (Section 6.2). That is, males of reduced viability, who are expected to sire low-viability offspring, are physically incapable of producing high-energy signals (cf. Section 3.2). Here, and under direct sexual selection for material benefits as described below, it is tacitly assumed that a high amount of metabolic energy input is necessary to produce a high-energy signal. This assumption, while reasonable, has been tested in only several cases, and it may be obscured by various complications. In the lesser wax moth, *Achroia grisella*, highly attractive males are distinguished by acoustic signals delivered at high intensity, a high pulse rate, and with lengthy "asynchrony intervals" between paired pulses, a specific temporal feature (Jang and Greenfield 1998; Figures 4.13 and 4.14). Signaling energy expenditure as measured respirometrically is related only to pulse rate, which makes high metabolic energy input necessary but insufficient for generating attractive signals (Reinhold et al. 1998; cf. Hoback and Wagner 1997). Signal intensity may be dependent on tymbal size and structure, which might be influenced by energy allocations made earlier during immature development (Section 4.2.9), but past energy input as such would be very difficult to confirm.

The basic genetic assumptions and relationships of the Fisherian and good-genes mechanisms of sexual selection—in which a discriminating female obtains only indirect, genetic, benefits—remain untested in most signaling systems, arthropod and vertebrate. Nonetheless, the evidence that male signal characters exhibit repeatable phenotypic variation, additive genetic variance, and heritability is growing (see Table 4.2), and in several cases signal attractiveness has been found to be associated with heritable viability (Table 4.2; Section 6.2; see Møller and Alatalo 1999). While indications that female preference functions exhibit comparable features are far more meager, largely due to the relative difficulty in assessing these functions experimentally (see Jennions and Petrie 1997), various studies are beginning to uncover this evidence as well (e.g., Wagner et al. 1995, Hedrick and Weber 1998, Bakker 1999, Jang and Greenfield 2000, Ritchie 2000). These findings suggest that it is probably safe to consider the Fisherian and good-genes processes as realistic hypotheses and potential explanations for the exaggerated male signal characters and female preference functions found in many species (see Jennions and Petrie 2000).

In addition to obtaining indirect, genetic benefits, discriminating females may also obtain direct, material (somatic) benefits by preferring high-energy signals. Importantly, the processes that lead to material and genetic benefits are not mutually exclusive (see Jennions and Petrie 2000). Given the requisite genetic

Table 4.2 Sexual Selection and Mechanical Communication in the Arthropoda: Review of Findings on Female Choice

Order / Family / Species	Signaling mode[a]	Target of female preference[b]	Signal variance[c]	Preference variance[d]	Viability/size indicator[e]	References[f]
Araneae						
Ctenidae						
Cupiennius getazi	vb-ss	IPI, PL; dir↑				41,42
Lycosidae						
Hygrolycosa rubrofasciata	vb-ss	I, PR; dir↑			PR↑: ♂ viability, offspr. viab.	1,26,27 28,31,33
Orthoptera						
Gryllidae						
Acheta domesticus	sd-ff	I, PN; dir↑			PN↑: ♂ viability	12,40
Gryllus campestris	sd-ff	f, dir↓			f↑: ♂ symmetry	44
Gryllus firmus	sd-ff		PR, g_a			53
Gryllus integer	sd-ff	PN, stab; BL, dir↑	PN, BL; g_a	PN, p_r		13,17,18,19,51
Gryllus lineaticeps	sd-ff	I, PN, CR; dir↑				49,50
Oecanthus nigricomis	sd-ff	f, dir↓			f↓: ♂ size	4
Teleogryllus oceanicus	sd-ff	BL, dir↑	BL, p_r			25
Gryllotalpidae						
Scapteriscus acletus	sd-ff	I, dir↑				52
Tettigoniidae						
Amblycorypha parvipennis	sd-ff	I, CL; dir↑; prec			I↑: ♂ size	10,11
Conocephalus nigropleurum	vb-ss	IPI, dir↓			IPI↓: ♂ size	7
Ephippiger ephippiger	sd-ff	PN, dir↑; cont↑; prec	PN, g_a	PN, g_a	cont↑: ♂ youth	15,34,35,38
Neoconocephalus spiza	sd-ff	I, CL; dir↑; prec				14,45
Phaneroptera nana	sd-ff	CL, dir↑; prec				47

(*continued*)

Table 4.2 (continued)

Order Family Species	Signaling mode[a]	Target of female preference[b]	Signal variance[c]	Preference variance[d]	Viability/size indicator[e]	References[f]
Requena verticalis	sd-ff	I, f; dir↑				2,3
Requena sp. 5	sd-ff	f, dir↓			f↓: ♂ size	54
Scudderia curvicauda	sd-ff	PN, dir↑			PN↑: ♂ size	48
Tettigonia cantans	sd-ff	f, dir↓; I, dir↑				30
zaprochiline x	sd-ff	f, dir↑				16
Acrididae						
Chorthippus biguttulus	sd-ff	cont↑				29
Chorthippus brunneus	sd-ff	PL, stab	PL, g_a	PL, g_a		5
Chorthippus dorsatus	sd-ff	comp↑				46
Ligurotettix coquilletti	sd-ff	I, dir↑; prec				15,43
Ligurotettix planum	sd-ff	prec				32
Omocestus viridulus	sd-ff	BL, dir↑; PL, stab				8,9
Coleotera						
Anobiidae						
Xestobium rufovillosum	vb-ss	PN, dir↑				55
Diptera						
Drosophilidae						
Drosophila melanogaster	sd-nf	IPI, f; stab	IPI, f; x			36,37
Drosophila montana	sd-nf	PL, dir↓; f, dir↑				20,39
Lepidoptera						
Pyralidae						
Achroia grisella	sd-ff	I, PR, gap; dir↑	I, PR; g_a	PR, g_a	PR↑: offspr. dev.; I↑: ♂ size, offspr. size	6,21,22 23,24

[a]Male signaling mode: sd-ff: airborne sound in the far field; sd-nf: airborne sound in the near field; vb-ss: substrate vibration.

[b]Target of female preference: I: male signal intensity; BL: bout length; CL: chirp length; CR: chirp rate; PL: pulse length; PN: number of pulses per chirp or trill; IPI: inter-pulse interval; f: carrier frequency; cont: signal continuity; comp: signal complexity; prec: precedence effect or leading signal; gap: silent interval within pulse pair; dir↑: directional selection for higher character value; dir↓: directional selection for lower character value; stab: stabilizing selection.

[c]Male signal variance: g_a: significant additive genetic variance present for targeted signal character(s); x: little additive genetic variance present; p_r: repeatable phenotypic variation in signal character(s).

[d]Female preference variance: g_a: significant additive genetic variance present for preference; p_r: repeatable phenotypic variance for preference.

[e]Male signal character as a viability/size indicator: offspr. viab.: offspring viability; dev.: developmental rate.

[f]References: 1: Alatalo et al. 1998; 2: Bailey and Yeoh 1988; 3: Bailey et al. 1990; 4: Brown et al. 1996; 5: Charalambous et al. 1994; 6: Collins et al. 1999; 7: DeLuca and Morris 1998; 8: Eiríksson 1993; 9: Eiríksson 1994; 10: Galliart and Shaw 1991b; 11: Galliart and Shaw 1996; 12: Gray 1997; 13: Gray and Cade 1999b; 14: Greenfield and Roizen 1993; 15: Greenfield et al. 1997; 16: Gwynne and Bailey 1988; 17: Hedrick 1986; 18: Hedrick 1988; 19: Hedrick and Weber 1998; 20: Hoikkala and Suvanto 1999; 21: Jang and Greenfield 1996; 22: Jang and Greenfield 2000; 23: Jang et al. 1997; 24: Jia and Greenfield 1997; 25: Kolluru 1999; 26: Kotiaho et al. 1996; 27: Kotiaho et al. 1998b; 28: Kotiaho et al. 1999; 29: Kriegbaum 1989; 30: Latimer and Sippel 1987; 31: Mappes et al. 1996; 32: Minckley and Greenfield 1995; 33: Parri et al. 1997; 34: Ritchie 1996; 35: Ritchie 2000; 36: Ritchie and Kyriacou 1994; 37: Ritchie et al. 1994; 38: Ritchie et al. 1995; 39: Ritchie et al. 1998; 40: Ryder and Siva-Jothy 2000; 41: Schmitt et al. 1993; 42: Schmitt et al. 1994; 43: Shelly and Greenfield 1991; 44: Simmons and Ritchie 1996; 45: Snedden and Greenfield 1998; 46: Stumpner and von Helversen 1992; 47: Tauber et al. 2001; 48: Tuckerman et al. 1993; 49: Wagner 1996; 50: Wagner and Reiser 2000; 51: Wagner et al. 1995; 52: Walker and Forrest 1989; 53: Webb and Roff 1992; 54: Wedell and Sandberg 1995; 55: White et al. 1993.

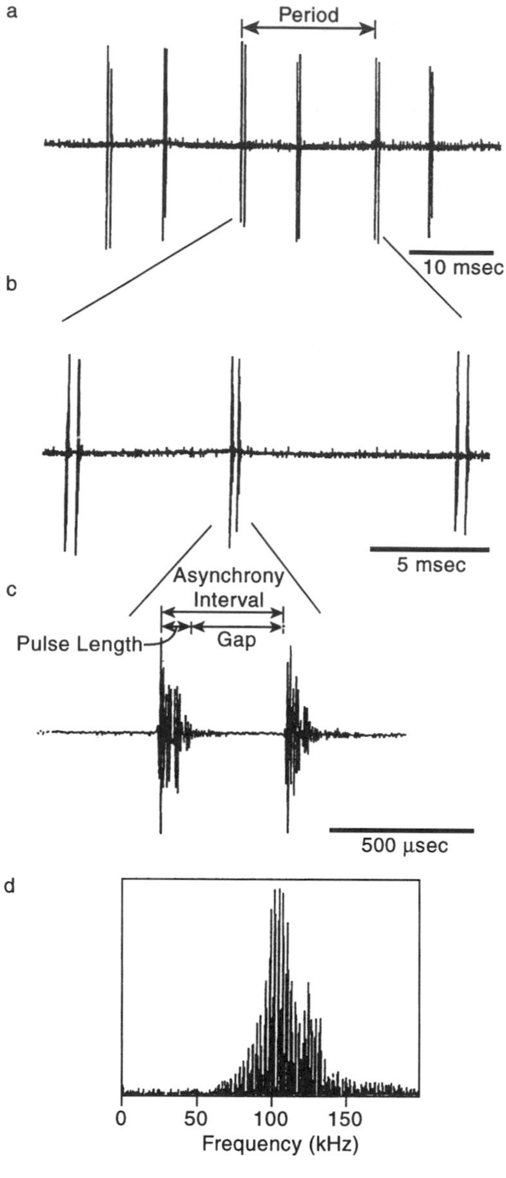

Figure 4.13 Male advertisement call in the lesser wax moth, *Achroia grisella* (Lepidoptera: Pyralidae). (a) Oscillogram showing paired ultrasonic pulses delivered at approximately 100 pulse-pairs·s^{-1}. (b) A pulse-pair is produced during both upstroke and downstroke of each ≈20-ms cycle of wing movement. (c) A brief silent gap, 100–600 μs in length, typically separates the two pulses of a pulse pair. The gap arises because of a slight asynchrony between left and right wing movement or tymbal action. (d) Frequency spectrogram of call; most acoustic energy is found between 70 and 130 kHz. (Reprinted from *Animal Behavior*, vol. 51, Y. Jang and M.D. Greenfield, "Ultrasonic communication and sexual selection in wax moths: female choice based on energy and asynchrony of male signals," pp. 1095–1106, 1996, by permission of the publisher Academic Press, London.)

variances and heritabilities in signaler and receiver traits, Fisherian or good-genes processes may be expected to function eventually (Section 6.2) whether or not direct, material benefits are found.

One potential way for a female to obtain material benefits by discriminating in favor of high-energy signals is via a nutritional spermatophore or specialized spermatophore attachment, a spermatophylax, transferred by the male (cf. Section 3.2). While the function and evolution of many of these extra-gametic materials passed with the ejaculate remain controversial (Wedell 1994), some evidence has indicated that they may be incorporated into somatic tissue or developing eggs and influence a female's fecundity (e.g., Gwynne 1984). Spermatophore or spermatophylax size may be related to a male's body size, and a female who receives and absorbs or ingests a larger amount of the material may produce more or larger eggs. Because certain signal features may correspond reliably with a male's size (Simmons and Zuk 1992, Tuckerman et al. 1993), a female's preference for male calls or vibrations could represent a means by which she maximizes the amount of male-derived material benefits passed to her offspring. This possibility may explain female preferences seen among various acoustic Orthoptera, many of which transfer sizeable spermatophores or attached spermatophylaces (e.g., Morris 1980).[48] However, in some species males who are limited by energy and should consequently transfer a small spermatophylax may reduce calling effort and thereby maintain a normal-sized transfer (Simmons et al. 1992). Is female preference for high-energy calling songs relaxed in these species? Possibly not, if benefits as described next apply.

Other subtler forms of material benefits that arthropod females may obtain by virtue of preference for high-energy signals include food resources, health, and reduced risk of predation. In territorial species, a male's signal characters may reflect his physical condition which, in turn, may be influenced by the quality of food resources at the defended site (Greenfield 1997c; cf. Wagner and Hoback 1999). Similarly, signal characters may reveal a male's infection by sexually transmittable parasites (see Simmons and Zuk 1992, Zuk et al. 1998). Thus, female preferences for signals bearing certain high-energy characteristics may represent "short-cuts" for locating high-quality food resources (Walker and Masaki 1989, Muller 1998; cf. Stamps 1987) or uninfected males (see Ryder and Siva-Jothy 2000). Such female preferences may also yield material benefits by affording an easier path, with fewer risks of predation, toward a male (Section 6.1). As with genetic benefits obtained via the Fisherian and good-genes processes of sexual selection, these material benefits are possible but nonetheless remain speculative in most cases.

Which signal characters would be expected to indicate a male's signal energy and condition most reliably? Signal rate and signal length would be least affected by distortion and degradation over the transmission path and should therefore accurately reflect unattenuated signal energy (Simmons 1988), and possibly the male's energy input and condition. A survey (Table 4.2) indicates that these

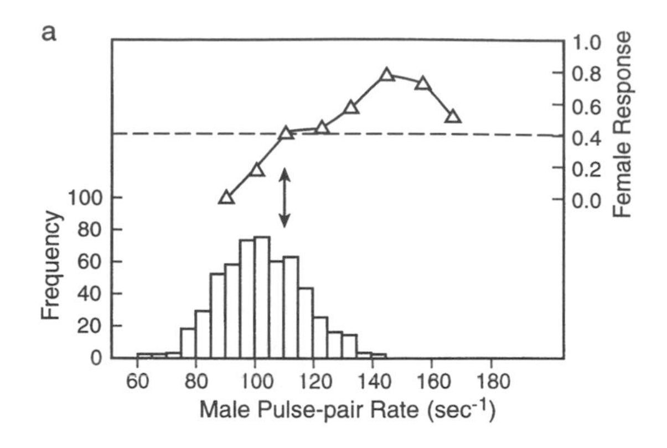
a
Frequency
0
20
40
60
80
100
Female Response
0.0
0.2
0.4
0.6
0.8
1.0
60
80
100
120
140
160
180
Male Pulse-pair Rate (sec-1)

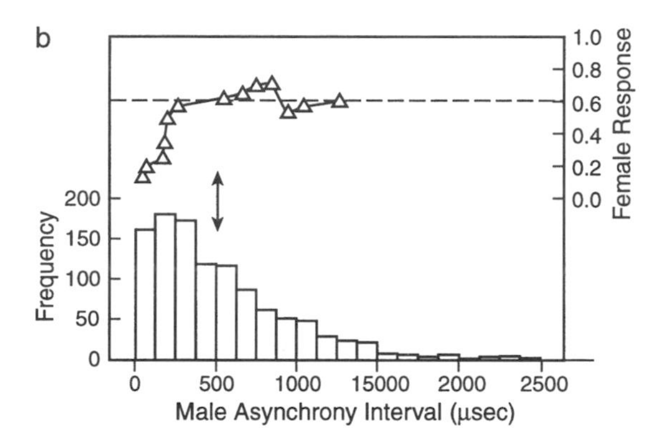
b
Frequency
0
50
100
150
200
Female Response
0.0
0.2
0.4
0.6
0.8
1.0
0
500
1000
15000
2000
2500
Male Asynchrony Interval (μsec)

c

temporal characters do influence female attraction in many arthropods, but signal intensity also ranks as an important factor. In the latter case, passive attraction may represent a potent evolutionary force. Despite the general inability of receivers to judge unattenuated (absolute) signal energy from perceived intensity, females who do respond to increased levels of perceived intensity may, on average, pair with males producing high-energy signals. They will also normally be rewarded with a shorter, and perhaps safer, journey to a male. Conversely, males may be under considerable selection pressure to generate intense signals because stronger broadcasts, having a greater radius of attraction, will, more often than not, reach a greater number of receptive females.

Female Preference for Spectral Features. Are other signal characters that are unrelated to energy also evaluated in the context of sexual selection? Carrier frequency plays a minor role in female choice in some acoustic Orthoptera, and various explanations for its evaluation may be proffered. In mole crickets, which have narrowly tuned ears, females may simply orient toward males whose call frequency spectra match their best frequencies most closely (Ulagaraj and Walker 1975): Females would perceive these calls as the most intense ones, and, from a mechanistic perspective, the choice does not entail any central processing of frequencies (cf. Section 5.3.1). In other insects, frequency-based preferences are directional, and an element of central frequency processing may exist. Females in some katydid species prefer calling songs whose spectra are skewed toward higher frequencies (Latimer and Sippel 1987, Bailey and Yeoh 1988, Bailey et al. 1990), which may represent a means of ranging and choosing nearby males, whose high-frequency energy would be attenuated relatively little (Gwynne and Bailey 1988). But, when females use an IID localization mechanism, they might simply orient toward high-frequency calls more easily (Section 4.3.5). These preferences for high frequencies may be confounded, however, by a general attractiveness of low-frequency sound energy (Latimer

---

Figure 4.14 Variation in male signal characters and female preference for exaggerated values of signal characters in the lesser wax moth, *Achroia grisella*. (a) Signal rate varies from 60 to 140 pulse-pairs·s^{-1} (see Figure 4.13) in the male population; histogram shows frequency of males in 5-s^{-1} bins. Females prefer faster signal rates within this range; triangles indicate female response levels. (b) Asynchrony interval (= time measured from onset of initial pulse to second pulse of a pulse pair; see Figure 4.13) varies from 0 to 1600 μs in the male population; histogram shows frequency of males in 125-μs bins. Females prefer asynchrony intervals >275 μs in length; triangles indicate female response levels. Vertical arrows indicate mean signal values; horizontal dashed lines indicate level of female response to mean signal value. (From Jang and Greenfield 1996 and Greenfield 1997b, reprinted from *Perspectives in Ethology*, vol. 12, with permission of Kluwer Press.) (c) Calling male *A. grisella*. Arrow indicates the ultrasound-producing tymbal on the tegula at the base of the forewing.

and Sippel 1987, Schatral 1990b). Females may benefit from mating with larger males for various reasons, and their heightened responses toward low call frequency and high call intensity may represent indirect means for pairing with preferred mates; both of these signal characters are often related to overall body size (Table 4.2; see also Sanborn and Phillips 1995).

Can Song Reveal Symmetry? Female acoustic insects may also evaluate various spectral and temporal calling song characters that reveal the extent of the signaler's fluctuating asymmetry. Because an individual's bilateral symmetry might reliably indicate its developmental homeostasis in the presence of environmental stress, females may be expected to evaluate this trait among potential mates (Møller and Swaddle 1997). Whereas arthropod visual acuity would seldom be adequate for evaluating minor departures from bilateral symmetry (cf. Swaddle 1999; Section 5.4; but, see Uetz and Smith 1999),[49] certain features of a male's mechanical signals might reveal his symmetry or lack thereof. Thus, the limited frequency analysis in cricket hearing may be sufficient for females to detect anomalies in the spectra of male calling songs that reflect morphological asymmetry between the left and right harps and wings (Simmons and Ritchie 1996). In the European grasshopper *Chorthippus biguttulus*, females discriminate against male calling songs that include silent gaps longer than 4 ms (Kriegbaum 1989). These gaps typically occur in calls produced by males who are missing a hindleg: intact males move their hindlegs slightly out of phase, and the brief silent interval during one leg's reversal of direction is masked by sound produced by the other leg. Because loss of hindlegs usually occurs during molting, these losses may result from developmental inadequacies (von Helversen and von Helversen 1994). In what may ultimately be a similar process, lesser wax moth females prefer male calls that include lengthy silent gaps between paired pulses (Jang and Greenfield 1996; Figures 4.13c and 4.14b). Here, the gaps occur because individual pulses are produced by tymbals activated by the forewings, and movement of the left and right forewings is not perfectly synchronous. Thus, a given upstroke or downstroke of the wings normally generates two distinct pulses. The proximate explanation for this female preference is that pulses separated by lengthy gaps mechanically excite the tympanum more than pulses separated by brief gaps or that are overlapping, despite equivalent acoustic signal energy. But, by favoring males who move their wings slightly out of phase (phase angle ($\cong$5–10°) and thereby produce separated rather than overlapping pulses, females may also be assured that their mates bear an intact tymbal mechanism on each side and can properly activate them.

Influences of Life History and Ecological Features of Signaling. Playback experiments have clearly demonstrated in numerous acoustic insects that females preferentially orient toward male calling songs that bear certain exaggerated features, which often correspond with signal energy and may indicate aspects of a male's condition (Table 4.2). Do these findings, generally obtained via two-choice trials conducted in rarefied laboratory settings, necessarily imply that directional selection operates and that males transmitting the desired calls enjoy elevated mating and reproductive success in the field? Calling song characters undoubtedly influence male fitness to some extent: otherwise, we would not expect to observe the strong directional preferences by females for exaggerated male signal characters and the allocation by males of substantial energy to transmitting such signals (Section 6.2). Nonetheless, various other factors must intervene in natural populations. Foremost among these factors are a male's reproductive lifespan, his phenology relative to that of sexually receptive females, the number of days during which he actively signals, the lengths of his daily signaling bouts, his location during signaling, and the activities of cheaters. Such factors are likely to combine multiplicatively with signal characters to yield a measure of overall reproductive success. However, monitoring male reproductive success in the field with any accuracy has proven nearly impossible in most arthropods,[50] and comparatively little data have been collected demonstrating the influence of signaling lifespan, phenology, bout length, and location on male success (cf. Clutton-Brock 1988). Moreover, these life history and ecological factors may be correlated with each other and with the signal characters, making it somewhat difficult to extract their specific influences from field data on variation in individual male mating or reproductive success. Consequently, I can only claim that signal characters influence some unknown portion of a male's fitness and are subject to some degree of directional selection.

The above reservations notwithstanding, field data on various acoustic insects have been invaluable for demonstrating the ways in which life history and ecological aspects of advertisement signaling may potentially influence fitness. For example, the number of days during which a male signals may correspond with the potential number of females encountered, provided that the signaling lifespan coincides with the period during which females are present and receptive (Greenfield and Shelly 1985, Wang et al. 1990; cf. Murphy 1994a,b). Similarly, the length of a male's daily signaling bouts may also correspond with the number of females he encounters (but, see Murphy 1999). In either situation, it is most unlikely that females would actively monitor the time intervals over which males have been signaling and then choose a mate on this basis. Rather, these influences probably operate in a passive fashion: stochastically, those males who signal the longest will be most likely to be signaling when any given female becomes receptive and begins to evaluate potential mates.

On both lifespan and daily bases, patterns of female receptivity may influence males to allocate their limited energy to signaling early during the season or day (or night) rather than late. Because females may become temporarily or permanently unreceptive following mating, males who begin signaling at or before the seasonal or daily onset of female activity will enjoy an advantage over males who fail to make such adjustments. Thus, many insects exhibit protandry, the appearance of adult males, and the onset of their signaling, several days to many weeks prior to females (Wang et al. 1990), and male signaling often peaks at the beginning of the day or night (cf. Murphy 1999). The latter patterns of daily activity can reflect a "gating phenomenon" in female activity: Females may enter maturity continuously over a 24-hour period, but those who mature prior to the onset of the normal day or night activity period await dawn or dusk before showing any responsiveness toward males (Walker 1983). Male signaling therefore tracks the increased availability of receptive females that follows dawn and dusk in diurnal and nocturnal species, respectively (Greenfield 1992; Figure 4.15).

Energy limitations and predation risks may lead to a negative correlation between the exaggeration of signal characters and signaling lifespan or the length of daily signaling bouts. In lesser wax moths, males who transmit attractive, energy-demanding ultrasonic signals that are delivered at high pulse rates offset this feature with shortened daily signaling bouts, possibly as a result of faster weight loss and depletion of energy reserves (see also Section 6.2.4 and endnote 10). In field crickets, males who call steadily are more likely to attract phonotactic parasitoids, tachinid flies, than relatively quieter males (Cade 1975, 1979). But, in some acoustic insects the males that transmit signals at higher rates and intensities also enjoy equivalent or even greater signaling lifespans (e.g., Greenfield and Shelly 1985). The positive correlations in these species suggest that individual males may differ greatly in overall energy reserves or the ability to replenish and use energy efficiently (see Jennions et al. 2001).

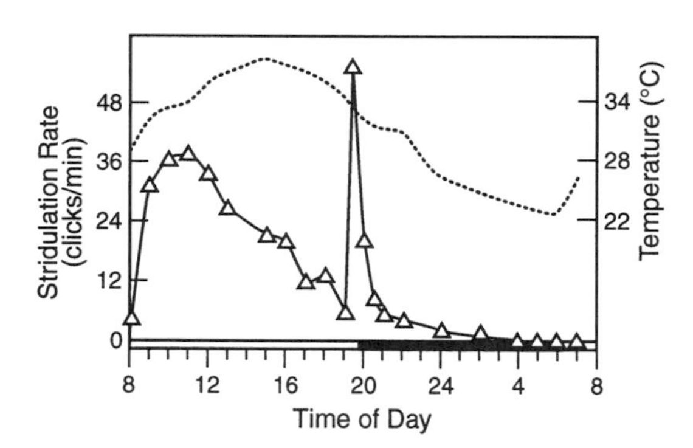

Figure 4.15 Morning and evening choruses in the desert clicker, *Ligurotettix coquilletti* (Orthoptera: Acrididae), as occurring within a local population of 25 males on a particular census day. Mean stridulation rate (triangles) of individual males rises to a broad plateau during morning hours and again to a sharp peak shortly after sunset. Timing of the evening chorus is set by light level (bar along x-axis) rather than temperature (dotted line) and coincides with the return of females to host shrubs at which males call. (From Greenfield 1992; recording from Deep Canyon, California, August 1990; reprinted from *Ethology*, vol. 91, p. 269, copyright 1992, by permission of Blackwell Wissenschafts-Verlag Berlin, GmbH.)

Owing to social and physical factors, the specific location from which a signaler advertises can exert a major influence on his success. If females are attracted by and move toward a congregation of male signalers from outside areas, males on the periphery of the congregation may have opportunities to induce arriving females to mate before they reach centrally located males (Forrest and Raspet 1994, cf. Farris et al. 1997). The signals of these males would simply be perceived as more intense, and passive attraction would draw a greater number of females to them. Unless females regularly move among the various males in the congregation and evaluate and compare their unattenuated signals prior to mating, a centrally located male may be disadvantaged regardless of the exaggeration of his signal characters. In other situations, however, most females may mature and mate within the area of a male congregation, and the centrally located signalers could hold an advantage. At present, these issues remain largely conjectural, as patterns of female movement and assessment in natural populations have received little scrutiny. But, in laboratory tests female field crickets have demonstrated the potential, when making a choice, for moving among males and remembering the signal characteristics of males visited earlier (Wiegmann 1999). This ability implies that the passive attraction models outlined above may not apply universally.

Unlike relative location within a male congregation, the substrate properties and environmental features at a signaler's location can affect signal transmission independently of his neighbors and may thereby exert a stronger, more direct influence on mating success. In mole crickets, for example, the calling song intensity perceived by a female flying above is influenced by the male's stridulatory apparatus and his burrow. The latter factor would include both burrow architecture and the male's ability to select and locate appropriate soil types in which to construct a burrow affording the greatest amplification. Here, female attraction toward more intense male signals may offer material (easier localization) as well as genetic (alleles associated with proficiency in signaling and burrow construction) benefits.

When valuable signaling locations are limited, male–male competition over those locations may arise. Such competition may be particularly strong when the locations owe their value to associated resources needed by both sexes. In *Ligurotettix* grasshoppers (Orthoptera: Acrididae), territorial species in which males defend host plants as mating territories, males vie to advertise from the most valuable shrubs, those whose foliage has a low titer of toxic compounds interfering with the conversion of digested food (Greenfield et al. 1989). Females attracted to a signaling male's shrub are most likely to remain as long-term residents and mate if the shrub is a high-quality one, which strongly selects for male abilities to locate, defend, and signal from these sites. Because prior residency affords a male an advantage in territorial conflicts, males who eclose early have a higher probability of establishing themselves as owners of the superior shrubs and encountering more females later in the season. Thus, protandry is quite marked in these species. The advantage to a male of eclosing early and then successfully defending and signaling from a superior shrub is strong enough to outweigh the possibility of dying before the female population peaks (Wang et al. 1990).

A certain amount of deception exists in the mating and social systems of most natural populations, and deceptive activity often has the potential to jeopardize the communication of legitimate signalers. In the context of advertisement signaling along the mechanical channel, individual males may indulge in deception by signaling at markedly reduced intensities and rates while strategically positioning themselves near superior signalers (e.g., Kiflawi and Gray 2000). Such cheaters, often termed satellite or silent males, may thereby intercept females attracted to the neighboring signaler, and they could reduce the general influence of exaggerated signal characters on mating success and fitness in the population. Because satellite males neither expend a large amount of energy on advertisement nor suffer predation from natural enemies attracted by signals (Cade 1979), their activity would appear to represent a rewarding alternative to regular signaling, particularly when population density is high such that there exist abundant signalers to "parasitize." Nonetheless, the few studies in which mating success of individual males has been monitored in natural populations do not support these possibilities. In *Ligurotettix* grasshoppers, satellites were consistently found to mate far less frequently than regular signalers, and their maintenance within populations is not clear (Greenfield and Shelly 1985, Shelly and Greenfield 1989). Satellite behavior in *L. coquilletti* is conditional on eclosion date (Wang et al. 1990), but this association does not explain why late-maturing males should accept their inferior status and reduced mating success.

For species that rely on acoustic or vibrational advertisement, a fundamental property of energy transport via wave action may reduce the potential impact of inferior signalers acting as satellites. When sound or vibrational waves broadcast from separate sources meet, the wave displacements add linearly at the point of intersection according to the superposition principle. However, the waves do not interfere with each other elsewhere, and they retain their original features. Thus, when acoustic or vibrational signals are transmitted from multiple sources within a limited area, a distant receiver may perceive the original signal features despite intersections with other signals en route. For humans, this principle helps us to concentrate on a particular person's voice in a crowded room containing many other conversations, the so-called "cocktail party problem" of psychoacoustics (Cherry 1953). For acoustic insects, such as crickets and katydids, that possess the neural ability to process different signals by the left and right ears, the principle may allow receivers to evaluate and localize a particular call in the presence of other calls broadcast simultaneously from different directions within a dense chorus (Pollack 1998, Römer and Krusch 2000; see also Schwartz and Gerhardt 1989 and Greenfield and Rand 2000 for treatment of the problem in anuran choruses). Consequently, a discriminating female may not be easily misled by an inferior signaler positioned near a superior one, and a superior signaler may expect to enjoy a female encounter rate that is commensurate with the expression of his signal characters.[51] This reliability may not exist along the olfactory channel, and potential susceptibility to cheating may be one reason why males seldom rely solely on pheromones for sexual advertisement (Section 3.1.6).

The Counteracting Influences of Natural Selection. In cases where females prefer exaggerated values of an energy-based male signal character, the mean value of that character seldom matches the female preference. Rather, natural selection normally counters the influences of sexual selection and holds the mean value in the male population at a level below that which females would most prefer (see Kotiaho et al. 1998b). This discrepancy is particularly evident when female preferences are essentially open-ended (e.g., Figure 4.14a), as opposed to focused on a specific character value (e.g., Figure 4.8), but it may also occur in the latter situation (e.g., Ritchie 2000). For signaling along the mechanical channel, energetic or neuromuscular limitations may prevent nearly all males from producing the most attractive signals. Additionally, many mechanical signals draw natural enemies (Burk 1982, Zuk and Kolluru 1998), who may be more strongly attracted by more intense or elaborate signals,[52] and acquiring the energy necessary to produce a signal preferred by females may incur higher risks of predation. Any of these factors could hold the mean exaggeration level of male signal characters to values below the mean female preference. Alternatively, males who can transmit highly exaggerated signals may continue to do so, but they compensate for their potential conspicuousness with heightened vigilance or other cautionary behavior (Hedrick 2000).

For acoustic insects, natural enemies attracted by male calling songs include parasitoid tachinid and sarcophagid flies and foliage-gleaning bats. Parasitoid tachinid flies are sufficiently threatening to some populations and species of crickets that they may have selected for the secondary loss of acoustic signaling (cf. Walker 1977). In other species of crickets, these parasitoids may be responsible for changes in song characteristics (Rotenberry et al. 1996) or a reduction in the length of daily signaling bouts or shifts of signaling to times of day when the parasitoids cannot forage for hosts efficiently (Cade 1984, Zuk et al. 1993). Katydids, many of which are arboreal, are more vulnerable to foliage-gleaning bats than crickets are (but, see Bailey and Haythornthwaite 1998). Especially in the Neotropics, these phonotactic predators may have selected for the sporadic production of low-duty cycle, very high-frequency calls found in many katydid species (Morris et al. 1994). Such temporal and spectral features reduce both the transmission distance and the likelihood that a female receiver perceives and recognizes the call, but the predators thereby suffer equal or greater difficulty in locating the male katydids. In some katydid species, tremulation partially supplants calling, presumably because bats cannot effectively eavesdrop on substrate vibrations (Belwood and Morris 1987, Belwood 1990). These signaling modifications are not observed among katydids found in dense vegetation, including secondary habitats and tall grass, where bats cannot maneuver.[53] They are also not conspicuous in the Palaeotropics, where bat faunas and foraging habits are different (Heller 1995).

Arthropods may experience as much or more risk of predation during orientation movements as during signaling (Section 3.1.5; Sakaluk and Belwood 1984). Consequently, various modifications in female preference and orientation, and even the entire pair-forming protocol (see below), may be expected in some species. In situations wherein the evaluation of potential mates exposes females to greater risks than would be experienced were they to orient and mate with the nearest or most localizable male, female choice may be relaxed. For example, when the perceived danger from predators is high, female field crickets may forgo the movements necessary to locate a male who would be preferred under ideal conditions and simply choose an easily located caller (Hedrick and Dill 1993).[52] Thus, natural selection acting on the receiver may reduce a discrepancy between the mean value of a signal character in the male population and the female preference. Ultimately, such natural selection might indirectly diminish the strength of sexual selection acting on the male signal.

For nocturnal receivers, aerial-hawking bats pose a major threat to those species that fly in search of and while orienting toward conspecific signals. Ultrasonic sensitivity and specialized evasive flight maneuvers in response to pulsed ultrasound are reported in five insect orders (Orthoptera, Mantodea, Coleoptera, Neuroptera, Lepidoptera; Hoy 1994), and these behaviors are generally believed to reflect selection pressure from bats. In many environments a major portion of the ultrasounds present at night are the echo-locations of foraging insectivorous bats (Fenton and Fullard 1981, Hoy 1992). Among Orthoptera, the anti-predatory responses evolved in taxa that were already relying on acoustic communication for pair-formation (Section 4.5). Crickets, which generally transmit advertisement calls below 7 kHz in frequency, may largely distinguish mating from predator signals by frequency and do so via categorical perception (Wyttenbach et al. 1996; cf. Section 3.5.2). Laboratory tests demonstrate that crickets in tethered flight veer toward appropriately pulsed low-frequency sound but away from ultrasonic pulses (Moiseff et al. 1978, Nolen and Hoy 1986). Katydids also exhibit negative phonotaxis toward bat echo-locations (Libersat and Hoy 1991, Schulze and Schul 2001), but most species use ultrasonic advertisement signals themselves, and it is not clear how mating and predator signals are distinguished. Here, receivers may evaluate tempo characters, which raises the possibility that selection pressure from bats has influenced the evolution of specific pulsing patterns in katydid calls (Sections 4.2.7 and 6.1). In both Lepidoptera and Orthoptera, the responses of females orienting toward male signals are modified when they perceive bat echo-location signals. A female's decision to initiate or continue orientation depends on the relative strengths of the male and predator signals (Acharya and McNeil1998, Farris et al. 1998).

In some acoustic insects, males switch from an advertisement call to a specialized courtship signal when an attracted female approaches to within close range and they detect her. Among crickets, courtship signals are normally higher in frequency and lower in intensity than advertisement calls. These signals may offer the female further opportunity to assess the male (Balakrishnan and Pollack 1996, Nelson and Nolen 1997, Wagner and Reiser 2000), but their lower intensity could be a specific adaptation to danger from phonotactic natural enemies: females might avoid making the final approach toward and remaining near a male calling intensely because the risk of being attacked themselves is increased.

Signaling Dialogues. When predation levels are particularly severe, we might expect the entire pair-forming protocol to be radically modified. Because females may be selected to avoid performing those pair-forming activities that incur greater risk (Section 3.1.5), males might assume both signaling and orienting roles if distant females can discreetly indicate their reception of and interest in a male's advertisement (Heller 1992). Signaling dialogues as such do occur in various arthropods using mechanical signals, and they are characterized by an intense male advertisement followed, during a critical latency measured from the end of the male signal, by a relatively weak female reply (Figure 4.16; cf. Section 5.2.1 for the photic analog). Among acoustic insects, these dialogues are particularly evident in certain acridid grasshoppers (e.g., von Helversen and von Helversen 1997) and bradyporine and phaneropterine katydids (e.g., Hartley 1993), where they are represented by male–female "duets."[54] In katydids, the stridulatory device with which females signal (Nickle and Carlysle 1975) is not homologous with the male one (Robinson 1990), and the carrier frequency of the female reply may be as much as 10 kHz higher than the male advertisement

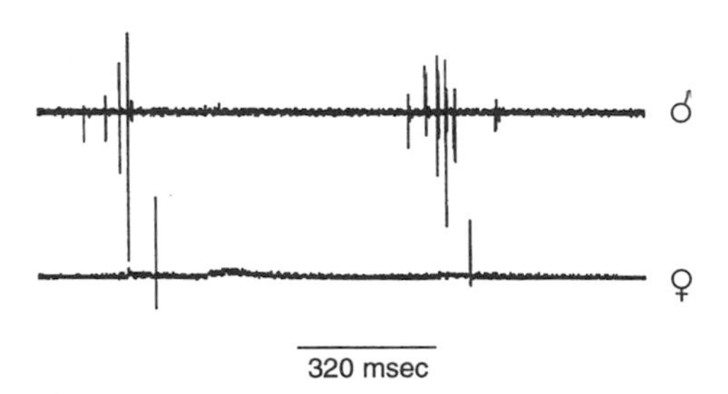

Figure 4.16 Example of male–female duetting in *Phaneroptera nana* (Tettigoniidae: Phaneropterinae). Oscillograms of male call (upper trace) and female reply (lower trace) indicate the higher amplitude and greater length of the male call and the occurrence of the female reply within a critical time window (60–100 ms) following the male call. The male and female are separated by ≈1 m (from Tauber et al. 2001).

call. The best frequencies in female and male hearing may differ accordingly in these cases (Dobler et al. 1994b, Heller et al. 1997a).

Signaling dialogues assume various formats among duetting species (Robinson 1990). In many duetting katydids, males whose advertisements elicit a perceptible female reply then localize the female and move the entire distance toward her. Duetting continues while the male approaches the female, who has shifted to her mate the potential risks associated with movement (see Heller 1992). In other species, however, the male may move only a partial distance toward a responding female, after which the female makes the final approach toward the male. Alternatively, responding females may move over most of the distance, with males making the final approach to the female, or both the male and female may move simultaneously toward each other.

The latency of a female's reply following the male's advertisement is generally species-specific, and males normally do not orient toward female replies that fall outside of the critical time window (Heller and von Helversen 1986, Robinson et al. 1986, Zimmermann et al. 1989, Dobler et al. 1994a). In some duetting species, females may reply indiscriminately to all local males (Heller et al. 1997b). These duets are often distinguished by reply latencies as brief as 15 ms, and it is likely that under this constraint the females do not even determine whether the male advertisement is a conspecific one. Possibly, species recognition and mate choice are performed later on the basis of acoustic and other information that can be assessed more accurately at close range. If males transfer material gifts at mating, the females may gain by attracting and mating with numerous suitors and receiving their nuptial offerings (Heller and von Helversen 1991). Afterward, they may practice post-copulatory mate choice by differentially using the sperm of only certain of the males to fertilize their eggs (cf. Eberhard 1996). Differential sperm use is accomplished by prematurely eating certain spermatophores, before they are completely empty of sperm.

In other duetting species, though, females may be quite discriminating in their choice of mates, particularly where several acoustic insect species co-occur. Females may choose to reply only to those conspecific males whose advertisements bear certain exaggerated signal characters by timing replies to follow, within the specific latency interval, calls of the preferred males (Tauber et al. 2001). Here, reply latencies are normally 50–100 ms, which may be sufficient for females to evaluate the male's signal characters. Provided that females can reliably evaluate male signals and associated attributes at a distance, a high level of female discrimination of mates may be expected in duetting species: Females would not incur higher energy costs and risks by being choosy, and they would avoid the conspicuousness and harassment that would ensue were they to reply to all perceived males and attract many of them.

Are males in signaling dialogues vulnerable to neighboring cheaters who eavesdrop and interlope in their duets and courtships? Neighboring males may occasionally insert their calls within duets, ostensibly as a means of diverting the female's attention away from her duetting partner and to themselves (Galliart and Shaw 1991a,b, Bailey and Field 2000; see Section 5.2.1). However, there are no reported cases of neighboring males acting as satellites by tracking the movements of duetting males toward replying females and then intervening in their courtship. Signaling dialogues may be inherently resistant to such cheating, because duetting continues until the male and female make physical contact. Thus, females have the opportunity to assess a male continually, and they may only mate with males with whom they have been duetting throughout the entire pairing process. A second male attempting to intervene at the last moment would normally be recognized as an interloper and rejected, which would select against satellite behavior. Moreover, advertising males may evaluate their attractiveness relative to that of neighbors and behave accordingly (Tauber et al. 2001; see Section 4.4.2): if inferior in signaling prowess, a male may forgo approaching local replying females, because such approaches are nearly always unrewarding.

Methodology and Cautionary Notes. Female responses to and preferences for male advertisement calls are usually tested via the presentation of recorded or synthetic stimuli in playback experiments. In duetting species wherein female responses are strictly acoustic, responses can be tested readily via single-stimulus presentations. The strength of a female's response to a given stimulus model might be assessed by the intensity or length of her reply and by the proportion of repeated presentations of the stimulus that elicit replies. In species wherein females orient toward advertising males, the single-stimulus testing protocol can also be used, with response strength assessed by latency or the precision and speed of orientation toward the stimulus. Locomotion compensation devices (Kramer 1975) are especially valuable for these assessments, because a tested female may continue her orientation for an extended time interval without experiencing an increase in stimulus intensity. The value of the single-stimulus approach lies in the ease and rapidity with which trials may be conducted (see Wagner 1998). Thus, an individual female may be tested with the full range of stimulus models, and each model may be tested in multiple trials to determine the repeatability of her responses.

Unfortunately, the ease of application that the single-stimulus approach affords is often acquired at the expense of sensitivity. In many cases, females may exhibit comparable responses to two different signals presented in separate trials, whereas a two-choice test would reveal a clear preference for one signal over the other (Doherty 1985c). But, testing all possible pairs of stimulus–signals would normally be a daunting task, and measures of preference repeatability may be precluded. An experimenter may therefore be forced to adopt a compromise and test only carefully selected pairs of stimuli, but determining a female preference function from such restricted data could be problematic. Tests presenting two or more choices may be adapted for species in which females either orient toward advertising males or signal their replies. In the latter case, the different choices would be presented in alternation so that the preferred stimuli, which female replies follow by a specific latency, are clearly indicated.

When designing stimuli for playback tests, should recorded or synthetic calls be used? Edited recorded calls offer the advantages of including a complete frequency spectrum, which may be difficult to generate in a synthetic call, and an accurate rendition of higher order temporal units and silent pauses and gaps between units, all of which may be critical to females (Schmitt et al. 1993). On the other hand, the recorded call used to generate stimuli may represent an outlier in the population, and anomalous results could ensue from presenting it in playback tests (Kroodsma 1989; but see Searcy 1989). While this latter problem might be more severe in testing the more complex calls of birds and mammals, especially where dialects occur, care should nonetheless be taken to ensure that the recorded call chosen for editing and stimulus playback bears intensity, spectral, and tempo characters resembling average values in the population. Alternatively, multiple exemplars of the call may be tested.

4.4.2 Intrasexual Interactions: Assessment, Competition, and Cooperation

Whereas the transmission of mechanical advertisement signals is usually restricted to one sex, the male, the reception of mechanical signals is nearly always found in both males and females. The sexes may differ in their sensitivities and best frequencies, but male signalers are normally at least as capable as female receivers of perceiving and evaluating the signals of neighboring males. These dual abilities also exist in chemical signaling, as males normally perceive, and may respond to, male courtship pheromones (Section 3.2). Females too may perceive female advertisement pheromones (Ljungberg et al. 1993, Schneider et al. 1998a), a generally overlooked (female) ability. But, along the mechanical channel, the dual abilities of males are often developed to a very high degree (e.g., Ulagaraj and Walker 1973). Males may rely on perception of neighboring male signals to adjust their spacing and maintain minimum nearest-neighbor distances, to form aggregations and to locate valuable resource patches more easily, and to assess their neighbors' motivation for and ability in aggression. In some species, males may switch to a specialized aggressive signal when interacting with neighboring males. These aggressive signals may be graded and thereby serve as reliable indicators of motivation and ability (Section 6.2). Possibly, the same physical factors that led many species to favor the mechanical channel for sexual advertisement and pair-formation have led them to favor it for intrasexual functions as well.

The distinction between a signal's function as an advertisement in the context of pair-formation versus a mediator of intrasexual interaction is often hazy, and many mechanical signals, and others,[55] may operate in both realms (e.g., Boake and Capranica 1982). For example, intensity or tempo of a male advertisement call may indicate both preferred traits to a female and aggressive prowess to a rival male. In other cases, males may monitor each other's signals and adjust certain signal characters so that they match or exceed their neighbors' values. Thus, males engage in signal competition to enhance their relative attractiveness to females. Signal competition may also entail aspects of relative timing that are independent of any intrinsic signal characters. For example, female receivers may be relatively insensitive to signals that immediately follow an identical, leading signal. This psychophysical effect may select for mechanisms by which interacting males adjust their signal rhythms in ways that decrease their chance of following and increase their chance of leading. Among acoustic insects, when males mutually apply these timing mechanisms, structured choruses can emerge in which the signals of neighbors occur in a regularly synchronizing or alternating pattern.

Most of the examples of intrasexual signal interactions provided below are from acoustic insects. This bias might reflect the ease with which airborne far-field sounds can be observed and studied. But, we might also inquire whether the apparent scarcity of interactions among other types of mechanical signals is real.

Spatial Adjustments. Because males can usually perceive the mechanical advertisement signals transmitted by neighboring males, they may be able to adjust the distances to their nearest neighbors. In some acoustic insects, the basic male response is to move away from a neighbor whose signal intensities as perceived exceed a threshold value (e.g., Thiele and Bailey 1980, Simmons and Bailey 1993). If all males signal at comparable intensities and amidst environments with equivalent attenuating properties, their mutual responses would maintain minimum distances between signaling neighbors, and a regular spatial distribution would emerge among the population. The threshold intensity that elicits negative phonotaxis is expected to be higher than the physiological hearing threshold (e.g., Römer and Bailey 1986) and to change with the level of background noise in the surrounding chorus. For example, if the density of signalers and their noise level are high, males may adapt (Section 2.3) and set their thresholds at a higher intensity. Thus, males would acquire exclusive signaling areas surrounding themselves that elastically contract or expand in accordance with the local density. This distribution of male signalers is "ideal free" (Fretwell 1972): assuming that all males are comparable in signaling prowess and the initial distribution of females within the habitat is random, no one male would suffer a lower female encounter rate than any other. Moreover, a male population that has arrived at this ideal free distribution will be "evolutionarily stable": an aberrant male who fails to respond to the appropriate threshold intensity by moving away from signaling neighbors would encounter fewer females than an average male in the population would. On the other hand, males who do respond in the expected fashion described above benefit because they avoid expending energy and time advertising close to signaling neighbors, males with whom they must necessarily share any nearby females (see Arak et al. 1990).

Despite the evolutionary stability of the ideal free distribution, its occurrence among acoustic insects, or in other species and situations, is only rarely reported. The most likely explanation for its failure to emerge dependably in the pure form above is that the requirements of comparable signaling prowess among males and a homogeneous habitat are seldom met. Under the more typical conditions of male and habitat variation, positive phonotaxis leading to aggregated spatial distribution may sometimes occur.

Within a population, males commonly differ in signal features that influence their attractiveness to females. Regardless whether such variation reflects age, chance events during immature development, or genetic differences, inferior signalers often position themselves adjacent to superior ones. But, inferior signaling in the context of satellite behavior may indicate to the superior signaler that his "parasitic" neighbor lacks motivation or proficiency in aggression. This indication may invite him to attack the inferior signaler, who may then depart and seek signaling opportunities elsewhere or abandon signaling altogether. In the latter case, by remaining silent, he reduces his conspicuousness and usually lessens attacks by the more aggressive neighbor. It is generally reasoned that an inferior signaler would achieve higher fitness through satellite, or even silent, behavior adjacent to an attractive signaler than he would by signaling weakly in a solitary location (cf. Arak 1988).

Aggregated distributions of signalers may also arise where habitat quality is not homogeneous. In the extreme case, preferred habitat is distributed across the landscape in the form of patches, which males may treat as economically defensible signaling stations and mating territories. Here, one might naïvely expect an even distribution of signaling males among patches as long as the patch : male ratio is equal to or greater than 1.0. However, various ecological and social factors may arise and generate rather uneven distributions, with some patches harboring aggregations of signalers while others remain vacant. How would these factors operate to circumvent the expected ideal free distribution?

Long-term field studies of *Ligurotettix* grasshoppers have suggested various ways in which the economics of resource requirements and availability might lead to signaler aggregations (Greenfield 1997c; Figure 4.17). First, patches themselves may vary in quality that could affect male signalers and the females drawn to the male advertisements. If the variable quality is an aspect of food resources in the patch, a male may transmit more intense, faster, or longer signals when he settles on a high-quality patch and has access to the food there. The high-energy signals that he produces may be sufficiently attractive

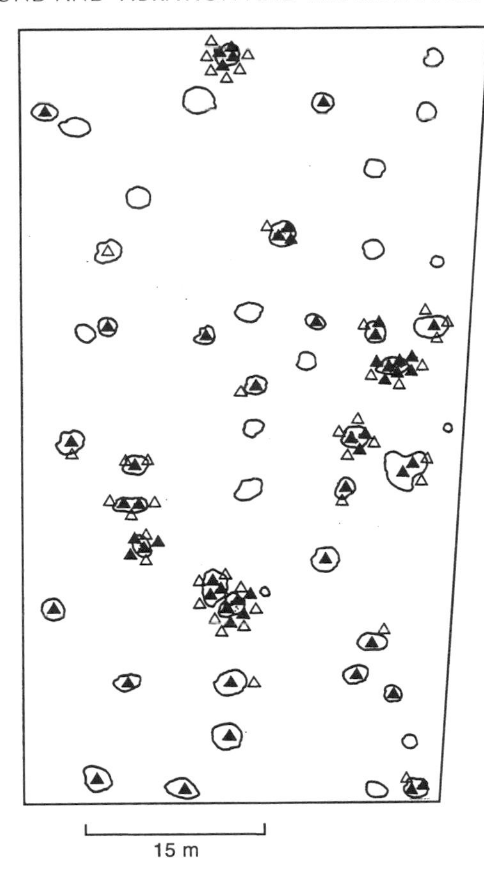

Figure 4.17 Aggregated dispersion of male and female desert clickers, *Ligurotettix coquilletti* (Orthoptera: Acrididae), as occurring on a particular census day. Solid triangles represent males, open triangles represent females, and irregular outlines represent host shrubs; most of the remaining area is bare ground. Approximately 2/3 of the males were actively calling. (Data from Deep Canyon, California, July 1983; reprinted from M.D. Greenfield, "Sexual selection in resource defense polygyny: lessons from territorial grasshoppers," p. 75 in *The Evolution of Mating Systems in Insects and Arachnids*, eds J.C. Choe and B.J. Crespi, 1997, with permission of Cambridge University Press.)

to other males that they too settle on the patch. These latter males may use the powerful conspecific signals of the initial male as a short-cut for locating the best food resources (Muller 1998; see Section 3.3.1). Should the value of the patch be high enough, the opportunity to partake of its food resources may outweigh the drawback of frequent aggressive encounters with the other males. Second, female preferences may reinforce aggregation behavior both directly and indirectly. If female receivers can evaluate male signalers more accurately by means of a simultaneous as opposed to a sequential comparison, they may only approach and mate with signalers found in aggregations, regardless of the strength of a lone signaler's transmission (but, see Real 1990, 1991, Wiegmann et al. 1996, Wiegmann 1999). Thus, all males would be forced to join aggregations, which may be considered "resource-based leks" (sensu Alexander 1975). And, like males, females may also rely on the powerful signals of males at high-quality patches, and of the male aggregations that may exist there, as short-cuts to the best food. Valuable resource patches may then become "sinks" at which signaling males continue to accumulate until the per capita female encounter and mating rates level off and eventually decline.

Acoustic Duels. It is quite common for males to assess their relative ability and motivation for aggression via a mutual comparison of graded signals. Among acoustic insects, these assessments may involve advertisement, courtship, or specialized aggressive signals. In a typical engagement, two males who have approached closer than an acceptable nearest-neighbor distance repeatedly exchange calls, which may be accompanied by visual displays if the males are very near one another. The encounter may result in various outcomes. Ordinarily, the two males adjust their spacing, with one continuing to signal at the contested site and the other retreating or downgrading itself to satellite status. Rarely, the encounter escalates to physical contact and overt fighting. In most cases, the males are contesting exclusive ownership of a territory, signaling station, or access to females.

Contests over territory ownership in the acridid grasshopper *Ligurotettix planum* are a particularly instructive example of intrasexual signal interactions, because males switch to a specialized aggressive call whose features are good predictors of the encounter's outcome (Greenfield and Minckley 1993). Here, two contesting males wage an acoustic duel wherein they alternate transmission of aggressive calls, which are longer, more intense, and more acoustically complex than the regular advertisement calls that attract females to a territory (Figure 4.18). In 85% of contests observed in the field that did not escalate to overt fighting, the male whose overall sound production, computed as the product of mean call length times call rate (calls·min^{-1}), was measurably higher won. The loser usually ceased calling within 1–10 min, retreated from the contest, and typically sought another patch on which to settle or remained as a satellite.[56] In 15% of all contests, however, the two males escalated to grappling, kicking, and biting. Males paired in escalated contests were distinguished by having nearly identical sound production, and they continued to exchange aggressive calls for extended periods, exceeding 1 hour in some cases. Eventually, overt aggression broke the signaling stalemate.

Overall sound production as computed probably reflects a male's energy expenditure during a contest. Moreover, from the perspective of a focal male's perception, a neighbor's overall sound production may be graded more reliably

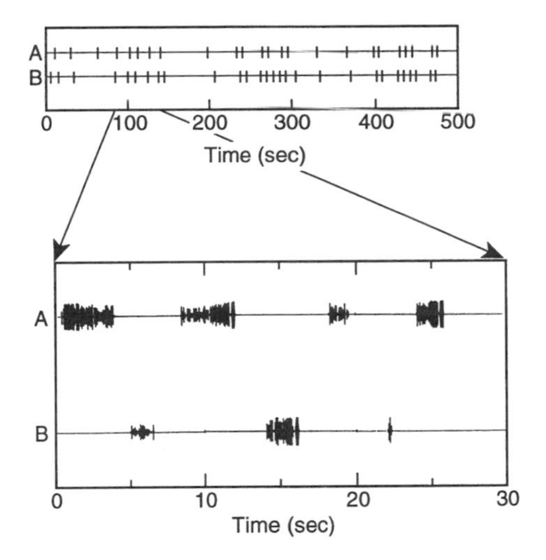

Figure 4.18 Example of acoustic dueling in male tarbush grasshoppers, *Ligurotettix planum* (Orthoptera: Acrididae). Oscillograms of the aggressive calls (see Figure 2.2d) of two males (upper and lower traces, A and B) indicate the typical pattern of alternation (from Greenfield 1997a,c). The two males were separated by <50 cm and were contesting ownership of a mating territory, a host shrub. (From Greenfield and Minckley 1993 and Greenfield 1997a,c; recording from Big Bend, Texas, August 1991; reprinted from *The Bionomics of Grasshoppers, Katydids, and their Kin*, p. 213, with permission of CAB International.)

with his energy expenditure than call intensity is (cf. Simmons 1988). But how can the focal male measure his neighbor's sound production and compare it with his own? Probably, the alternating exchange of calls is the critical factor that enables such seemingly complex measurements. Behavioral observations indicate that *L. planum* males suffer a reduced hearing sensitivity during call production (Section 4.3.6), and the alternating protocol is likely a means by which a focal male both improves his ability to monitor the neighbor and allows the neighbor to monitor the focal male's calls. A basic rule, wherein a male does not initiate a call while his neighbor is transmitting one, enforces the alternating protocol, and a byproduct of call alternation is that via a simple algorithm a focal male might accurately judge his sound production relative to his neighbor's and respond appropriately. The focal male remains in the contest as long as he is able to call following every call of the neighbor, but he departs once the neighbor calls two or more times before his next call. Similar rules and algorithms appear to govern ritualized contests among vertebrates (e.g., Clutton-Brock and Albon 1979), and they may be the most effective solutions to various problems in signal perception. Nonetheless, in *L. planum* acoustic duels the key signal feature, overall sound production, is influenced by call length as well as by call rate, and it is not clear how males might assess length.

It is assumed that males generally obey rules as outlined above because certain signal characteristics, namely features of endurance or activity that demand a high expenditure of energy, are reliably correlated with aggressive ability or motivation (e.g., Mason 1996, Hack 1997a,b). Thus, a male whose signals are noticeably inferior to his neighbor's but who nonetheless remains in a contest would be inviting defeat in an energetically costly and potentially injurious fight. Ritualized signal interactions may be a basic means of avoiding the elevated costs and risks of fighting and arriving at the same outcome that a fight would provide. But these advantages could require that the contestants have access to relatively complete information about each other, that their signals are honest and not overly susceptible to cheating (Section 2.6), and that interactive signaling itself is not so costly that a male's energy reserves and fitness suffer from the very act of participating. We return to these theoretical concerns in Section 6.2.

Because males may use their advertisements for both pair-formation and resolving contests with other males, both inter- and intrasexual selection may act on these signals. Under such circumstances, the two influences might combine additively and favor the evolution of signal exaggeration. But it is also possible for these influences to act in opposite directions on a signal feature or to favor separate signal features (Cremer and Greenfield 1998), which may be a significant factor maintaining signal variants (see Section 3.2; Moore and Moore 1999) or complex signals in a population. For example, male advertisement calls in the Polynesian cricket *Teleogryllus oceanicus* include both chirps, which are attractive to females, and recurring trills, which influence male spacing and aggregation (Pollack 1982).

Signal Modifications. Female choice may also drive males to modify critical features of their advertisement signals so that they match or exceed their neighbors' signals and compete more effectively in the mating arena (see Gerhardt et al. 2000). Signal modification as such may be expected whenever females assess local males on a relative basis before orienting toward and choosing a mate. Modification might be restricted, however, because males may not have direct control over all signal features. Additionally, males might be unable to assess many of their neighbors' signal features or compare them with their own.

Among acoustic insects, the most basic way in which males modify their signals when within earshot of their neighbors is by initiating calling bouts as soon as neighbors do (e.g. Schatral and Bailey 1991). These responses can generate an impressive chorus in which nearly all local males start calling within mere seconds of each other. The collective onsets of calling may occur once at the beginning of the day or night activity period or repeatedly during these periods (Greenfield 1994a, cf. Schwartz 1991).

On a finer scale, males may modify various signal characters when responding to the onsets of their neighbors' calling bouts. In the lesser wax moth, males adjust one of their key signal characters, pulse rate, to higher, more attractive levels when a neighbor begins to call within 20 cm (Jia et al. 2001). No other signal characters are modified, and the 5–15% elevation of pulse rate only lasts for approximately 10 min. Probably, males are unable to control other signal characters, such as intensity, which are determined by overall body size and the dimensions and structure of the tymbals. The limited duration of signal modification may reflect constraints on energy expenditure, which are severe in this insect that neither feeds nor drinks as an adult and must rely entirely on energy reserves acquired during larval development. Consequently, any signal modification that demands greater energy must be accomplished economically. Because the elevated pulse rates require an 11% increase in metabolic rate on average (Reinhold et al. 1998), the 10-min duration of modification may be all that a male can afford. Nonetheless, males strategically time their brief signal modifications to coincide with occasions when females are most likely to be attracted, the beginning of the night and the onsets of signaling bouts (see Section 4.4.1; Walker 1983). Perhaps allocating limited energy reserves to a significant increase in attractiveness at these times would be more worthwhile than spreading them out over the course of many hours.

When males modify their signals in a competitive context, do they scale the adjustments relative to features of their neighbors' signals? While such scaling might be expected on economic grounds, perceptual limitations may often prevent it. Lesser wax moth males show no evidence of elevating their pulse rates by greater or lesser amounts depending on the neighbor's pulse rate, and it is most unlikely that they could effect the necessary assessments to do so. But by simply elevating their signal rate above its normal, basal level, they increase the probability that they will match or exceed the neighbor's level and remain attractive to local females.

Chorusing: Synchrony and Alternation as Outcomes of Competitive Signal Jamming. Particularly among acoustic insect species that signal rhythmically, sexual selection may drive males to adjust the timing of their advertisements on a fine scale. Some of these signal adjustments are made in response to (female) receiver preferences for leading signals, and the responses of individual signalers may collectively generate the emergence of temporally structured choruses of synchronized or alternating signals. In other cases, however, it is conceivable that these collective signaling events per se are adaptive and not mere emergent properties arising from the separate activities of myriad individuals.

Receiver biases toward leading acoustic signals are known among mammals, anurans (Dyson and Passmore 1988), and several groups of insects (Wyttenbach and Hoy 1993, Greenfield 1994b). Analogous responses also occur among arthropod species using visual signals relying on bioluminescence (Vencl and Carlson 1998) and reflected light (Backwell et al. 1998; see Section 5.2). The responses were first reported in humans (Wallach et al. 1949), where they were termed the precedence effect. This term has been subject to various interpretations and defined in different ways over the years. Here, I shall adhere to the broad definition in Zurek's 1987 review chapter:

> When two binaural sounds are presented with a brief delay between them, and are perceived as a single auditory event, the localization of that event is determined largely by the directional cues carried by the earlier sound. This observation is known as the precedence effect in sound localization.

For many years, precedence effects did not attract the attention of scientists outside the confines of mammalian psychoacoustics, but investigations of sexual selection in acoustic insects and anurans eventually revealed that various of these species were also influenced. Females presented with two or more identical calls broadcast from different azimuths would orient toward, or otherwise respond to, the first call only. The effects begin at the onset of the leading call and extend for varying durations, ranging from as little as 70 ms in some katydid species to approximately 2 s in an acridid grasshopper. Following calls initiated during these time intervals do not attract females, unless they are at least 6–10 dB higher than the leader (Snedden and Greenfield 1998). Moreover, the female's preference for the leader does not result from simple physical masking of the following call's onset by the leading call's acoustic energy. The preference holds even when the following call begins well after the end of the leading one. In the Malaysian katydid *Mecopoda* sp. (Römer et al. 1997), neurophysiological investigation suggests that a form of forward masking (see Sobel and Tank 1994), in which the receiver's contralateral ear is desensitized for a short time interval beginning with stimulation of its ipsilateral ear, is responsible for the precedence effect.

Why do precedence effects in sensory perception exist? Because they occur in both arthropods and chordates, groups whose most recent common ancestor undoubtedly had just a simple nervous system with limited sensory capability (Knoll and Carroll 1999; Chapter 1), an adaptive explanation reflecting convergent or parallel evolution is appealing. Thus, some workers have suggested that precedence effects prevent animals from inappropriately responding to echoes of signals, but the 2-s duration of precedence effects in some species argues against this particular interpretation. It is difficult to imagine that any echoes distracting to an acoustic insect would remain after a delay of that length. Alternatively, we might consider that precedence effects reflect constraints in neural design and the central mechanisms involved in signal localization. The number of ways to construct a neural device for signal recognition and localization may be quite limited (see Dumont and Robertson 1986 on the conservative nature of nervous systems), and circuitry responsible for precedence effects may be imbedded within some of the effective designs. Provided that they are not selected against strongly, the effects remain as byproducts. Nonetheless, precedence effects do not occur in all acoustic species,[57] and, where the effects do occur, they may lead to various signal interactions and collective signaling displays. If these interactions and displays are not selectively neutral, the underlying precedence effects may be subject to selection as well.

Investigations of the coneheaded katydid *Neoconocephalus spiza* (Tettigoniidae: Conocephalinae) showed how precedence effects may generate both signal interactions between advertising males and collective chorusing events (Greenfield and Roizen 1993; Box 4.1). Male *N. spiza* produce rhythmic advertisement signals that females up to 30 m distant can hear. The signals are approximately 50-ms chirps that males deliver at rates varying from 1.8 to 4.1 chirps s^{-1}. A solo male usually maintains a given chirp rate, his free-running rhythm, for several minutes, but he may increase his rate slightly if neighbors begin to call. In addition to calling at a slightly faster chirp rate, he invariably adjusts the phase of his rhythm such that his chirps coincide in time with his neighbors'. Thus, a synchronous chorus emerges (Figure 4.19), but it is an imperfect synchrony: Neighbors' calls only overlap each other but do not necessarily begin at precisely the same instant in time, and participants periodically drop out of the chorus for a cycle or two. When they re-enter, however, they are again in phase with their neighbors' rhythm.

Box 4.1 Inhibitory-resetting, Chorusing, and Precedence

Chorusing in many rhythmically signaling acoustic insects, and similar phenomena in bioluminescent species, may be modeled by assuming that free-running rhythms are maintained by a central nervous system oscillator, which periodically rises from a basal to a peak (trigger) level and then returns to the basal level. Signals are triggered when the oscillator attains the peak level, but their onsets do not occur until a brief interval *t*, an effector delay, has elapsed. When a signaler perceives a stimulus, such as a neighbor's call, that begins after delay *d* relative to the onset of his last call, the oscillator descends immediately to the basal level and remains inhibited until the stimulus ends and is no longer perceived. Following the end of the stimulus, the oscillator ascends to the peak level at a rate equivalent to or faster than that during its free-running mode. Such inhibitory-resetting lengthens the call period from T to T' when the stimulus occurs during the oscillator's ascent, but a stimulus occurring during its descent shortens the next period while leaving the concurrent period unaffected. As described, inhibitory-resetting is a phase-delay mechanism that affects one call period only (i.e., the free-running rhythm resumes immediately following the modified period).

continued

Box 4.1 (*continued*)

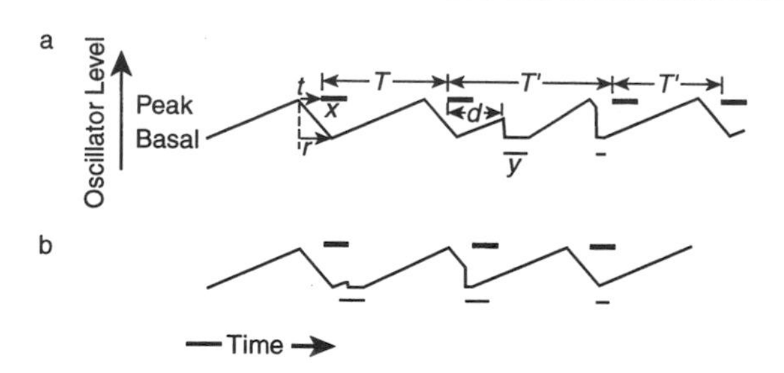

(a) Sawtooth oscillator model for inhibitory-resetting mechanism. (b) Sawtooth oscillator under the assumption that the signaler does not hear during its own call (from Greenfield et al. 1997; reprinted with permission of the Royal Society of London).

The graphical depiction of inhibitory-resetting shown above may be transformed into a linear model following determination of a phase response curve (PRC), which regresses response phase, $((T - T')/T) \times 360°$, against stimulus phase, $(d/T) \times 360°$ (Figure 4.22). Given that s = PRC slope, a measure of how fast the oscillator rebounds from inhibition, v = velocity of signal (sound) transmission over the stimulus-focal individual distance l, y and x are respective lengths of the stimulus and focal individual's call, and ε is a stochastic element in each call period (see Buck et al. 1981b),

$$T' = s \times [(d + l/v) - (r - t)] + (T + \varepsilon) + (y - x) \qquad (4.8)$$

Equation 4.8 is modified slightly if it is assumed that signalers do not perceive stimuli during their own calls (Greenfield et al. 1997).

A Monte-Carlo simulation model of mutually interacting male signalers whose calling is described by equation 4.8 shows that an imperfect synchrony (e.g., Figure 4.19) typically emerges when the PRC slope, s, is less than 0.7, whereas alternation (e.g., Figure 4.21) emerges at shallower slopes—which reflect a rapid rebound by the oscillator following inhibition. Provided that female perception is influenced by precedence or a similar effect in which more orientation is directed toward leading calls, inhibitory-resetting will be selected for in males. Inhibitory-resetters produce fewer following calls than males who signal regardless of their neighbors, and they are more attractive to females than these "regardless" callers as long as (1) both males and females selectively attend to their nearest neighbors only (Figures 4.20 and 4.21) and (2) the PRC intersects the y-axis below the origin (Figure 4.22b; descent, r, to the basal level by the free-running oscillator is relatively long). This second condition offers a "relativity adjustment" which ensures that an inhibitory-resetter produces calls that are leading as perceived by local females who may be situated several meters distant. An examination of various acoustic insects has revealed inhibitory-resetting mechanisms that meet these conditions.

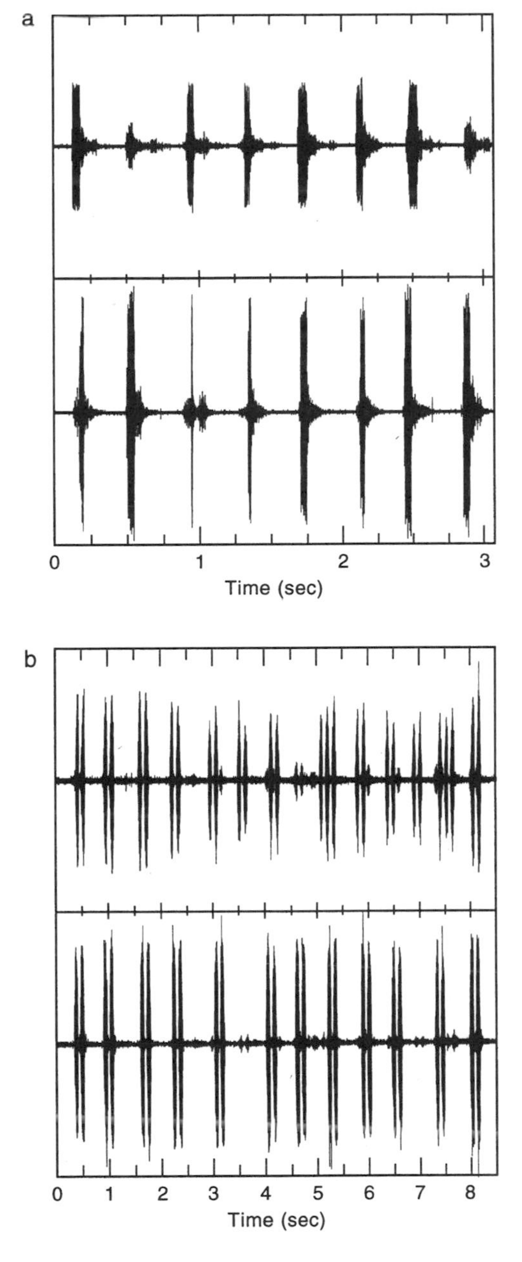

Figure 4.19 Imperfect synchrony between calls of two neighboring males (oscillograms in upper and lower traces) in (a) *Neoconocephalus spiza* (Tettigoniidae: Conocephalinae) and (b) *Sphyrometopa femorata* (Tettigoniidae: Agraeciinae) (from Greenfield 1994b; *N. spiza* recording from Gamboa, Panama, December 1992; *S. femorata* recording from Monteverde, Costa Rica, May 1991). A run of synchrony may be interrupted briefly when an individual drops out of the chorus for one call cycle and re-enters in phase. (Reprinted from the *American Zoologist*, vol. 34, p. 607, with permission of the Society of Integrative and Comparative Biologists.)

The responses of *N. spiza* males to single, isolated stimulus chirps presented to them in playback experiments demonstrated that their phase adjustments, and, ultimately, their synchronous choruses, are produced by an "inhibitory-resetting mechanism." This mechanism works as follows: upon hearing an experimental stimulus or a neighbor's chirp that begins *d* ms (the stimulus delay) after his last chirp, a male lengthens his concurrent chirp period by an increment (the response delay) slightly less than *d* ms for one cycle, unless the stimulus begins just before he is about to generate his next chirp. In the latter case, his concurrent chirp period is unaffected, but he shortens his subsequent one by a small decrement (cf. Walker 1969). These adjustments do not extend beyond the period affected, and the male returns to his previous rhythm in the next period. Thus, inhibitory-resetting is a type of "phase-delay mechanism" (Hanson et al. 1971, Hanson 1978, Buck et al. 1981a). When two or more neighboring males whose free-running chirp rhythms are comparable all adhere to the inhibitory-resetting mechanism, synchrony arises by default. But, when the males' free-running rhythms differ markedly, the faster individual generates most of the chirps. This result probably occurs because the slower male is repeatedly inhibited from calling by the faster one's chirps before his own chirps are triggered.

Laboratory phonotaxis trials indicate that female *N. spiza* strongly prefer leading calls and ignore following calls whose onsets begin within 75 ms of the leader's onset, unless they are considerably more intense than the leader (Snedden and Greenfield 1998). This receiver preference would select for signal timing mechanisms in males that reduce the production of following calls and (particularly in *N. spiza*) improve the chance of producing leading ones; these two objectives are not identical. Males who practice inhibitory-resetting, as opposed to those whose call rhythms are unaffected by their neighbors' signaling, appear to achieve both objectives. A Monte-Carlo simulation model shows that inhibitory-resetters would produce fewer following calls and more leading ones, and that they would attract more females, provided that that their timing mechanism includes a "relativity adjustment" for the velocity of signal transmission and "selective attention" (Figures 4.20 and 4.21) toward the nearest calling neighbors (Greenfield et al. 1997). Selective attention would be necessary in an effective inhibitory-resetting mechanism because signalers who adjust their rhythms in response to all audible neighbors may be inhibited so often that they rarely call. However, males who only make adjustments in response to the intense calls of nearby neighbors, their strongest competitors for local females, may produce a significant number of leading calls, relative to these neighbors, and still maintain a high chirp rate (Snedden et al. 1998). Hearing in many acoustic insects exhibits sensory adaptation to ambient noise levels (Römer 1998), and selective attention may arise as an aspect of this general neural feature. Moreover, some species, namely ensiferan Orthoptera, process sounds received by the left

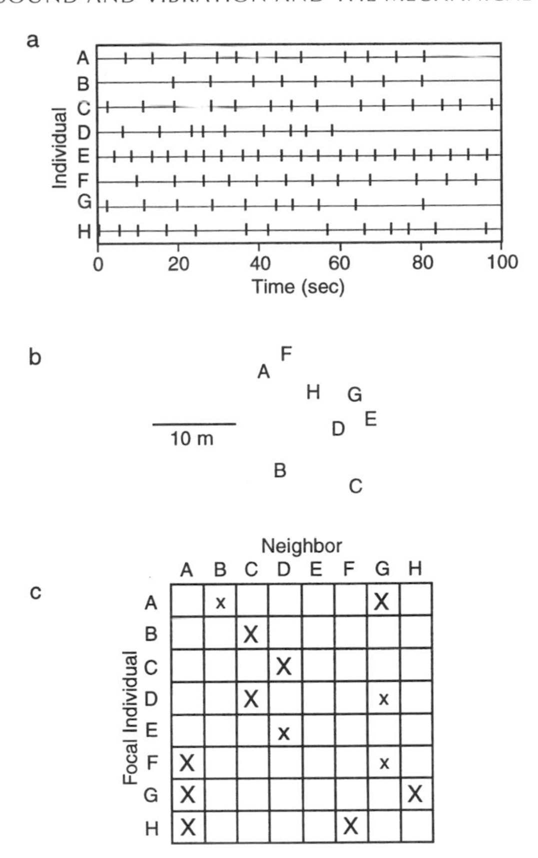

Figure 4.20 Selective attention in the tarbush grasshopper, *Ligurotettix planum* (Orthoptera: Acrididae). (a) Traces indicate the timing of advertisement calls by each of eight males, A–H (see Figure 4.19). (b) Relative positions of the eight calling males depicted in (a). (c) Each focal male avoids producing calls immediately following the calls of one or two neighbors (**X**: strong avoidance; **x**: weak avoidance) and ignores the remainder. In general, ignored neighbors are more distant than the attended ones (indicated by **X** or **x**) (recording from Portal, Arizona; August 1997).

and right ears separately and can readily attend to a specific neighbor (Pollack 1988). The inhibitory-resetting mechanism in *N. spiza* incorporates both requirements, relativity adjustment and selective attention, and it is evolutionarily stable. The simulation model predicts that in a population of inhibitory-resetters, an individual with identical signal characteristics but who forgoes this timing mechanism would be less attractive to females than the average inhibitory-resetter.

As described above, synchronous chorusing in *N. spiza* is an epiphenomenon that emerges as a byproduct of individual males who are competing to jam each other's signals. This epiphenomenon arises solely from a precedence effect and

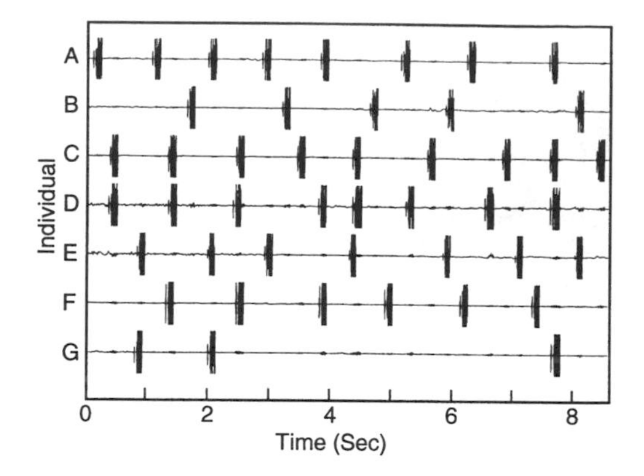

Figure 4.21 Alternation between calls among seven neighboring males, A-G, in *Ephippiger ephippiger* (monosyllabic "race"; Tettigoniidae: Bradyporinae). Each of the seven traces represents the oscillogram of calls produced by a different individual. Males typically alternate with one or two of their neighbors (e.g., A alternates with B only) and ignore the others (see Figure 4.20). Runs of alternation are typically interrupted by one to several cycles of synchrony (e.g., C and D) (data courtesy of Andy Snedden; recording from St. Jean de Buèges, Hérault, France; July 2000).

selective attention in perception and a resettable neural oscillator controlling a rhythmic signal. When these receiver and signaler features occur, an inhibitory-resetting mechanism in signalers would be evolutionarily stable and may be selected for strongly.

Playback experiments indicate that precedence effects and inhibitory-resetting mechanisms may be responsible for chorusing phenomena in a diverse set of acoustic insects (Greenfield et al. 1997). Importantly, they are as likely to generate alternating choruses (Figure 4.21) as synchronizing ones. The Monte-Carlo simulation model predicts that the specific outcome of signal interactions—a synchronous or an alternating[58] chorus—depends solely on the parameters of the signalers' inhibitory-resetting mechanism. Here, the ratio of a male's response delay to the stimulus delay he experiences is most critical: when the response delay is similar to the stimulus delay, as in *N. spiza*, synchrony emerges (Figure 4.19), but when the response delay is considerably shorter, alternation is predicted (cf. Jones 1966). These predictions are corroborated by data from various acoustic orthopterans for which both observations of chorusing and measurements of inhibitory-resetting parameters are available (Figure 4.22). Synchrony and alternation do not have distinct phylogenetic affinities, and the only apparent generalization is that species whose free-running rhythms are faster than 1 call·s^{-1} tend to have high response delay : stimulus delay ratios and generate synchronous choruses. Given the above perspective, is it possible that our interest in the format of chorusing originates largely in human perceptual sensibilities? That is, whether a chorus is synchronous, alternating, or has any other regular temporal structure may not matter to the signalers and receivers who generate and listen to the event!

Whereas chorus structure may be inconsequential in many acoustic species, sexual selection could favor the underlying precedence effects owing to benefits afforded female receivers. On the one hand, precedence effects differ fundamentally from all female preferences discussed previously in that intrinsic physical characteristics do not necessarily distinguish the signals preferred under this influence. Rather, the preferred signals merely occur first in a sequence, and it is not immediately clear how attractive males would differ from avoided ones. However, in the competitive milieu wherein female choice normally occurs, a precedence effect could enhance the ability of females to assess male signal characters indicative of energy expenditure. For example, females may prefer signals delivered at faster call rates (cf. Backwell et al. 1999 for an example in optical signaling), but among a dense aggregation of male signalers, a female's assessment might be impaired. If a precedence effect exists in receivers and male signalers have evolved the practice of inhibitory-resetting, though, females may then assess the relative call rates of local males more easily. The faster males would do most of the calling, while the slower ones would seldom call owing to a cruel bind: forgoing inhibitory-resetting would lead to the production of ineffective following calls, while practicing it would lead to repeated inhibition. Thus, the precedence effect would ultimately magnify small differences in an energy-based signal character assessed by females. This possibility illustrates how a receiver trait that is at first selectively neutral in a particular context could become favored via a feedback loop that ensures its maintenance.

Chorusing: Synchrony and Alternation as Cooperative Phenomena. Alternative views of chorusing in acoustic insects maintain that collective synchrony and alternation are cooperative phenomena that benefit the participating signalers. What adaptive explanations have been proffered, and can they be reconciled with observations and experimental findings? First, it has been proposed that synchronous chorusing preserves species-specific call rhythms that females must recognize before they localize any one signaler. Under this selection pressure, any male who calls out of phase with his neighbors would be committing a spiteful act, which is generally not expected. Second, if male signalers cluster in aggregations due to social factors or because preferred habitat is patchily distributed, individual signalers might cooperate to maximize the number of females attracted to their group as a whole. One way for them to boost group attractiveness would be through synchronizing their calls so that peak sound pressure levels are as high as possible. Provided that receivers assess call energy by integrating sound over a fairly short time interval, more females would orient toward groups that synchronize than toward those that do not. Thus, synchrony

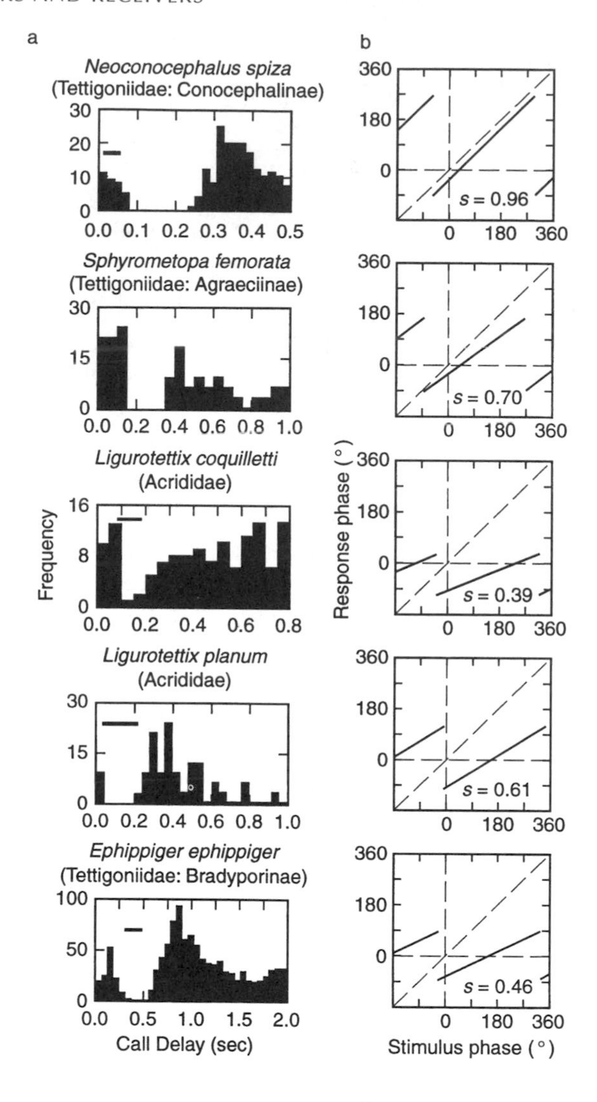
a
Neoconocephalus spiza
(Tettigoniidae: Conocephalinae)
Sphyrometopa femorata
(Tettigoniidae: Agraeciinae)
Ligurotettix coquilletti
(Acrididae)
Ligurotettix planum
(Acrididae)
Ephippiger ephippiger
(Tettigoniidae: Bradyporinae)
Frequency
Call Delay (sec)
b
s = 0.96
s = 0.70
s = 0.39
s = 0.61
s = 0.46
Response phase (°)
Stimulus phase (°)

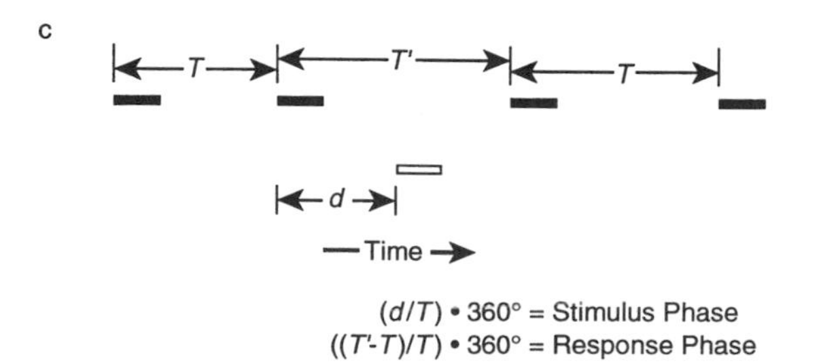
c
T
T'
d
Time
(d/T) • 360° = Stimulus Phase
((T'-T)/T) • 360° = Response Phase

may be selected for by inter-group competition in situations where such competition is a potent force. However, once females have arrived at a group, signal competition between individual members of the group should take priority and may override the cooperatively driven synchrony. Third, synchrony might render localization of any one signaler by phonotactic natural enemies difficult. But, we might consider that localization should not be made so difficult that female receivers too cannot easily orient toward the male signalers. Fourth, in duetting species, males may find it easier to perceive a female's replies if they synchronize the silent gaps between their advertisement calls with the gaps in their neighbors' calls.

While apparently sound, there exist few observations among acoustic insects that may be construed as supportive of the above proposals. In many species, call rhythms at a given temperature vary substantially among individuals, making the rhythm preservation hypothesis unlikely. And even where rhythms are relatively invariant, selective attention mechanisms may allow a female receiver to focus on and recognize one or two male signalers while signalers in other directions call out of phase. Specific experiments have not tested the potential influences of inter-group competition and natural enemies on synchronous chorusing,[59] but these impacts, should they occur, may not be strong enough to counteract signal interactions driven by competition between individuals.

Figure 4.22 Inhibitory-resetting in five species of acoustic insects (from Greenfield 1997a, Greenfield et al. 1997). (a) Call-delay histograms showing the frequency distribution of call onsets of a focal male following playback of a single, randomly timed stimulus call whose onset is at time 0. In each species there exists an interval beginning shortly after the onset of the stimulus call during which the focal male calls little or not at all. Horizontal bar indicates the typical duration of the precedence effect in the species: Females hearing two identical, spatially separated calls that respectively begin at time 0 and at any later time prior to the end of the bar are likely to orient toward the first call. (Reprinted from *The Bionomics of Grasshoppers, Katydids, and their Kin*, p. 213, with permission of CAB International; reprinted from *Perspectives in Ethology*, vol. 12, with permission of Kluwer Press.) (b) Phase-response curves showing the relationship between response phase of the focal male's calls and stimulus phase of a single, randomly timed stimulus call. In synchronizing species (*Neoconocephalus spiza, Sphyrometopa femorata*), response phase is relatively large for a given stimulus phase, and PRC slope (s) is ≥ 0.7; in alternating species (*Ephippiger ephippiger, Ligurotettix coquilletti, L. planum*), response phase is relatively small for a given stimulus phase, and PRC slope is <0.7 (Greenfield et al. 1997). (c) Response phase is the proportional increase or decrease ($\times 360°$) in the male's call period (T) concurrent with playback of the stimulus call (open horizontal bar). Stimulus phase represents the timing of the stimulus call during the focal male's concurrent call period, T'. (Reprinted with permission from the *Annual Review of Ecology and Systematics*, vol. 34 © 1994 by Annual Reviews www.AnnualReviews.org.)

Nonetheless, synchronous events based on cooperation undoubtedly occur among natural phenomena. The beating of the heart, which is a composite of many separate rhythmic elements, and the timed applause of concert audiences in certain countries[60] are two examples taken from human experiences. The synchronous events in arthropod signaling that are most likely to represent cooperative events are those in which synchrony does not arise by default, as via phase advance mechanisms, but is achieved by mutual rhythmic adjustment. This type of adjustment, which has been modeled by the physical construct known as "coupled oscillators" (Strogatz and Stewart 1993), accurately describes the synchrony found in certain fireflies (see Section 5.2.1), but its occurrence among acoustic insects is uncertain. The most likely candidates are species that can maintain nearly perfect synchrony over an extended time interval (Figure 4.23), but the mechanisms and adaptive value of these choruses have not been investigated.[61]

Non-competitive explanations for signal alternation mirror those proposed for synchrony: (1) preserving call features, as opposed to rhythms, that female receivers must hear and recognize before they localize any one male signaler and (2) maximizing the acoustic energy, integrated over a relatively long interval, emanating from a cluster of male signalers. Again, specific experiments testing these hypotheses have not been conducted, and most alternating choruses among acoustic insects appear to arise by default, when the participants happen to maintain comparable call rates.

Vibratory Chorusing? Because vibratory signals are transmitted rapidly and often retain specific rhythms, we might expect to find analogues of chorusing among species in which males produce advertisement vibrations. A review of the

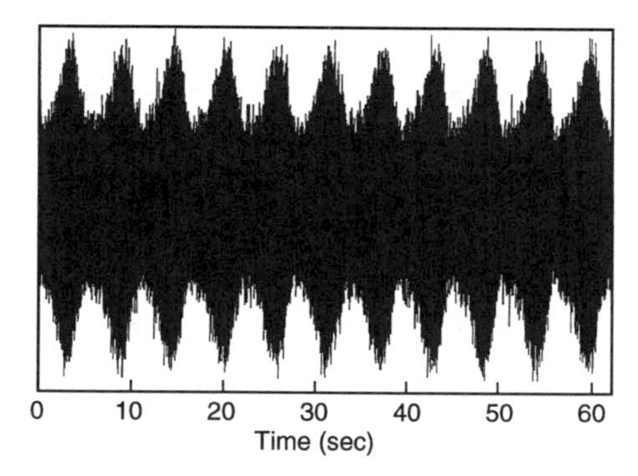

Figure 4.23 Synchrony between calls of an undetermined number of chorusing males in the periodical cicada *Magicicada cassini*. The overall pattern of sound pressure level in the chorus as revealed by the oscillogram indicates that call periods of individual males are approximately 5 s in length and that many individuals maintain nearly perfect synchrony with their neighbors (recording from Jefferson County, Kansas; June 1998).

literature does not reveal such events, however, and we may ask whether fundamental differences between acoustic and vibratory signaling are responsible. One major difference between these two forms of mechanical signaling is that substrate vibrations are often restricted to spreading in one dimension only. For example, when male signalers transmit vibrations along plant stems, a receiver may only detect the signals produced by males located on the same stem. Consequently, there may exist few opportunities or situations in which collective interactions among multiple vibrators might arise.[62]

4.4.3 Social Behavior: Group Formation and Organization

Mechanical signaling is not the most common means by which arthropods maintain group cohesion or social structure and hierarchy. But, from simple mutual attraction establishing local aggregations to specialized messages between and within castes in advanced eusocial species, mechanical signals do play various roles in social behavior. Generally, tactile, vibratory, or near-field sound signals rather than far-field sounds are used for social functions. These signals have more limited ranges than far-field sounds do, which would make them more appropriate for most social contexts, wherein receivers are close. The limited ranges of tactile, vibratory, and near-field sound signals would also reduce attraction of natural enemies. Unlike sexual advertisement, where the high stakes may lead males to disregard dangers inherent in signaling, transmitters of social signals may not be expected to take unnecessary risks.

Subsocial Aggregation: Group Cohesion and Parent-Offspring Contact. During October 1985, I witnessed the unfolding of a remarkable series of signal interactions among subsocial tortoise beetle larvae (*Stolas* sp.; Chrysomelidae: Cassidinae) at a secondary forest site in central Panama (Figure 4.24). The larvae, which represented several late instars, were feeding on the foliage of *Ipomoea* (Convolvulaceae) and formed loosely knit aggregations that extended among adjacent leaves. On several successive occasions, from one to several individuals began to vibrate visibly, after which the behavior quickly spread to surrounding larvae. Larvae as far as several leaves distant responded by moving toward the initial vibrators. In some cases, responding larvae would walk onto the stem from their leaf, start moving in the direction opposite the vibrators, and then reverse course. At the location of the initial vibrators, the larvae arranged themselves in a circular formation (cycloalexy), with their heads at the center and their abdomens directed radially outward. Normally, the formation was completed within one minute of the initial vibrations. These larvae are "trash-bearers," retaining old exuviae from previous molts on their abdomens, and their postures within the concentric formation probably protect them from

a

b

Figure 4.24 Defensive behavior facilitated by vibratory communication in cassidine beetle larvae (*Stolas* sp.). (a) Larvae form a cycloalexic aggregation in response to each other's vibratory signals, which are elicited when one or more vigilant individuals detect a natural enemy. (b) Predatory pentatomid bug feeding on a lone *Stolas* larva who failed to join an aggregation (photographs taken in Gamboa, Panama; October 1985).

attacks by some arthropod enemies. I observed that vibration and the aggregation response could be elicited by my gentle disturbance of the leaves or stem. Following one naturally occurring aggregation, a solitary larva, who either failed to respond to the vibrations or to localize the circular formation, was attacked and killed by a predatory pentatomid bug. Presumably, the alarm signal in these beetles is transmitted as a bending wave through the leaves and stem, and joining the aggregation offers a larva "safety in numbers" (Hamilton 1971) and physical protection of the more vulnerable parts of its body. Thus, the signaling would be cooperative, as it appears to benefit both the signaler and receiver. While the larvae may be siblings, their genetic relationship would not be critical for this group alarm to operate.

The above theme of vibratory alarm signaling arises in various subsocial arthropods found on surfaces conducive to the transmission of bending waves and similar disturbances. A major variation on this theme occurs when parents become involved and signals are exchanged between the parent and her offspring as well as among the offspring (Cocroft 1996). In the treehopper *Umbonia crassiconis* (Homoptera: Membracidae), a mother remains in the vicinity of her nymphs while they feed on stems of the host plant (Cocroft 1999a). Visual or mechanical disturbances elicit vibratory signals from the nymphs and mother, both of whom are fitted with specialized tymbals; adults otherwise use these tymbals to generate their mating signals (Section 4.2.4). Nymphs also signal when they perceive the signals of other nymphs, which spreads the alarm response rapidly throughout the group. Such nymph–nymph communication would reduce the cost of vigilance, because a given individual would not be required to remain continuously alert for the approach of natural enemies. The nymphal signals are synchronized, and this feature evokes a stronger response from their attending mother than randomly timed signals would (Cocroft 1999b). Synchronized nymphal signals draw the mother closer if the nymphs are the first to recognize potential danger, but, more importantly, they serve to retain the mother's close presence for a longer interval. Similarly, the mother's signals alert the nymphs if she is the first to recognize danger, and they may also enhance cohesion of the nymphal aggregation. It is assumed that the soft-bodied nymphs, who are rather vulnerable to attack by parasitoid wasps and predatory hemipterans, rely on their strongly sclerotized mother as a shield and general deterrent of natural enemies, which they can do most readily if they form a cohesive group.

Eusocial Colony Organization: Collective Defense, Alarm, and Recruitment. Many eusocial Hymenoptera and Isoptera also rely on mechanical signals to alert colony members of impending danger and to generate a collective defense response (e.g., Kirchner et al. 1994). Additionally, individual members of eusocial colonies may use mechanical signals to elicit help from nestmates when they are in physical danger, to request a food donation, or to attract local assistance in exploiting a rich food supply. These messages are either tactile or transmitted via substrate vibration, and some are accompanied by olfactory signals or cues (see Kirchner 1997 for review).

Eusocial Colony Organization: Dance Languages as Acoustic Recruitment Signaling. The most prominent role for acoustic signals among eusocial arthropods is in the recruitment of foragers to distant food sources. Acoustic recruitment signaling, also known as the dance language, occurs in certain aerial species, where pheromone trails (Section 3.7) would not be feasible. In the honeybee *Apis mellifera*, the dance language operates by means of a returning scout vibrating her wings while performing a series of abdominal waggling maneuvers on the nest surface. These activities convey information to potential recruits concerning the value, distance, and direction of the floral food resource she had just visited. A scout adjusts the vigor, duration, and orientation of her wing-vibrating and waggling maneuvers in accordance with the profitability of that food source, the visually perceived distance she has flown on her return trip (Srinivasan et al. 2000, Esch et al. 2001), and the horizontal angle, viewed from the perspective of the nest, between the sun and the food source.[63] These correlations between a scout's wing-vibrating maneuvers and the food source were revealed through a series of painstaking observations and field tests made over a 50-year period by Karl von Frisch and his colleagues at the University of Munich (von Frisch 1967). Subsequently, several elegant laboratory experiments confirmed the field observations by demonstrating that a scout's wing-vibrating maneuvers are at least partly responsible for directing a recruit to the food source (Gould et al. 1985; but see Wenner 1990, who challenges the communicative role of these maneuvers). Nonetheless, the actual channel along which the information was transmitted from the wing-vibrating, maneuvering scout to the attending recruit remained a mystery. Because *A. mellifera* perform the dance within the darkened confines of the cavity containing their nest, the visual channel could be ruled out, and the apparent absence of tympanal organs in bees convinced researchers that airborne sound too was unlikely.

A major breakthrough in understanding recruitment communication in *A. mellifera* occurred in the late 1980s when several workers reconsidered the possibility of transmission along the acoustic channel (Towne 1985, Michelsen et al. 1986, 1987, Towne and Kirchner 1989). This time, however, they focused on near-field sound. Acoustic particle velocity recordings of dancing scouts revealed that their wing vibrations produce 250–300 Hz sound with characteristic pulse patterns. Although rapidly attentuated, these low-frequency pulses are extremely intense within several millimeters of the bee's wings (Section 4.2.2). Moreover, the Johnston's organ in the bee antenna is sufficiently sensitive to detect these sounds within 1 cm of the dancer (Dreller and Kirchner 1993a,b, Kirchner 1994), and potential recruits do position themselves such that their antennae would perceive them. But, the unequivocal support for the role of near-field sound came from experiments with a "robot bee" whose waggling maneuvers and associated sounds could recruit an attending bee to a predictable direction and distance (Michelsen et al. 1989). Because the waggling maneuvers of the scout are an integral part of the dance code (Figure 4.25), these findings also suggest that a fourth basic character, movement, may define a mechanical signal. That is, in addition to evaluating intensity, carrier frequency,[64] and tempo of the near-field sounds transmitted by wing vibration, honeybee receivers may also monitor the location and movement of the signal source (cf. Chapter 5). Possibly, a receiver

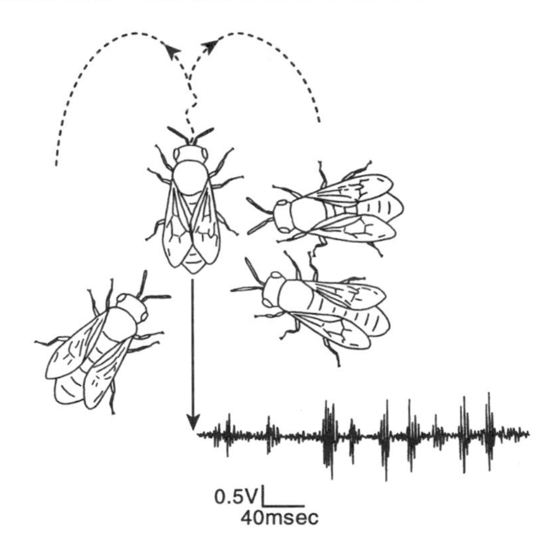

Figure 4.25 Acoustic transmission of the honeybee (*Apis mellifera*) dance language. Three potential recruits attend a dancer during the straight run of her waggle dance. Oscillogram depicts near-field sounds recorded 2 mm dorsal to the dancer's wings. (Reprinted from "The acoustic near field of a dancing honeybee," *Journal of Comparative Physiology A*, vol. 161, p. 633, figs 2a and 5, A. Michelsen, W.F. Towne, W.H. Kirchner, and P. Kryger, copyright 1987 with permission of Springer-Verlag.) The dance is normally performed on the vertical comb surface within the nest, and the angle between the straight run and a vertical line codes for the horizontal angle between the sun and the direction to the food source.

monitors the length and orientation of the scout's waggling maneuvers by evaluating relative changes in sound intensity perceived at her two antennae, but more likely she evaluates the movement via tactile and vibratory information (see Rohrseitz and Tautz 1999, Nieh and Tautz 2000).

Like the sophisticated olfactory communication that eusocial insects use for nestmate recognition, defense, recruitment, etc. (Chapter 3), the *Apis mellifera* dance language is a remarkable evolutionary achievement. But, as with these olfactory signaling systems, the evolutionary origin of the dance language is somewhat obscured. The dance language appeared to represent a pinnacle of complexity in signal coding, standing alone among the far simpler systems in most bees and other eusocial insects. In the absence of intermediate and alternative levels of development in recruitment signaling among various species, any suggested evolutionary pathways could only be highly speculative. During the early 1990s, though, biologists began to probe these pathways by investigating two related groups of eusocial bees reputed to exhibit recruitment signaling similar to *A. mellifera*, the various Asian species of *Apis* and the more distantly related stingless bees (Apidae: Meliponinae) (see Lindauer 1961, Esch et al. 1965). Because the acoustic mechanism with which *A. mellifera* transmit and perceive dance language signals was known by this time, the parallel investigations of Asian honeybees and stingless bees could be more focused than they had been in earlier studies.

Investigations of four Asian species of *Apis* revealed that two of them, *A. cerana* (Indian honeybee) and *A. dorsata* (giant honeybee), use acoustic recruitment signals similar to those of *A. mellifera*. The two other species investigated, *A. florea* (dwarf honeybee) and *A. laboriosa* (Himalayan honeybee), recruit foragers via dance maneuvers that are silent (Kirchner and Dreller 1993, Dreller and Kirchner 1994, Kirchner et al. 1996). Because *A. cerana* and *A. dorsata* are cavity nesters and partially nocturnal, respectively, whereas *A. florea* and *A. laboriosa* are diurnal and nest in open locations, it is inferred that the need to communicate in darkness selected for use of a non-visual channel. Specifically, near-field sound signals may have originated by exploiting or coevolving with ancestral receiver biases (cf. Section 6.1) in *Apis*, as *A. florea*, a silent dancer, possesses the ability to perceive low-frequency sounds with its antennae. Near-field sound may also offer some adaptive advantages, because its rapid attenuation would allow multiple scouts to signal simultaneously on the nest without interfering with each other. Were they to use substrate vibrations transmitted through the nest comb, the signals of a given scout might be perceived by potential recruits attending neighboring scouts transmitting information regarding other food sources. Such interference, however, apparently does not affect the messages that recruits transmit back to dancing scouts: after perceiving a scout's dance for a sufficient interval, a recruit sends a vibratory "stop signal" through the nest comb. The scout then normally stops dancing and offers samples from the food source. Presumably, the sample's odor helps the recruit to pinpoint the source once she arrives in its general vicinity.

A comparison of recruitment signaling among *Apis* species suggests the following evolutionary sequence: Ancestral species[65] evolved dancing maneuvers, performed on the horizontal comb surfaces of open nests, that represented the distance and horizontal direction toward a food source. The critical information in these maneuvers was most probably transmitted and received visually. As species evolved either crepuscular/nocturnal activity or nesting within dark cavities, visual dance signals no longer sufficed, and acoustic accompaniment arose. Additionally, where comb surfaces in cavity nesters became vertical, direction could no longer be represented by dance movements oriented toward the food source. Thus, the more abstract representation of direction by the dance's vertical angle evolved (Figure 4.25).[66]

Even so, the visually transmitted, horizontal dances of ancestral *Apis* species are also a sophisticated means of recruitment signaling, and we might ask how the basic dance convention originated. One possibility is that bees returning from a trip to a distant food source unintentionally performed maneuvers on the nest surface commensurate with the length of their trip and its direction. Waggling movements may have been performed to dissipate heat accumulated in the body, and more intense or lengthier waggling would have been necessary following a longer trip. And, bees preparing to return to the food source were it valuable may have aligned their waggling movements with the direction in which they would take off again. Did nestmates ever follow returning bees back to the food source by simple visual tracking? If so, their success in tracking and arriving at the distant source would have been markedly improved by associating characteristics of the unintentional waggling with the source location. As with the various chemical signals of eusocial insects that may have originated as inadvertent cues or autocommunicative information (Chapter 3), more definitive waggling maneuvers that accurately represented the trip's distance and direction would have benefited both the dancer and recruits. Such selection pressures may have led to the specialized representational dances seen among *Apis* honeybees.

Investigations of the neotropical stingless bee *Melipona panamica* showed that dance language has evolved either convergently or in parallel in another eusocial group that forages on distant floral resources (Nieh and Roubik 1995, 1998, Nieh 1998a, 1999; cf. Esch et al. 1965). Like *Apis* honeybees, returning foragers in *M. panamica* perform a representational dance at the nest. *M. panamica* nests are built within cavities, and their dances occur in darkness and include pulsed acoustic signals. Both distance to the food source and its elevation above ground are encoded in temporal features of these pulsed sounds. Horizontal direction to the source, however, appears to be indicated by other activities and movements that the scout bee performs outside the nest as well as odor cues (Nieh 1998b). Do *M. panamica* dancers indicate source elevation, which *Apis* bees are not known to do, because this parameter is critical in the highly stratified Central American rainforests that they inhabit? Does their directional communication outside of the nest represent a less abstract—and more primitive—coding of a food source's azimuth than that exhibited by *Apis* bees? *Melipona* and other stingless bees have been more difficult to observe and study than *Apis* bees, and few comparative data from other species, which might help to resolve these questions, are available.[67]

4.5 Origins and Limitations

The perception of mechanical stimuli may be as pervasive among arthropods as olfaction. Most, if not all, arthropods possess some ability to detect substrate vibrations, local air or water currents, and other external inertial forces directly impinging upon the cuticle. It is also assumed that they normally receive and process various types of proprioceptive information concerning posture of the body and appendages. Nonetheless, reliance on the mechanical channel for long-range sexual communication, either advertisement, courtship, or competitive (intrasexual), is relatively uncommon (Table 4.1; Figures 4.26 and 4.27, 4.28). Why should this be so?

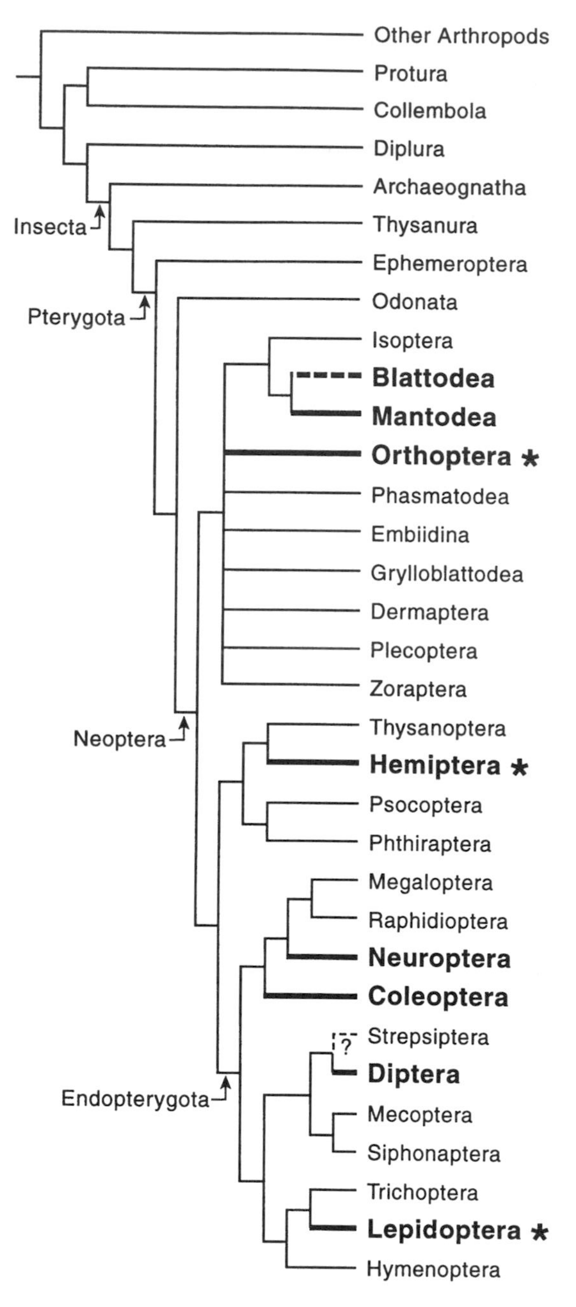

Figure 4.26 Phylogeny of arthropods, focusing on Hexapoda (cf. Figure 1.2 and Table 1.1) and showing known occurrence of far-field hearing (thickened lines and bold type) in seven (or eight: Homoptera and Hemiptera are grouped together under Hemiptera here; far-field hearing in Blattodea is uncertain and indicated by dashed, thickened line) insect orders and acoustic communication (asterisk) in only three (phylogeny from Gullan and Cranston 2000).

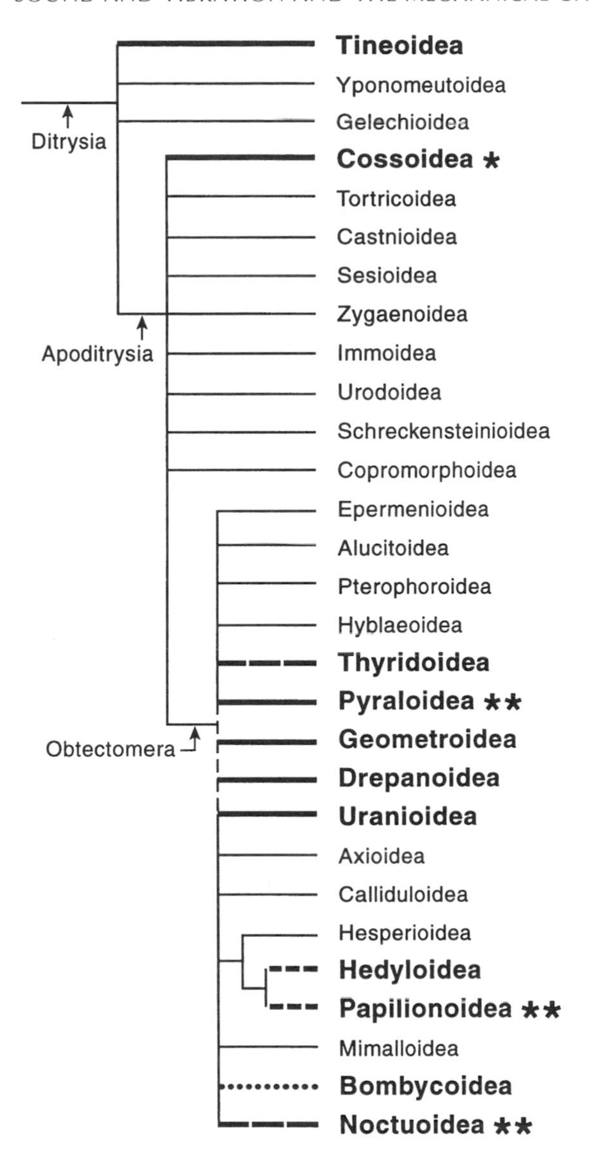

Figure 4.27 Phylogeny of the Lepidoptera, Ditrysia (cf. Figure 3.12), showing known occurrence of far-field hearing (thickened lines and bold type) in 11 superfamilies (Scoble 1992) and of acoustic communication (double asterisk: confirmed case; single asterisk: possible case) in only three (phylogeny from Nielsen 1989). Of the 11 superfamilies with reported far-field hearing, only Papilionoidea are diurnal. Papilionoid ears are not tuned to ultrasound, and male acoustic advertisements found in this group are low-frequency (≈2.4 kHz) sounds (Monge-Najera et al. 1998). Solid thickened lines: abdominal tympanal organs; long dashes: thoracic organs; short dashes: organs located at the wing bases; dots: organs located in the mouthparts.

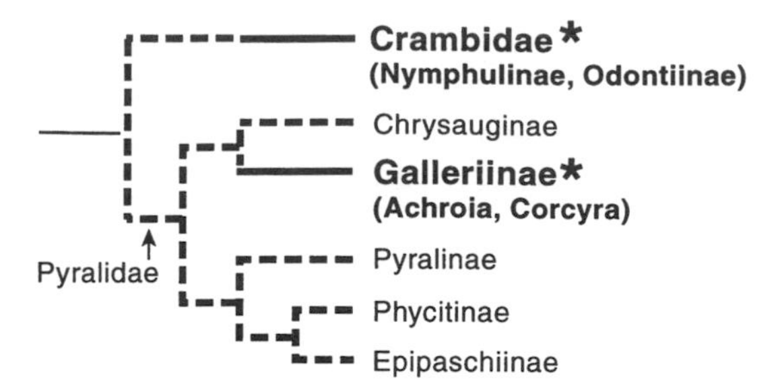

Figure 4.28 Phylogeny of the Pyraloidea (cf. Figure 4.27), showing distribution of far-field hearing (thickened lines) throughout the group (Minet 1983, 1985) and the known or suspected occurrence of acoustic communication (solid lines, bold type, asterisk) in only two genera among two subfamilies of the Crambidae and two genera of the pyralid subfamily Galleriinae (phylogeny from Solis and Mitter 1992).

4.5.1 Substrate and Body Size Restrictions?

Vibratory sexual communication may be restricted to relatively few groups because local substrates and body size often conspire to constrain signal production. Most arthropods would have to remain on substrates conducive to the transmission of surface or bending waves in order to generate vibrations that could represent effective signals over any but the shortest distances. Consequently, only species found on the water surface and certain types of vegetation and similar materials, or that construct silk webs or wax or paper nests, may readily generate effective vibrations that conspecific receivers might detect. Those arthropods that send vibratory signals through other types of substrates are typically rather large, robust species, such as shore-dwelling crabs which can transmit Rayleigh and Love wave signals in damp sand by rapping it with an enlarged leg (Salmon 1971, Crane 1975, Salmon and Hyatt 1979, Aicher and Tautz 1990; cf. Field 2001 for similar behavior in ground- and tree-dwelling stenopelmatoids). Because mechanoreception is so common whereas mechanical signaling is quite rare, we might infer that the evolution of many cases of vibratory sexual communication among arthropods involved exploitation of a pre-existing perceptual ability found in both sexes and originally used in a non-sexual context (Section 6.1). That is, male signals that were simply most easily perceived and localized by female receivers evolved, and they did so without any corresponding evolutionary change in the receiver at the time they originated (e.g., Proctor 1991, 1992).

4.5.2 On the Rarity of Acoustic Perception and Communication

Acoustic sexual communication among arthropods is even more restricted than vibration is, which may come as a surprise given the advantage of rapid communication over long distances—and in all directions—that sound affords. Additionally, effective acoustic communication would not be limited by a rarity of required substrates or other stringent environmental conditions. However, certain anatomical factors involved in receiver function may have limited the evolution of acoustic sexual communication to a relatively small number of groups. Among arthropods, far-field acoustic sexual communication is essentially an insect phenomenon (Figure 4.26). The production of audible sound by various spiders and crustaceans is somewhat problematic, because sensitive pressure receptors insulated from substrate vibration (Section 4.3.4) may not be present in many of these species (but see Popper et al. 2001). Perhaps their sounds act in the near field only,[67] are just byproducts of vibrational signals transmitted through the substrate, or, in the case of some spiders, are transmitted through the air and cause a receiver's substrate to vibrate. Alternatively, some sounds may be intended for heterospecific receivers or recipients.[68]

In the far field, most insect hearing relies on tympanal organs that are either part of or coopted from the tracheal system. The expanded tracheal air sacs formed of thin cuticle that are found in insects appear to be ideal for use as pressure or pressure-difference receivers, and these systems and structures may represent critical preadaptations for hearing.[69] Were well-developed tracheal systems with thin cuticular linings then a prerequisite for arthropod hearing, and does their absence or different construction and composition in the chelicerates, myriapods, and crustaceans explain why hearing and acoustic communication are largely restricted to insects? Possibly, but far-field hearing is far from ubiquitous among insects regardless of the development of their tracheal system. With the exception of grasshoppers and cicadas, it mostly occurs among nocturnal species, where an inability to rely on visual perception to detect critical environmental stimuli, and the specialized predation pressure from insectivorous bats, may have been strong agents of selection. Does this restricted distribution of hearing imply that most insects could not afford to divert a major part of their tracheal system to a non-respiratory function unless it was absolutely essential for them to do so? Among insects, the most active aerial groups (e.g., Odonata, Diptera, and Hymenoptera) include very few species with tympanal hearing (Figure 4.26). On the one hand, their general lack of tympanal organs may reflect the fact that highly active insects, sustaining high metabolic rates, were least able to relinquish respiratory function for acoustic perception.[70] On the other hand, these active species are also largely diurnal and rely heavily on visual perception. Thus, selection pressure influencing them to evolve tympanal hearing may have been relatively weak, particularly if hearing would have necessitated a substantial diversion of brain function away from visual processing (Chapter 5).

Morphological and neural analyses suggest that insect tympanal organs evolved as modifications of chordotonal organs monitoring vibration (Lakes-Harlan et al. 1991, 1999) or offering proprioceptive information on the movement or position of the body and its appendages. For example, the tympana in eared moths may be homologous to organs that monitor wing activity in related atympanate species (Yack and Fullard 1990). And, in various acoustic insects the scolopidia and interneurons associated with tympana appear serially homologous to features of chordotonal organs in adjacent body segments (Meier and Reichert 1990, Boyan 1993). Proprioception and detection of substrate vibration are very common, though, and this origin alone cannot account for the restricted distribution of acoustic perception among insects. However, physiological and anatomical constraints in conjunction with predation and other selection pressures may explain why specialized tympanal ears evolved from generalized chordotonal precursors in certain groups but not in most.

4.5.3 Acoustic Signaling as Exploitation of Receiver Bias?

Whatever the reason, tympanal organs and acoustic perception for far-field sound occur in relatively few insect species (Figure 4.26).[71] Rarity of hearing function would have limited the opportunities for acoustic sexual communication to evolve via exploitation of pre-existing perceptual ability (Section 6.1), and a survey of acoustic insects suggests that this origin might have been a common one: in most insect groups that include tympanate species, hearing is much more widespread than sound production and normally occurs in both sexes when present.[72] While the absence of sound production can be ascribed to a secondary loss in isolated cases of mute species in those taxa that otherwise call,[73] such loss is unlikely to account for the general muteness observed in the majority of tympanate groups. Moreover, when acoustic advertisement is present, it is normally restricted to males. Where both sexes do produce sounds, the transmissions often function as disturbance or startle messages intended for vertebrate predators (e.g., Møhl and Miller 1976, Masters 1979, 1980); in many cases, the species producing these sounds lack hearing. One inference that may be drawn from these general patterns is that hearing evolved in various insect groups as a means of detecting the sounds, inadvertent or intentional, made by natural enemies. Male advertisement and courtship signaling later exploited basic acoustic perception in some of these groups, although coevolutionary mechanisms of sexual selection probably influenced subsequent modifications of female perception, and male signaling, in most cases (see Section 6.1 and Endler and Basolo 1998).

Acoustic communication in acridoid grasshoppers and moths is consistent with the evolutionary scenario presented above. Among acridoid grasshoppers, tympanal hearing organs and thoracic auditory interneurons of similar design are widespread (Riede et al. 1990), whereas acoustic sexual signaling only occurs in a few, unrelated groups and involves non-homologous stridulatory and crepitating structures (Otte 1970, Riede 1987). Presumably, hearing evolved once and was a response to predation.

Ears are also widespread among nocturnally active Lepidoptera, where they occur in no fewer than 10 superfamilies (see Scoble 1992; Figure 4.27). Furthermore, hearing organs are very common in four of these superfamilies: Pyraloidea, Hedyloidea, Geometroidea, and Noctuoidea. Lepidopteran ears, however, are neither uniform nor homologous. Tympanal organs are located on various body segments and appendages and probably evolved independently six or more times (Yack and Fullard 2000). Because lepidopteran ears are tuned to ultrasonic frequencies, it is assumed that they evolved specifically in response to predation pressure imposed by echo-locating, insectivorous bats. This assumption is supported by the general occurrence of specialized evasive behaviors elicited by bat echo-locations (Rydell et al. 1997) and by the reduction or absence of hearing in closely related species from regions devoid of bats (Fullard 1994; but see Rydell et al. 2000) or active at times of the day or season when bats are not (Fullard et al. 1997, Surlykke et al. 1998).[74] Male acoustic (ultrasonic) signaling occurs in several of the tympanate lepidopteran superfamilies, Pyraloidea (Figure 4.28) and Noctuoidea (Spangler 1988a, Conner 1999), but it is far less common than hearing. Again, we may infer that acoustic advertisement and courtship evolved via exploiting ultrasonic hearing in these unusual moth species.

Elsewhere among insects, ultrasound-sensitive tympanal ears are found in the Mantodea, two families of Coleoptera (Cicindelidae, Scarabaeidae), and Neuroptera (Chrysopidae) (Figure 4.26). Acoustic sexual signaling is unknown in these groups, though, which rely on pheromones, visual information, and substrate vibration for pair-formation.

Overall, predation pressure, especially from echo-locating bats, probably led to the evolution of tympanal ears on multiple occasions (Hoy 1992, Hoy and Robert 1996), but acoustic signaling is comparatively rare. Exploitation of receiver bias (Ryan et al. 1990, Shaw 1995, Endler and Basolo 1998) offers a plausible explanation for this uneven distribution of acoustic perception and signaling, and the process may have contributed to the origin of acoustic sexual advertisement and courtship in various insect groups. For example, were hearing already present and males performing a stereotypical leg or wing movement for visual display or evaporation of pheromones, sexual selection may have favored stridulatory, tymbal, or percussional devices by which the movements yielded sound (see Bailey 1991). But the receiver bias model on its own cannot readily explain how perception used in the context of detecting natural enemies, which would normally elicit a negative motor response, might be exploited by a male signaler for attracting females or influencing them to respond positively in other ways. Thus, male signal evolution appears to demand a leap of faith, unless the actions of other mechanisms are considered. We return to this perplexing issue in Chapter 6, Sexual Selection and the Evolution of Signals.

The acoustic insect groups in which communication is least consistent with exploitation of receiver bias are the ensiferan Orthoptera (crickets, katydids, and their relatives) and the cicadas. Both hearing and calling are nearly ubiquitous among three of the four ensiferan superfamilies, Tettigonioidea, Hagloidea, and Grylloidea, and hearing occurs in a major section of the fourth, Stenopelmatoidea. Throughout the ensiferan lineage, where hearing is present it is common to both sexes and accomplished by tympana, and associated crista acusticae, located in the tibiae of the forelegs. In the Tettigonioidea, Hagloidea, and Grylloidea, males call with a file and scraper stridulatory device at the bases of the forewings, and carrier frequencies of calls usually match best frequencies in hearing (e.g., Hill and Boyan 1977, Hill and Oldfield 1981, Lin et al. 1993, Dobler et al. 1994b, Meyer and Elsner 1996). Moreover, fossil evidence indicates equivalent antiquity, approximately 250 Mya (Sharov 1971), for origins of foreleg tympanal ears and forewing stridulatory devices in the Grylloidea. Therefore, the sensitivity and negative responses to ultrasound known in various crickets, and katydids, probably evolved as a sensory modification at a much later date,[75] subsequent to the appearance of echolocating bats during the Paleocene (60 million years ago; Novacek 1985). Whereas the phylogeny of the Ensifera is presently unclear (Flook et al. 1999), and both monophyletic (Alexander 1962) and polyphyletic origins (Gwynne 1995) of sound production in the suborder have been proposed, no analyses claim that forewing stridulation evolved many ($\gg 2$) times. Acoustic signaling and perception in these insects are ancient, appear to be tightly coupled, and most likely coevolved over many aeons. Do these features imply that male advertisement and courtship calling in crickets, katydids, and their relatives can be attributed to coevolutionary mechanisms of sexual selection, either Fisherian or good-genes selection? Here, the bulk of evidence suggests that these coevolutionary sexual selection mechanisms have had a major impact on the evolution of acoustic communication, but I cannot claim with certainty that Fisherian or good-genes selection were responsible for its origin.

In fact, acoustic communication in some ensiferans bears features that could be interpreted as favoring exploitation of receiver bias. The hump-winged cricket *Cyphoderris monstrosa* (Hagloidea: Haglidae) exhibits a striking mismatch between its best frequency in hearing ($\approx$2 kHz) and the carrier frequency of male calls ($\approx$12 kHz).[76] This mismatch arises because their tympana have two categories of scolopidial units, one specialized for low frequencies and one that responds to a broad frequency range extending above 12 kHz (Mason et al. 1999). Possibly, low-frequency hearing originated in response to general predation pressure, and male advertisement calling later evolved as an exploitation of this perceptual ability. But male song, being restricted to higher frequencies by physical limitations (Section 4.2.5), might have been poorly perceived, and scolopidial units more sensitive to these frequencies could have been selected for and coevolved with the signal. Comparative neural data from other members of this relict family, augmented by better resolution of ensiferan phylogeny (cf. Flook et al. 1999), may afford more definitive conclusions. The weta (Stenopelmatoidea) of New Zealand represent a different anomaly. Four types of stridulatory devices occur among these large wingless species, and the ancestral (plesiomorphic) condition, rubbing of abdominal and hindleg pegs, was probably used for defensive purposes (Field 1993). Calling and courtship stridulations appear to be a derived (apomorphic) condition and occur in only two genera, *Hemideina* and *Deinacrida*. Because hearing occurs throughout the seven genera of weta, sexual signaling in these two may represent a modification of interspecific signaling that exploited an ancestral perceptual ability.[77]

Like ensiferan Orthoptera, most species of cicadas (Homoptera: Cicadidae) rely on acoustic communication for sexual advertisement and courtship. Tympanal ears located ventrally on the anterior abdominal segment are found in both males and females throughout the family, and males of most species bear sound--producing tymbals above the tympana and specialized air sacs that amplify the calls. While other acoustic devices, including stridulation and percussion by wing clapping or banging on the substrate, occur in various cicadas (e.g., *Platypedia*; Sanborn and Phillips 1999), and in some groups even replace the tymbals, these alternatives most likely represent character states derived from tymbal-bearing ancestors. Thus, cicadas may represent the strongest case among the insects for acoustic communication originating and evolving under a coevolutionary sexual selection mechanism. Several evolutionary and ecological particulars bolster this view. Tympanal hearing appears to have evolved within the Cicadidae, as its relict sister group, the Tettigarctidae, lack these organs; tettigarctids communicate sexually via substrate vibrations generated by weakly developed tymbals, a trait believed to be primitive in the Cicadoidea (Claridge et al. 1999; Figure 4.29). Additionally, cicadas, being either diurnal or crepuscular, may primarily rely on vision rather than hearing to detect natural enemies. No lines of evidence indicate that the intense and incessant advertisement calls that male cicadas typically produce while at conspicuous perches have exploited an ancestral perceptual ability that originated in a non-sexual context. More so than in crickets and katydids, acoustic communication in cicadas appears to be a highly coevolved system of male signaling and female (and male) perception.

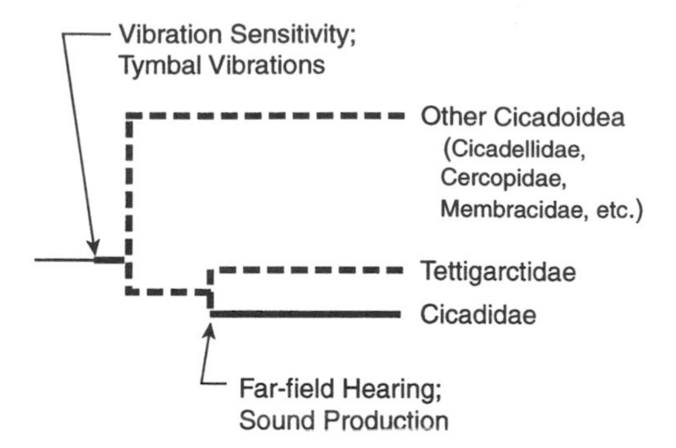

Figure 4.29 Phylogeny of the Cicadoidea, indicating restriction of far-field hearing and acoustic communication to the Cicadidae (see Claridge et al. 1999). Both hearing and sound production are widespread throughout the Cicadidae.

4.5.4 Limited Utility for Social Communication

Recent findings on recruitment signaling in honeybees and stingless bees clearly demonstrate that acoustic communication does play a critical role in the behavior of some eusocial insects. Overall, however, the non-contact forms of mechanical signaling—sound and vibration—are less important than tactile communication and far less consequential than chemical signaling in mediating the behavior of eusocial species. For example, maintenance of reproductive hierarchy and recognition of nestmates are wholly accomplished by chemical signaling in the advanced eusocial insects, and most alarm and recruitment signaling is chemical as well. In cases where signals must be transmitted continuously over a long time interval and reach a multitude of receivers, it is understandable that olfaction would be the channel of preference: pheromones linger in the environment (Section 3.1.6) and can thus maintain a queen's status, regulate reproductive activities by the various castes, etc. But, social functions such as nestmate, family, or group recognition, alarm, and recruitment are often accomplished acoustically in vertebrates, even in mammals possessing well-developed olfactory perception. What makes the eusocial insects different?

First, tympanal organs appear to have never evolved within the lineages of the main eusocial groups, Hymenoptera and Isoptera. In the absence of preexisting receptors, would acoustic social signals have readily evolved given that other channels (e.g., olfaction) were already functioning? Second, the rudimentary frequency processing in insect hearing may severely limit transmission capacity of the acoustic channel. That is, acoustic receivers may only distinguish signals by temporal qualities such as pulse rate or length, which may restrict the utility of sounds for encoding specific messages or recognizing nestmates. For example, in honeybee dance languages distance to a resource patch and its quality are communicated acoustically via simple, graded signals, but direction, which may require more information bits to encode (Section 2.5), is transmitted by tactile or visual signals. Other messages such as colony signatures may require an even greater quantity of information bits. Although a social insect need only distinguish its nestmates from outsiders, the transmission capacity of the channel used to encode colony signatures would have to be great enough to minimize the possibility that any two nearby colonies bear identical signatures (see Beecher 1989). In light of the receiver characteristics typically observed in insect perception, this possibility would be minimized by using the olfactory channel: A receiver who could detect five different odors via five different labeled lines (x) and differentiate five different concentration levels (y) of each odor would, in theory, distinguish among 3125 ($= y^x$) different chemical signatures. Were the acoustic channel used, the expected number of distinguishable signatures would be orders of magnitude less.

Reliability may also limit colony signatures of social insects to the olfactory channel. If environmental factors are critical determinants of signatures, cuticular odors would be a most reliable indication that another individual has or has not matured and lived in the exact same environment. In this context, mechanical signals might be much more vulnerable to cheating. Could a foreign individual penetrate a colony and learn and copy the local dialect of sound or vibration signals? Possibly, and she might do so before being recognized as an outsider. Odors impregnated in the cuticle would be difficult to absorb rapidly, however, and most invaders may be detected before acquiring insider status.

5

Bioluminescence and Reflected Light and the Visual Channel

Optical signaling among arthropods combines features common to both chemical and acoustic/vibratory communication. Like chemical signaling and perception, the visual channel is nearly ubiquitous among adult arthropods. Most species have some form of image-forming eyes, and an organism's very presence normally represents a visual stimulus.[1] Thus, many specialized optical signals may have evolved readily via simple elaboration or refinement of inadvertent cues. Like sound and vibration, optical signals are transmitted as waves that travel over long distances at high—infinite, for all intents and purposes—velocities, and they include intensity, spectral, and tempo features. But unlike sound, they do not pass behind solid (opaque) objects, and, with the exception of bioluminescent signals, they cannot be transmitted in darkness. The visual channel may also be fraught with more environmental noise than other channels, and achieving a critical signal : noise ratio may be difficult without arousing the attention of natural enemies. Arthropods generally cannot attain a private visual channel for communicative purposes as easily as they can with pheromones, sound (e.g., ultrasound or near-field sound), or substrate vibration.

The characteristics of visual information and stimuli make optical signals potentially more complex than chemical and most mechanical signals. Visual stimuli are characterized by quantitative (intensity) and qualitative (color spectra) features, and they usually include a specific tempo and movement pattern. Moreover, many visual stimuli represent a spatial arrangement of interlocking components, each with quantitative and qualitative features that may change and move over time in a stereotyped manner. To process even a portion of these characters, a receiver would need extensive neural apparatus, and such exaggeration is generally found (e.g., Strausfeld 1976). The optic lobes and other visual neuropil of the arthropod brain are very often its most extensive sector, even in species noted for olfactory or acoustic sensitivity. This heavy reliance on supporting neural circuitry may explain why image-forming eyes, alone among arthropod sensory receptors, are always located on the head, adjacent to the brain.[2]

In this chapter, we examine the nature of the visual channel and the ways in which optical signals are generated and transmitted along it. We then consider the devices with which arthropods perceive visual signals and the information that they can extract from them. At every step, the limitations and trade-offs that challenge arthropod communication along this channel are striking (and also quite reminiscent of those encountered in photography). Equally striking is how little we know about optical communication in arthropods compared with communication along the chemical and mechanical channels. It has been most difficult to determine the components of optical stimuli that receivers actually perceive and interpret as signals, and the application of digital technology to refining these determinations has begun only recently and in few species.

5.1 The Nature of Light

When an electric field at a given point changes in magnitude or polarity, as caused by the vibration of a charged particle, for example, it induces a changing magnetic field, which in turn induces another changing electric field, and so ad infinitum. These electric and magnetic fields are oriented at right angles to one another and propagate away from the source in the form of a train of electromagnetic waves. The wavetrain propagates in a direction perpendicular to the axis of the initial electric field's movement or vibration, and it travels at the velocity of light ($c \cong 3.0 \times 10^8$ m·s^{-1}, in a vacuum, and infinitesimally slower in air). Frequency (f) and wavelength (λ) of the electromagnetic wave are related by $f = c/\lambda$ (cf. equation 4.1), with values of f ranging from $< 10^5$ Hz (radio and other long waves) to $> 10^{23}$ Hz (X-rays). This range is known as the electromagnetic spectrum, of which light comprises a very small, but critical, intermediate band.

5.1.1 *Frequency and Wavelength*

The character of an electromagnetic wave is mainly determined by its frequency, which is directly proportional to the wave's energy (Section 5.1.3). Because the frequency values of electromagnetic waves are so high, however, we normally characterize a wave by the wavelength it sustains when transmitted through a vacuum (λ_{vac}). For example, the frequency value for an electromagnetic wave that we perceive as a particular hue of yellow light is 5.26×10^{14} Hz, and the frequency value of this wave would be invariant as it is transmitted through different media. But we normally describe this particular hue as having a λ_{vac} of 570 nm, even though its wavelength would decrease to 425 nm in water, where light travels considerably slower ($c \cong 2.24 \times 10^8$ m·s^{-1}, for $\lambda_{vac} = 570$ nm).

That band of the electromagnetic spectrum ranging from 430 to 690 nm in wavelength (λ_{vac}) is known as visible light. This designation arises because the molecules serving as human visual pigments can absorb waves of these lengths relatively efficiently, are thereby excited to electronically activated states and undergo reversible chemical transformations, and can evoke neural messages in the process. That is, our eyes perceive these waves but not ones with shorter or longer wavelengths.[3] It is generally accepted that this narrow intermediate range of wavelengths perceived visually is adaptive: waves of shorter length, and hence higher frequency and energy, are potentially damaging to cells, and they need to be screened out of the body rather than admitted to peripheral receptors. On the other hand, longer waves are so low in energy that any molecules serving as potential visual pigments for them would also be regularly excited by thermal agitation, generating excessive internal noise along the visual channel. It is also critical that most of the electromagnetic energy available near the earth's surface lies between 300 and 1100 nm in wavelength.

Whether for obtaining energy, as in photosynthesis, or information, as in visual perception, the range of the electromagnetic spectrum used by most organisms parallels what humans perceive as visible light. Among arthropods, the width of the visible spectrum is similar ($\approx$300 nm), but it is typically shifted to shorter wavelengths. A majority of arthropods can perceive ultraviolet (UV) light ($\lambda = 300$–400 nm) but remain relatively blind to red light ($\lambda = 640$–740 nm). This shift may be, in part, a consequence of the optics of compound eyes, which are limited by diffraction and would therefore perceive images formed of shorter wavelengths more effectively than longer ones (Section 5.3.1).[4] At the other end of the arthropod visible spectrum, the relative insensitivity to red light is often exploited by biologists: nocturnally active species may be observed with minimal disturbance by illuminating them with dim red light. Some arthropods, notably certain butterflies, fireflies (Coleoptera: Lampyridae), and dragonflies (Odonata), exhibit the typical visible spectrum shift at the UV end but also perceive some red light (e.g., Arikawa et al. 1987). These sensitivity ranges, among the widest visible spectra known in animals, may be critical for signal reception and communication in these highly visual species.

5.1.2 *Planes of Oscillation and Polarization*

Light and other electromagnetic phenomena are propagated as transverse waves in that the constituent electric and magnetic fields oscillate perpendicular to the direction in which the wave propagates. Relationships between these fields may be viewed from various spatial and temporal perspectives (Figure 5.1). At a given instant in time and at a given point in space, the vectors of the electric (**E**) and magnetic (**B**) fields are always perpendicular to one another and to the direction

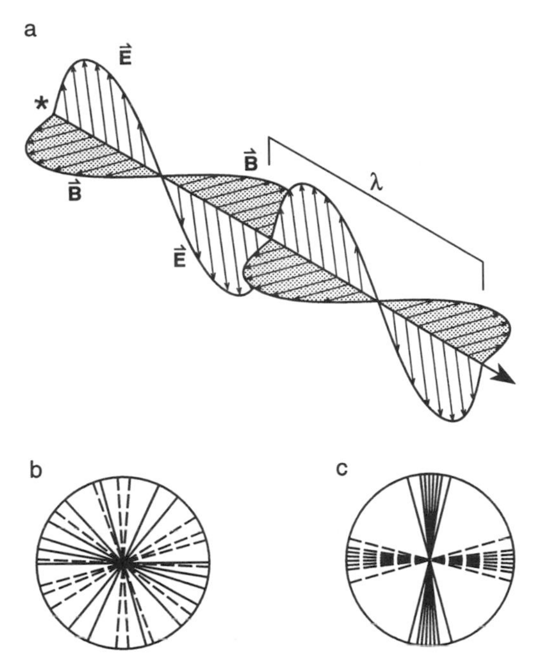

Figure 5.1 Orientations of electric and magnetic fields in electromagnetic waves, including light. (a) Wavetrain originating at asterisk (source) and propagating toward the right in direction of arrow. The electric (**E**) and magnetic (**B**) fields are oriented perpendicular to each other and transverse to the direction of wave propagation. (b) Orientation of **E** (solid lines passing through center) and **B** (dashed lines passing through center) fields of successive wavetrains propagating from an unpolarized source at the center of the circle. View shows wavetrains that are propagating directly out of the page; **E** fields of successive wavetrains are oriented randomly with respect to one another. (c) Orientation of **E** (solid) and **B** (dashed) fields of successive wavetrains propagating from a highly polarized source at the center of the circle. View shows wavetrains that are propagating directly out of the page, as in (b); **E** fields of successive wavetrains are mostly oriented in the same plane.

of wave propagation (Figure 5.1a). The **E** and **B** vectors of preceding and following waves in that wavetrain will also lie in the same pair of perpendicular planes at that instant, and earlier or later when they pass by that point. However, when a subsequent wavetrain passes by the same point approximately 10^{-8} s later, the **E** and **B** vectors of its component waves will almost certainly lie in a different pair of planes, although these two planes too will remain perpendicular to one another and to the direction of wave propagation. Such electromagnetic waves are considered unpolarized. From the perspective of an observer viewing the approach of unpolarized waves from a given source over a period of time, the **E** and **B** vectors will be seen cumulatively to lie in a random distribution of planes, all of which include the axis along which the approaching waves are propagated (Figure 5.1b).

Under certain circumstances, the oscillations of the electrical and magnetic fields of electromagnetic waves are restricted such that the **E** and **B** vectors of succeeding wavetrains are not randomly distributed. Instead, an observer would see that waves approaching from a given source are much more likely to have some **E** vector orientations than others (Figure 5.1c). Such electromagnetic waves are considered polarized, and the plane that includes the **E** vector of the majority of these waves is termed the plane of polarization. This convention is followed because most electromagnetic receivers, including the visual receptors of animals, perceive the electric field only. Polarization often occurs because particles in a medium may differentially scatter transmitted waves in some directions more than in others. Differential reflection of waves at a boundary between two media such as air and water represents another important source of naturally occurring polarization.

At an angular separation >30–45° from the sun, light in the sky tends to be polarized in a plane perpendicular to the line extending from the sun to that point in the sky. The percentage of light waves that are polarized in this plane increases at greater angular separation from the sun until it reaches a maximum $\cong$ 75% in an arc of the sky 90° from the sun. Polarization, like any type of scattering, is wavelength-dependent (Section 5.1.4), and a much higher percentage of UV light than longer wavelengths in the sky is polarized. Light transmitted through and scattered by water is also polarized, as is sunlight reflected from the water/air boundary. Many vertebrates, including humans, have some ability to perceive the plane in which light is polarized, and various species even use it for orientation. The sensitivity of arthropods to polarized light, however, is far more widespread and generally much sharper. Some species possess special sensitivity to polarized light aligned in a particular plane, while others are more versatile and can determine the plane of polarization of light in the sky that is 10% polarized or more. This latter ability has important consequences for communication in social species whose recruitment signals include information used for orientation. In the honeybee *Apis mellifera*, where the vertical orientation of a returning scout's dance indicates the azimuth angle between the sun and the food source she has just visited (Section 4.4.3), both signalers (scouts) and receivers (recruits) infer the position of the sun from the pattern of polarization planes in the sky (von Frisch 1949, Rossel and Wehner 1984). Thus, recruitment may proceed on partly overcast days (Gould et al. 1985).

5.1.3 Quantification of Light

Electromagnetic phenomena such as light have a dual nature. They are transmitted as waves and are normally described from this perspective, but they may also behave as a stream of quantized particles. For light, these particles are termed photons, and different wavelengths of light are comprised of photons of different energy content, E, calculated as

$$E = h \cdot f \tag{5.1}$$

where $h = 6.626 \times 10^{-34}\,\mathrm{J{\cdot}Hz^{-1}}$ (Planck's constant) and f is the frequency yielding that wavelength. It is often more convenient, and relevant, to describe the intensity of light as a measure of the number of photons emanating from a source or streaming across a given planar area per unit time than as a measure of wave energy (joules) or power (watts).

Intensity can be measured as energy (or photons) radiated by a light source or as energy (or photons) illuminating a surface, and either measurement may be global or local. These various measurements are expressed in different units, which must be specified carefully. The total amount of light emitted in all directions by a source, the light organ of a male firefly, for example, is a measure of global or radiant flux and is expressed in (quantal) units of photons$\cdot$s^{-1}. Because light sources do not necessarily transmit energy equally in all directions, it is often useful to measure the amount of light radiated within a particular solid angle. This measurement of local flux or radiant intensity is expressed in units of photons$\cdot$sr$^{-1}\cdot$s^{-1} (sr is the abbreviation for steradian). The amount of light illuminating a surface, the eye of a female firefly, for example, is a measure of flux density or irradiance and is expressed in units of photons$\cdot$m$^{-2}\cdot$s^{-1}. Because it may be critical to measure the amount of illumination received from a particular direction or source, a male firefly, for example, we may also determine the flux density per solid angle or radiance, expressed in units of photons$\cdot$m$^{-2}\cdot$sr$^{-1}\cdot$s^{-1}. A receiver's sensitivity to light may be measured as the radiant flux absorbed per photoreceptor for a given radiance (Land 1981, 1989) or as the radiance that would be needed to stimulate each photoreceptor with 1 photon$\cdot$s^{-1}.

The above measures lump all wavelengths of light together, which would not be appropriate where illumination is not spectrally uniform. Thus, each of the four measures may also be determined for a particular spectral band. For example, irradiance may be expressed as a spectral density in units of photons$\cdot$m$^{-2}\cdot$s$^{-1}\cdot$nm^{-1} for a band centered around a given wavelength.

5.1.4 *Scattering, Absorption, Reflection, Refraction, Diffraction, and Interference*

As a consequence of its wave nature, transmitted light bears various features in common with sound. Several of these features are critical for understanding the manner in which arthropods and other animals send and receive optical signals. When light encounters matter in its path, the waves are transmitted through or around it, absorbed, and reflected in varying proportions that depend on the nature of the light and the matter (Figure 5.2). Very small particles (diameter $< \lambda/10$) reflect intercepted light waves in multiple random directions, a phe-

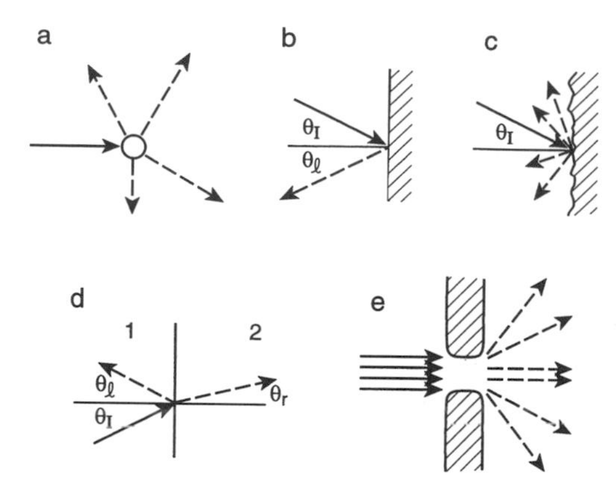

Figure 5.2 Interactions between light and matter. (a) Rayleigh scattering of light wave (solid arrow) by small particle (diameter $<\lambda/10$). (b) Reflection from a smooth surface. (c) Diffuse reflection from a microscopically rough surface. (d) Refraction by a medium (2) in which light propagates at a reduced velocity (see equation 5.2). (e) Diffraction of light waves passing through a small aperture (diameter $\cong\lambda$).

nomenon termed Rayleigh scattering (Figure 5.2a). As with sound, light waves of shorter wavelengths are scattered (and also absorbed and then generally converted to heat) more strongly than longer ones. Larger particles ($\lambda/10 <$ diameter $< 25\cdot\lambda$) redirect light waves in a more complex fashion, termed Mie scattering, and objects of yet greater dimensions may reflect light in a single direction. Here, light will be reflected from a smooth mirror-like surface at an angle equal to its angle of incidence (Figure 5.2b), but a microscopically rough surface will reflect light at multiple angles; this phenomenon is termed diffuse reflection (Figure 5.2c).

The color, or hue, that a receiver perceives from a non-radiant object results from light incident on the object that is neither absorbed nor transmitted through it (Figure 5.3). When light is not transmitted through an object, the hue perceived is simply determined by the spectrum of the incident light, the absorption spectra of surface molecules (those wavelengths that excite the molecules to electronically activated states), and the absorption spectra of the receiver's visual receptors. Waves of incident light that are not absorbed by surface molecules are reflected (cf. equation 4.7 for the acoustic analogy), and it is this reflectance spectrum that evokes the sensation of hue when perceived by a receiver. Thus, color, hue, and coloration are strictly psychophysical percepts that only exist after light has been perceived and processed by a receiver. This meaning is implicit wherever these terms are used in this chapter.

When transmitted light passes from one medium into another, it is normally redirected via refraction. As with sound, light is refracted toward the medium in which it travels more slowly, and the bending of light waves at the boundary between the two media is determined by

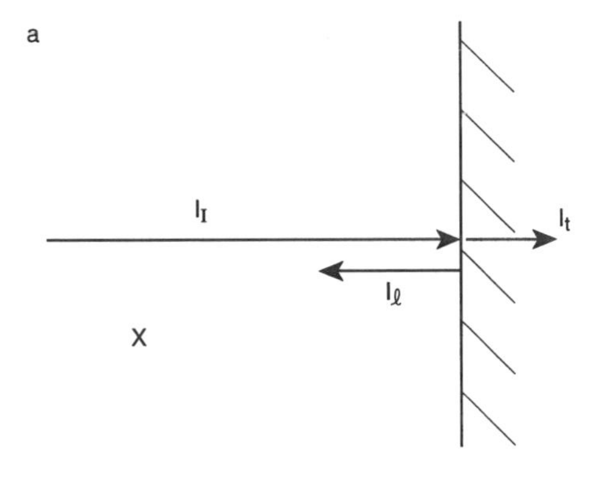

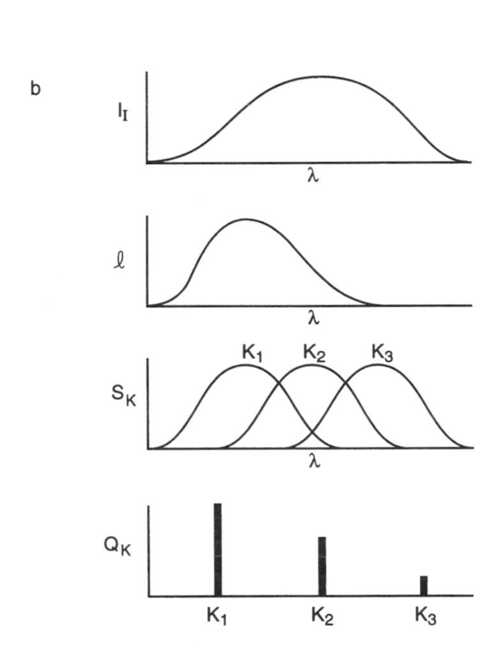

Figure 5.3 Absorption, reflection, and photoreceptor capture of light incident on (irradiating) the surface of an object. (a) Absorption = Incident light (I_{I}) – reflected light (I_ℓ) – transmitted light (I_{t}). (b) For a receptor of category K situated at point X and viewing the object surface in (a), the number of photons captured is calculated by

$$Q_k = \iint I_{\text{I}}(\lambda, t) \cdot \ell(\lambda) \cdot S_k(\lambda)\, \mathrm{d}\lambda \mathrm{d}t \tag{5.3}$$

where $I_{\text{I}}(\lambda)$ is irradiance (photons·m^{-2}·s^{-1}) of the object surface by light of wavelength λ at time t, $\ell(\lambda)$ is surface reflectance ($I_\ell \cdot \text{sr}^{-1} \div I_{\text{I}}$) of light of wavelength λ as received at X, and S_k is the spectral sensitivity (radiant flux absorbed per unit radiance) of a category K receptor. A receiver's sensation of color in a functional visual unit (e.g., an ommatidium) is determined by the combination of all Q_k values in that unit.

$$\sin\theta_I / \sin\theta_r = n_2/n_1 \tag{5.2}$$

where θ_I is the angle of incidence in the first medium, θ_r is the angle of refraction in the second, and n_1 and n_2 are the refractive indices of the two media (Figure 5.2d); this formula is known as Snell's Law.[5] Refractive indices are computed as the ratio of the velocity of light in a vacuum to its velocity in the given medium. All media slow the transmission of light to some extent, and refractive indices range from 1.0003 (air) to 1.5 and higher (e.g., glass). Moreover, light of shorter wavelengths is slowed more than longer wavelengths of light, and this effect increases in media whose refractive indices are high. While the speed of violet light is only 0.0008% slower than that of red light in air, it is more than 1% slower in glass. Hence, when the refractive index for a given medium is listed, the wavelength (λ_{vac}) should be specified. The refraction of light is critical for both the production of signals and their perception. Hues that observers perceive from many arthropods do not result from simple surface reflection, but rather originate from complex mechanisms involving reflection from and refraction through thin films or layers of material (Section 5.2.2). Lenses of compound eyes typically have a refractive index $\cong 1.5$ (for $\lambda_{vac} = 589$ nm), which yields a low *f*-number and hence a bright image (Section 5.3.1).

As light waves pass by the edge of an object whose refractive index differs from the surrounding medium, they are diffracted in new directions, as are sound waves when they flow around a barrier. Diffraction becomes a significant phenomenon when light waves pass the edge of a small opaque object or through a small aperture in an opaque object (Figure 5.2e). Generally, the diameter of the object or aperture must approach, or fall below, λ in order to generate substantial diffraction. Thus, an increasing proportion of the light waves entering individual ommatidial units of arthropod compound eyes would be diffracted were these units to become excessively small, and visual perception and eye structure may be subject to constraints imposed by this optical effect.

Two intersecting light waves interfere with one another in accordance with the principle of superposition (Section 4.4.1). Consequently, light waves may add to or cancel each other in certain circumstances. Such constructive and destructive interference is critical for the generation of arthropod iridescence, which results from combined reflection and refraction by thin films (Section 5.2.2).

5.2 Signaling Along the Visual Channel

Organisms generate optical signals by radiating, reflecting, or, in the case of various translucent freshwater and marine species, transmitting light that contrasts with the local visual background that receivers perceive. The contrast may be in intensity, spectrum, or polarization, and it may occur between the surface generating the signal and the rest of the organism or between the organism and its visual environment. For example, some male butterflies have patches of scales on their wings that reflect UV light, which is perceived by conspecific females during courtship. Patrolling male butterflies may also advertise to distant females by overall reflectance from their entire wings (Rutowski 2001). Energy-demanding activity often amplifies or modifies the signal, bestowing it with a tempo, changing its features over time, or moving it against the visual background (e.g., Jennions and Backwell 1998). Those UV-reflecting patches may be alternately revealed and concealed as the male opens and closes his wings or moves them in some other stereotypical fashion.

5.2.1 Bioluminescence

Bioluminescent signals of arthropods include some of the most spectacular animal displays known. The massive aggregations of *Pteroptyx* spp. fireflies in the Indo-Malayan region, in which thousands of males cluster in single trees along the banks of waterways and flash in nearly perfect synchrony, have attracted the attention of explorers and natural historians for centuries (Buck and Buck 1976, 1978). Less acclaimed, but equally fascinating, are the multi-colored lights of glowworm larvae and adults (Coleoptera: Phengodidae; Viviani and Bechara 1997), the collective luminescence of cave-dwelling fungus gnat larvae (Diptera: Mycetophilidae) in Australia and New Zealand (Sivinski 1998), and the temporally patterned flashing of marine crustaceans (Ostracoda), fittingly dubbed "firefleas of the sea" (Morin 1986). Because their signals are clearly defined in time and space, bioluminescent arthropods are particularly valuable subjects for the study of visual communication. Measurements from photic recordings show that the flashes of various bioluminescent arthropods bear the same regular temporal features commonly found in acoustic and vibratory signals (Section 4.2.7).

Chemical Mechanisms. Bioluminescence is produced by various forms of a chemical process, chemiluminescence, and has evolved independently on at least 30 occasions in organisms distributed among four kingdoms (Hastings and Morin 1991). Among metazoans, it occurs in nine phyla and is most prevalent in marine forms. The bioluminescence of several groups of fishes and cephalopods is generated by symbiotic bacteria—the actual light producers—that inhabit chambers fitted with translucent windows, but arthropod bioluminescence is produced intrinsically via chemiluminescent reactions occurring or originating in specialized light organs.

From a chemical perspective, bioluminescent processes are noteworthy for the radiation of light that neither depends on high temperature of the radiating substance nor on its prior absorption of light. These features set bioluminescence apart from more familiar luminous processes such as incandescence and fluorescence, and the emission of bioluminescent light has intrigued scientists and laymen alike. Aided by material collected from literally millions of fireflies, various biochemistry laboratories conducted major research efforts during the 1960s and 1970s that ultimately determined the basic mechanisms of bioluminescent processes in several organisms.

The chemiluminescent reactions that generate bioluminescence all involve the oxidation of a substrate, generically termed luciferin, that requires catalysis by an enzyme, luciferase. Molecular oxygen normally serves as the oxidant. Additional substances or factors (e.g., ATP, Mg^{2+}, seawater, or a change in pH) are generally needed to complete the reaction. Owing to these requirements, bioluminescent reactants are used in sensitive bioassays for ATP and oxygen in a wide array of biomedical applications.

Comparative analyses show that luciferins are heat-stable compounds of relatively low molecular weight. They may be derived from pigment molecules and are distinguished by the property of emitting a photon when oxidized to an electronically activated state. Firefly luciferin is a benzothiazole, luciferyl adenylate (Figure 5.4), which is oxidized and then modified to a carbonyl compound that emits light while still enzyme-bound. Luciferases are much more variable among bioluminescent organisms than luciferins are[6] and tend to be species-specific. Their structure and activity suggest that they are derived from mixed-function oxidases operating in basic metabolic processes. The specific structure and configuration of an organism's luciferase may influence the energy of the photon released, and hence the spectrum of its bioluminescence. Light organ

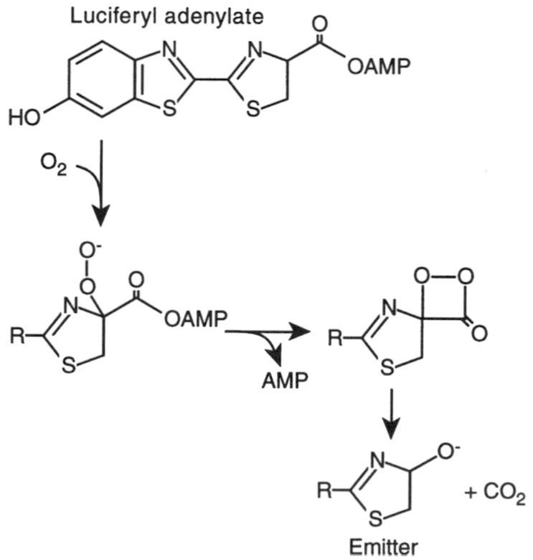

Figure 5.4 Bioluminescent reaction found in *Photinus* and *Photuris* fireflies (from Hastings and Morin 1991). Luciferyl adenylate, the luciferin substrate, is oxidized and converted to the carbonyl emitter. Both ATP and Mg^{2+}, as well as the specific luciferase, are required to complete the reaction.

filters, secondary emitters (byproducts of oxidation), and pH may also influence bioluminescent spectra.

Fireflies: Physiology and Behavior. Fireflies unquestionably include the most thoroughly studied examples of arthropod bioluminescence (Lloyd 1971). Both adult males and females in most nocturnal species of this beetle family produce bioluminescent flashes, and larvae in many species also glow (Carlson and Copeland 1985). The adult light organ consists of the terminal segments of the abdomen, and it typically emits light whose peak radiant flux falls within a narrow spectral range from 550 to 580 nm (yellow-green to yellow). Minor spectral variations do occur among firefly species, and the spectrum that a species emits may be that which contrasts best against the visual background normally present at the time and location of activity (Lall et al. 1980, Seliger et al. 1982). Adult flashing largely functions as an advertisement and courtship signal in pair-formation, whereas larval bioluminescence may represent a startle or aposematic signal that can deter nocturnal natural enemies (DeCock and Matthysen 1999). Fireflies are chemically defended (Meinwald et al. 1979, Eisner et al. 1997) and bear black, yellow, and red patterns commonly used in warning coloration.

Firefly light organs, commonly termed lanterns, consist of photocytes, specialized cells containing the bioluminescent reactants, that are stacked in dorsoventral columns arranged in rosette fashion around tracheae (Case and Strause 1978, Ghiradella 1998). The entire organ includes several thousand of these rosette units. Tracheoles branch horizontally off the tracheae and project between the photocytes, and neural axons, which enter the organ along the tracheae, terminate at the tracheolar cells (adults) or at the photocytes themselves (larvae). The temporal pattern of flashing, including flash length (25–500 msec) and rhythm (0.1–10 Hz), is controlled by both oxygen (Ghiradella 1977a,b), supplied via the tracheoles, and neural discharges (Christensen and Carlson 1981, 1982), which may be organized to synchronize activity in all of the organ's photocytes. However, it has not been clear how O_2 can diffuse from the tracheoles to the photocytes rapidly enough to effect the sudden onsets of light that occur in high flash rates and in signal interactions such as synchrony between neighboring males and male–female dialogues (see below). Similarly, the pathway by which neural messages travel from axonal discharges in tracheolar cells to the bioluminescent reactants, concentrated in peroxisomes in the center of photocytes, has defied identification. The most likely mechanism suggested to date implicates nitric oxide (NO) as a neurotransmitter. Following axonal discharges, NO is quickly manufactured and diffuses into the adjacent photocytes where it inhibits mitochondrial respiration, freeing O_2 for completing bioluminescent reactions in the peroxisomes (Trimmer et al. 2001). Because light can suppress the inhibitory action of NO, negative feedback from the flash leads to its extinction shortly after onset.

Pair-formation in many North American firefly species is a photic dialogue that partly resembles the duetting observed in some acoustic Orthoptera (e.g., phaneropterine katydids; cf. Section 4.4.1, Figure 4.16), and more closely that in various cicadas (e.g., Gwynne 1987) and oedipodine grasshoppers. Roving males emit regular advertisement flashes while in flight, which may elicit response flashes from stationary females (Lloyd 1997). The female replies are typically less intense and elaborate than the male advertisement and are given after a characteristic latency measured from the end of the male signal. If the male detects a female reply during his species-specific post-advertisement time window, he approaches while maintaining the dialogue. Upon landing near the female, the male may emit a different flash pattern as well as pheromones that mediate courtship. Other pair-forming protocols found in firefly species include stationary male advertisement attractive to flying females (e.g., *Pteroptyx*) and male flight toward glowing, stationary females.

Female fireflies largely evaluate male signals on the basis of temporal parameters (Case 1984, Carlson and Copeland 1985, Lloyd 1997), although flash intensity and movement during signaling may also influence their responses. Male firefly advertisements are typically species-specific flash patterns: flashes may be simply repeated at a characteristic rate, but in some species they are grouped into flash trains comprising from 2 to 10 flashes which are themselves repeated at a characteristic flash train rate. Thus, females may potentially distinguish male signals by their flash rate, flash length, inter-flash interval, duty cycle, number of flashes per train, and even the rate at which light intensity rises and decays during flashes (flash envelope; Figure 5.5); these characters are analogous to the pulses, chirps, and associated features in acoustic insects (Section 4.2.7). Moreover, males may fly in a characteristic ascending pattern during emission of a flash or flash train, and the resulting spatial pattern of light emission too may be critical (but see Section 5.4). Spectral parameters are claimed to be less important (Case 1984), although receivers do have the perceptual ability to evaluate color (Section 5.3.1) and spectral sensitivities in firefly vision may closely match spectral peaks in the conspecific flash signal (Cronin et al. 2000).[7] Thus, peripheral filtering may yield some female preference for male flash signals based on spectral features, as it does for carrier frequency among acoustic species (Section 4.3.5).

As with rhythmic mechanical signals in arthropods (Section 4.2.7), firefly flash rates are temperature dependent. Within a median range, flash rate, or the flash rate with a train, increases linearly with temperature (Carlson et al. 1976, Branham 1995),[8] and the evaluation of male flash rate by female receivers may be concomitantly modified (Carlson et al. 1976; cf. Section 4.2.7 on temperature coupling in acoustic signaling systems).

Understandably, most initial investigations of firefly flash patterns focused on their role in species recognition (cf. Section 3.2). But female fireflies probably evaluate male quality in addition to species identity, and it is likely that signals

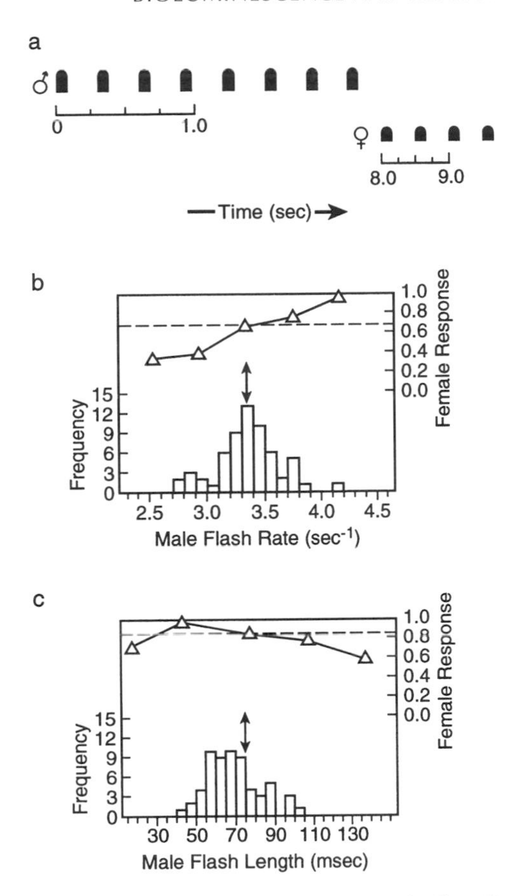

Figure 5.5 Bioluminescent communication in the firefly *Photinus consimilis* (Coleoptera: Lampyridae). (a) Male advertisement flash train and female reply. Males produce approx. 3 flash trains·min^{-1}, each train including from 4 to 12 70-ms flashes delivered at 3–4 flashes·s^{-1}. Female replies begin approx. 8 s after the male flash train and include from 1 to 7 flashes delivered at approx. 2 flashes·s^{-1}. (b) Variation in male flash rate (from 2.7 to 4.1 flashes·s^{-1}) and female preference for faster flash rates (flash length held constant at 70 ms). Histogram shows frequency of males in 0.1-s^{-1} bins; triangles indicate female response level. (c) Variation in male flash length (from 40 to 105 ms) and relative female indifference to flash length (flash rate held constant at 3.3 s^{-1}). Histogram shows frequency of males in 5-ms bins; triangles indicate female response level. Vertical arrows indicate mean signal values; horizontal dashed lines indicate relative level of female response to mean signal value. (From Branham 1995, Branham and Greenfield 1996, and Greenfield 1997b; reprinted from *Perspectives in Ethology*, vol. 12, with permission of Kluwer Press.)

containing greater photic energy are favored (cf. Section 4.4.1). One study (Figure 5.5) has shown that, at a given temperature, males of the North American firefly *Photinus consimilis* vary considerably in their within-train flash rates and that females are more inclined to reply, and also to reply more strongly, to fast rather than slow flash rates (Branham and Greenfield 1996).

These findings were obtained from single-stimulus photic playback experiments. When presented with a choice, as must often occur in the field when several males flash locally, females might indicate their preference by inserting their replies within the time windows of specific males or by beaming their replies toward those males. Stochastic factors too may play an important role in the efficacy of firefly signals. For example, a roving male emitting flash trains at a higher rate may have a greater chance of being seen by or stimulating a given female and consequently will more likely elicit her reply than a male flashing more slowly.

In many firefly species, males respond to conspecific male flashes at least as strongly as females do. The male responses typically entail adjustments in flash rhythm that lead to local synchrony and are thus reminiscent of the chorusing observed in certain acoustic insects (Section 4.4.2). In species with roving males who flash while in flight, an individual may temporarily enter into synchrony with several others who happen to be passing by at that time (Rau 1932, Buck 1938), but in various *Pteroptyx* spp. of the Indo-Malayan region, wherein stationary males flash while perched, thousands of individuals may gather in a dense aggregation and maintain precise synchrony over an extended period (Buck and Buck 1976). These various levels of synchrony result from different mechanisms (e.g., Moiseff and Copeland 2000) and selection pressures. Photic playback experiments show that synchrony in some firefly species results from a simple phase-delay mechanism (Hanson et al. 1971, Hanson 1978, Buck et al. 1981a,b) very similar to the inhibitory-resetting described in acoustic Orthoptera. Because (female) receivers in fireflies may be influenced by psychophysical precedence effects in which the earlier of two or more flashes or flash trains draws more attention (cf. Section 4.4.2), it is possible that by adjusting their flash rhythms, males both improve their stimulation of females and jam their neighbors' flashes (Vencl and Carlson 1998).[9] In such cases, synchrony may represent only an epiphenomenon, a mere byproduct of competitive signal interactions between neighboring males. But in species that maintain precise synchrony over an extended period, the interaction arises from a more complex mechanism in which males adjust and match their actual free-running rhythms to those of their neighbors (Ermentrout 1991, Strogatz and Stewart 1993; cf. Figure 4.23). Rather than a byproduct, synchrony in these fireflies may be a cooperative phenomenon selected for its maximization of an aggregation's collective photic energy (Buck and Buck 1978, Buck 1988; cf. Section 4.4.2). Presently, however, this and other explanations (Otte and Smiley 1977) for the massive synchronous displays seen in fireflies must be treated as speculative. Evolutionary and ecological factors underlying these most spectacular of bioluminescent phenomena remain mostly unstudied (see Lloyd et al. 1989).

Bioluminescent Diversity among Arthropods. Outside of the fireflies, arthropod bioluminescence is known from a very small number of species. Nonetheless, these species are distributed among a diverse range of taxa: four other families of beetles (Phengodidae, Elateridae, Throscidae, Staphylinidae), two additional insect orders (Collembola, Diptera), several orders of Diplopoda, Chilopoda, Chelicerata, and various marine Crustacea (Copepoda, Ostracoda, Malacostraca). The physiology and behavior of light production in these arthropods vary greatly, even among closely related groups (Hastings and Morin 1991, Wood 1996). For example, phengodids (glowworms), which are assigned to the same beetle superfamily (Cantharoidea) as fireflies, may bear organs on several parts of the body that emit light of different spectra. The most elaborate species, found in the genus *Phrixothrix*, have rows of lateral green organs, several yellow dorsal organs, and red organs on the head; they are known as railroad worms (Viviani and Bechara 1997). As suggested by their common name, phengodids tend to emit light continuously. Males, larviform females, and larvae may all emit these glows, which are believed to function in defense or for illumination. Bioluminescent elaterids (click beetles) are more similar to fireflies in their habits, but they bear paired yellow-green light organs dorsally on the prothorax and a large yellow-orange one ventrally on the abdomen (Colepicolo-Neto et al. 1986). Males emit lengthy (>1 s) flashes that probably function in mating, but behavioral analyses have not been reported.

Few of the other known occurrences of arthropod bioluminescence have been studied in sufficient detail to afford an understanding of the behavioral role of light emission. In the various species of bioluminescent mycetophilid flies (fungus gnats), light emission is generally a larval trait that may function to lure phototropic arthropod prey (Sivinski 1998). But, in some species adults also glow, and males may locate receptive females by orienting toward their light. Among bioluminescent marine crustaceans, ostracodes in the genus *Vargula* have been studied most thoroughly (Morin 1986, Cohen and Morin 1990). Whereas their bioluminescence is extracellular, the reactants being secreted out of the body and mixed with seawater to emit light, they are behaviorally similar to certain fireflies. Swimming males emit pulses of light in characteristic temporal and spatial patterns that attract non-luminous females. In other contexts, *Vargula* light may also function to startle natural enemies.

Evolution: Multiple Origins but Overall Rarity. Bioluminescence apparently can result from a variety of fundamental biochemical processes, and the many independent origins suggest that it is not phylogenetically constrained. Moreover, bioluminescent flashes should be a most effective means for advertisement and courtship signaling. These signals may be generated and perceived at night, in which case they bear a relatively high signal : noise ratio, and the potential for both temporal and spatial patterning implies that the channel has the capacity to transmit considerable information bits (Section 2.5). Thus, the overall rarity of bioluminescent communication among arthropods is surprising.[10]

What factors may have impeded the spread, or origin, of bioluminescence beyond the few species in which it occurs? Following, we consider three possibilities: energy, risk, and perceptual limitation. Because the emission of bioluminescent flashes does not necessarily require locomotion or other forms of exaggerated muscular activity, they seem energetically cheap to produce. However, the energetics of light emission have not been measured in any bioluminescent arthropod, and it would be premature to eliminate this possibility entirely. Additionally, hidden costs may be incurred in synthesizing and sequestering the reactants or in developing and maintaining the light organ (cf. Section 4.2.9).

A first analysis also suggests that bioluminescent flashes do not expose a signaler to excessive risks of predation or parasitoid attack. Adult fireflies do not attract insectivorous bats with their flashing (see Lloyd 1989), and the primary natural enemies that do exploit firefly flashes in order to attack the signalers are other firefly species[11] (Lloyd and Wing 1983, Lloyd 1997). Many of these species are aggressive mimics (*Photuris* spp.) in which females may reply to heterospecific (*Photinus* spp.) male flashes to lure and then consume an unwary male (Lloyd 1975); such femmes fatales may operate by altering their responsiveness to flash patterns as they graduate from a mating behavioral state to a feeding one (Vencl et al. 1994). But fireflies may be a poor example with which to argue that bioluminescence is a relatively safe form of signaling. Fireflies owe some of their freedom from natural enemies to chemical noxiousness, and their bioluminescence may actually reinforce any learned avoidance that predators acquire for these insects (see DeCock and Matthysen 1999). Among other bioluminescent arthropods, some reports do suggest that hymenopteran parasitoids of mycetophilid fungus gnats can locate their hosts phototactically. Based on currently available information, any verdict on the relative safety of bioluminescent signaling would be tenuous at best.

Perceptual limitation would seem to be the least likely reason for the rarity of bioluminescence. Because the flashes of many firefly and luminescent click beetle species (e.g., *Pyrophorus* spp.) are so conspicuous to human observers, it is hard to imagine that conspecific receivers might fail to detect them. However, bioluminescent flashes may only be effective in relatively open habitats or in species where both signalers and receivers emerge to conspicuous locations during the nightly activity period. More critically, the flashes may appear barely larger than points of light subtending extremely small angles of arc at the receiver's eyes (but see Section 5.3.1, Image Formation, Resolution, and Field of View), save when the transmission distance is quite short. In Section 5.4, Perceptual Influences on Optical Signal Design, we consider why many arthropods, particularly the majority that view their world through compound eyes, may not reliably perceive or extract information from such images.

5.2.2 Reflected Light

Other than bioluminescence, the only means with which arthropods can communicate along the visual channel is by reflecting light that originates from astronomical sources, mainly the sun. Arthropods can generate reflected light signals by using pigment substances, or with specialized structural features that reflect and refract incident light via some remarkably elaborate mechanisms. In some environments restrictions on the potential reflectance spectra of signals may exist because the spectrum of available light is modified or limited (Lythgoe 1979). These restrictions are most pronounced in freshwater and marine habitats, where longer wavelengths of light are severely attenuated at depths greater than several meters below the surface. Consequently, red or orange signals cannot be generated there via reflected light. In terrestrial habitats, less severe restrictions may arise at dawn and dusk and in forested areas (Endler 1993b). Under the dense shade of tropical broadleaf forests, most red and blue light is filtered out because of absorption by chlorophyll in foliage. Thus, yellow and green light predominate in the available spectrum. Nonetheless, an organism might exploit that reduced amount of red light that does occur near the forest floor to generate a signal bearing high spectral contrast against the green background.

Biological Pigments. Pigments are molecules that have specific absorption spectra for light and thereby reflect photons whose wavelengths differ markedly from the peaks in those spectra. Thus, pigments act as filters that give an object its reflectance spectrum and, ultimately, its perceived hue. A pigment with an absorption peak of 420 nm may cause an object illuminated by sunlight to reflect primarily orange or red light, the specific reflectance spectrum depending on the width of the absorption peak and the spectrum of incident illumination, while a 650-nm pigment may reflect mostly blue or violet light. Green reflectance spectra might be created by pigments with two absorption peaks (e.g., 420 and 650 nm).

The biological pigments, sometimes termed biochromes or zoochromes, that are responsible for the reflectance spectra of many arthropods include a vast array of compounds (Needham 1974). Many are synthesized de novo, often as byproducts of basic metabolic processes, but others, such as carotenoids, may require specific dietary precursors. In general, pigments yielding green, blue, or violet reflectance spectra are rarer than those yielding spectra composed of longer wavelengths. It may be noted that the general rarity in nature of pigments yielding blue or violet spectra is reflected by the special value accorded them in ancient Mediterranean societies and by the association of purple with European royalty.

Among the diurnal Lepidoptera (butterflies and various moth taxa) whose reflectance spectra have been analyzed most intensively, blacks, grays, tans, and browns are often produced by melanins incorporated in the scales, whereas reds, oranges, yellows, and some whites are produced by pterins. Additionally, flavones, carotenoids, and ommochromes may be responsible for yellows and browns in some cases (Nijhout 1981, 1991).

Did synthesis and deployment of the specific compounds serving as pigments in Lepidoptera and other arthropods originate under selection for generating optical stimuli? Perhaps in the case of cryptic coloration affording protection from natural enemies, but, in general, the creation of optical stimuli was more likely a byproduct of compounds selected for their other attributes. Because most pigments absorb relatively short wavelengths, these compounds screen out potentially harmful UV radiation and prevent it from penetrating internal organs. Pigments may also serve in thermoregulation for either heating or cooling. And some pigment compounds are noxious defensive substances, while other extracellular pigments are merely derivatives of nitrogenous wastes that are excreted in a form requiring minimal water loss. There exists rather little evidence suggesting that pigments were primarily selected as optical signals for sexual advertisement or courtship. Sexual dimorphism is seldom observed for the general presence of pigmentation or its basic chemistry. Rather, sexual dimorphism in coloration, and the optical signals that function in pair-formation, usually entail a specific color pattern or hue, and these features often cover only local body regions. It appears more likely that such optical signals evolved as modifications of general pigmentation that had originated in non-communicatory contexts.

Structural Color. Interference, diffraction, and scattering (Section 5.1.4) may all generate reflectance spectra in the absence of pigment, or they may supplement reflectance derived via pigmentation (Srinivasarao 1999). Animal coloration derived in this fashion is termed structural color, and it is prevalent in the emission of UV, violet, blue, green, white, and variable spectra. Again, structural color mechanisms have been analyzed most thoroughly in butterflies. Nanoscale investigations of their wing scales reveal an amazingly ornate and precise ultrastructure that can generate brilliant colors via the interaction between light and cuticle (Figure 5.6a).

The common mechanism by which the interaction of incident light waves with scales generates structural color is the constructive interference of directly reflected waves and waves that are first refracted through and then reflected back across the scale, or one of its component modules (Ghiradella et al. 1972). Thus, light of wavelength λ can be reflected by an unpigmented scale, or module thereof, whose thickness is $\gtrsim \lambda/(4n_2)$, where n_2 is the refractive index of the scale (Figure 5.6b). When incident light waves reach the outer surface of the scale, they are either directly reflected or transmitted on through the scale. Because the scale's refractive index is normally high ($\cong 1.5$), the transmitted waves are usually slowed and bent (if the angle of incidence $\neq 0°$). When these refracted waves reach the inner surface of the scale, they again are either reflected back toward the outer surface of the scale or continue transmission into the air space behind the scale. The waves reflected back then cross the outer surface of the scale, where they interfere with direct reflections of incident waves arriving at that instant. Letting the scale thickness = d, waves of length $\lesssim 4d \cdot n_2$ crossing back out of the scale will be in phase with incident waves of that length arriving and reflected directly by the scale surface. These waves will constructively add, and the intensity of their wavelength will be reinforced in reflection from the scale. Waves of other lengths will be out of phase, and some will destructively cancel, attenuating or eliminating those spectra from the scale's reflection.

The above description is the classical physics explanation for generation of structural color by interference from a thin transparent film. However, butterfly scales are far more intricate in design than thin films, and their brilliance cannot be fully understood without examining their finer points. Electron microscopy reveals precise arrangements of adjoining scales and very elaborate ultrastructure of individual scales.

In overall arrangement, butterfly scales are normally layered as shingles or tiles on a roof. Six or more layers of scales, with five or more interdigitating layers of air between the scales, are found in some species. Incident light waves will pass through these layers as they are transmitted toward the wing and then reflected away from it. The process of reflection and refraction is repeated in each of the scales through which the waves pass, and each scale affords an opportunity to recapture additional incident light for reflection. Thus, the percentage of incident light that is ultimately reflected away from the insect's wing via constructive interference is greatly increased. In some butterflies, this percentage may approach 40%.

Whether generated by a single layer of scales or augmented by many, the structural color that an observer perceives should depend on the angle from which the wing is viewed (Figure 5.6b). When viewed from an oblique angle, the reflected waves seen by an observer have travelled farther while refracted through the scales than they have when viewed perpendicular to the wing surface. Consequently, the wavelengths that constructively interfere will be shorter, possibly making the wing appear violet rather than indigo or blue. This phenomenon, a form of iridescence, implies that an observer will perceive a variable color that depends on his/her relative position and the tilt of the butterfly's wings at that moment (see Ghiradella et al. 1972; see Vukosic et al. 2001 for a variation on this theme). However, some structural colors in butterfly wings are not iridescent but appear as the same hue regardless of the viewing angle. How are these non-iridescent hues produced? Via coherent scattering, a matrix of periodically spaced regions of altered optical density can, in principle, yield the same reflectance spectrum from any viewing angle. While butterfly scales are generally flat in overall profile, their internal anatomy is exceedingly ornate and includes regularly spaced ribs, vanes, pores, and lamellae (Figure 5.6a; Ghiradella 1984) that present numerous geometrical possibilities

a

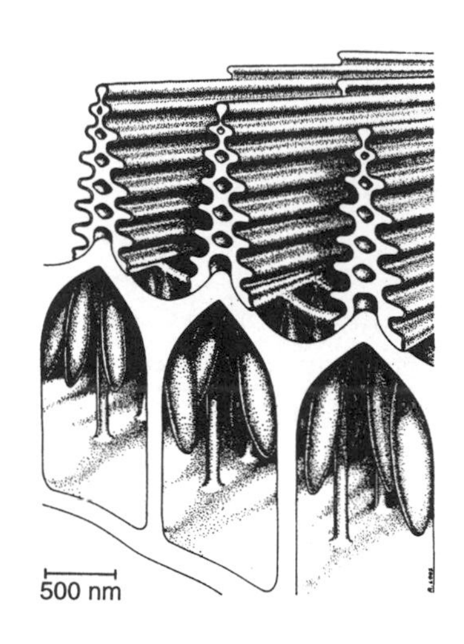

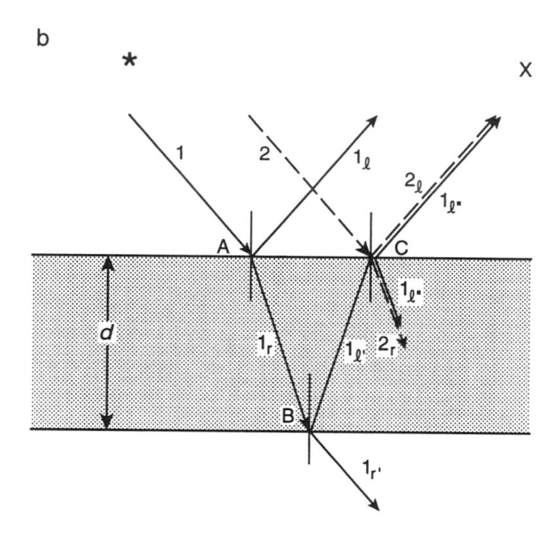

c

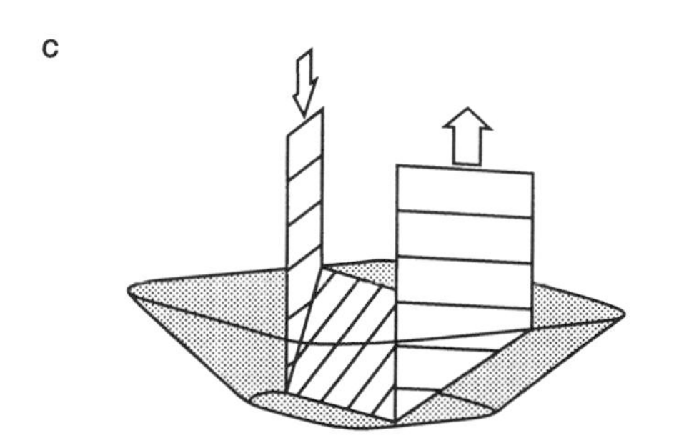

for maintaining a relatively constant spectrum of constructive interference at all viewing angles.[12] Scales may also bear perforated granules and other miniature structures that can generate a particular hue via combined diffraction and scattering (Morris 1975). Understanding how these ultrastructural features contribute to coloration is only beginning and will require sophisticated physical measurement and modeling.

In many butterflies, two or more distinct types of scales are found. A common arrangement is to have an inner tier(s) of pigmented "ground" scales overlaid with an outer tier(s) of transparent "glass" scales. The inner ground scales absorb longer wavelengths of incident light such that light reflected away from the wing is mostly violet or blue, while the outer glass scales can act as a diffraction grating for the reflected light and increase the angle of arc over which it spreads. Measurements taken of *Morpho* butterflies show that ultra-

---

Figure 5.6 Morphological features and mechanisms by which butterfly scales generate structural color. (a) Nanostructure of scale of *Eurema lisa* (Lycaenidae), showing parallel vaned lamellae and ellipsoid granules as revealed by scanning electron microscopy. (Reprinted with permission from "Ultraviolet reflection of a male butterfly: interference color caused by thin-layer elaboration of wing scales," H. Ghiradella, T. Eisner, R.E. Silberglied, D. Aneshansley, and H.E. Hinton, *Science*, vol. 178, 1215, copyright 1972 American Association for the Advancement of Science.) (b) Generation of structural color via constructive interference of light reflected and refracted by a thin film (shaded region) (e.g., a scale lamella) $\leq \lambda \cdot m/4$ in thickness, where $m = 1, 3, 5$, etc. When incident wave 1 reaches the scale surface at A, it is reflected (1_ℓ) and refracted (1_r) in accordance with its wavelength λ and the reflective and refractive properties of the scale. The refracted wave 1_r is reflected ($1_{\ell'}$) by the opposite surface of the scale at B and then passes back out of the scale at C($1_{\ell''}$). For incident light waves originating at *, an observer at X will mostly perceive those waves whose wavelengths in air are $\lesssim 4d \cdot n_2$, where d is the scale thickness and n_2 is the refractive index of the scale. This phenomenon arises because transverse waves reflected at a boundary between two media experience a 180° phase change when wave propagation is toward a medium in which wave velocity is slower (e.g., junction A), but they do not change phase when propagation is toward a medium in which velocity is faster (e.g., junction B). Consequently, reflected waves $1_{\ell''}$ and 2_ℓ (direct reflection of incident wave at C) will be in phase and constructively add. An observer viewing the scale surface at different angles will perceive structural color of longer or shorter wavelengths, depending on the viewing angle. In general, $\lambda = 4d \cdot n_2[(n_2^2 - \sin^2\theta_1)/(n_2\sqrt{(n_2^2 - \sin^2\theta_1)}]$, where θ_1 is the angle of incidence of waves whose reflections are observed at X. (c) Generation of composite structural color in *Papilio palinurus* (Papilionidae). Concave pigmented scales directly reflect yellow light from the center and retroreflectively yield blue light (open arrows and ruled planes) from the slanted margins; the angle of the slant and the scale width restrict retroreflection to waves of a specific wavelength and oriented as shown. The lateral spatial separation between the yellow and blue waves reflected by a given scale is sufficiently small that only their combination, green, can be perceived. (Reprinted from *Nature*, vol. 404, p. 457, "Structural colour: colour mixing in wing scales of a butterfly," P. Vukusic, J.R. Sambles, and C.R. Lawrence, copyright 2000 by permission of Macmillan Magazines Ltd. and Dr P. Vukusic, University of Exeter.)

structural features within the individual scales can generate dispersion of reflected light over an angle greater than 100° in one plane and greater than 15° in an orthogonal one. Diffraction by the outer tier of glass scales increases the angles of overall reflection from the wing still further (Vukusic et al. 1999).

Distinct regions within individual scales may also have profound influences on structural color. The dazzling green coloration observed in the Indonesian butterfly *Papilio palinurus* is generated via mixing light reflected by yellow pigment in the central region with light reflected by specialized retro-reflective surfaces in the outer region of each scale (Figure 5.6c). Because light waves reflected from the two parts of the scale are so close, an observer perceives the yellow and blue light as a single hue, their spectral combination of green (Vukusic et al. 2000). This mechanism of color generation is also found in the cuticle of beetles (Schultz and Bernard 1989), and humans have used the effect to create images in pointillistic painting and on television screens.

Intricate coloration incorporating a wide range of hues could also be generated by pigment granules in epidermal cells or by structural features of the cuticle,[13] and we may ask why scales have evolved so extensively as specialized emitters of color in the diurnal Lepidoptera. The above findings on *Morpho* and *Papilio* butterflies suggest that lightweight scales enhance the visibility and variety of visual advertisement signals more than alternative devices may. Layering of scales maximizes the capture of light for reflection, which can greatly increase the distance over which a female, or male, receiver may perceive signals from a fluttering wing, and, diffraction by outer glass scales would expand the angle of arc within which these signals are visible. Scales can produce either specific hues, including those which are not readily generated by pigment, or iridescent ones. The latter may be particularly valuable for adding yet another character, temporal pattern of hue, to a visual signal. A fluttering butterfly could present a regularly changing color spectrum to a receiver viewing from a fixed point (e.g., Vukusic et al. 2001).

Anatomically, scales are modified setae that have evolved in several insect orders in addition to the Lepidoptera. Like pigment compounds, they may serve various functions, including thermoregulation, aerodynamic enhancement, protection, facilitation of escape from entrapment by predators or their webbing, as well as coloration. The elaborate morphology of lepidopteran scales, especially the precise dimensions and periodic spacing of morphological features with respect to wavelengths of light, though, argues strongly that these structures have been modified specifically for generating color. Following, we examine the large-scale modifications of the distribution of scales on lepidopteran wings, features which establish the spatial patterns of their visual signals.

5.2.3 *Optical Stimuli versus Optical Signals*

A chemical or physical stimulus per se cannot be perceived as a signal unless it is discriminated from background noise (Section 2.4). Provided that receivers are fitted with the requisite detection apparatus and stimuli are sufficiently intense, however, discrimination may not be especially difficult to achieve along either the olfactory or mechanical channel. A signaler's pheromone is unlikely to be matched in chemical characteristics by background odors (but see Section 3.1.3), and a calling song usually differs from ambient sound in carrier frequency and tempo and exceeds it in intensity. However, visual environments in many habitats are exceedingly varied and complex, and merely displaying an optical stimulus of a particular light intensity and hue will not ensure a signaler's distinction from the optical background. Consequently, achieving a sufficient signal : noise ratio is a major problem in communication along the visual channel, and various characters of optical advertisement signaling may be devoted to it (see Endler 1992).

Bioluminescence is one way for an arthropod to generate a conspicuous optical signal. Although normally restricted to nocturnal or crepuscular activity and possibly limited by other unidentified factors, bioluminescent stimuli have the advantage of attaining a high signal : noise ratio with respect to the physical environment by just being turned on. Thus, some bioluminescent arthropods can communicate by emitting simple continuous glows. Nonetheless, noise in the biotic (bioluminescent) environment may select for specialization in additional signal characters. Species identity and mate quality and motivation are normally indicated and recognized by various temporal characters, as they are along the mechanical channel (Chapter 4).

Spatial Patterns. For diurnal arthropods, the task of generating a conspicuous optical signal is much more difficult and generally entails specific spatial patterns of hue and intensity. Moreover, these spatial patterns may be enhanced by movement of all or part of the stimulus as well as by a specific tempo. The diurnal Lepidoptera offer myriad examples for observing how spatial patterns in optical stimuli might be created via the distribution of scales bearing different pigments and generating various structural colors.[14]

Common schemes by which butterflies and diurnal moths generate spatial patterns enhancing optical stimuli include wing outlining and novel color patterns. Wings are often bordered by a hue that differs markedly from their interior, a pattern that may increase overall visibility of the signaler. And novel color patterns that contrast with the wing background may increase the likelihood that the signaler displays features that do not occur in its visual environment (Hailman 1977). Stripes, curves, polygons, circles, and other geometric patterns may all function as unique, visible markers. So diverse are these patterns among butterflies that a sample of 36 has been used to form a "butterfly alphabet and set of numerals" on a popular poster promoting Lepidoptery! Nonetheless, the functions of such patterns remain unchecked, and the potential errors resulting from assuming that human responses to stimuli parallel those of intended receivers should be avoided. As in other signals, trade-offs are expected between effective communication and concealment from unintended observers, and certain aspects of these optical stimulus patterns may actually improve a butterfly's camouflage as perceived by its natural enemies (Wickler 1968). Alternatively, they may direct a predator who has detected the butterfly to attack its wing margins rather than body (Blest 1957). In many butterflies upper and lower wing surfaces bear markedly different coloration, with the upper surfaces more conspicuous (to human observers). This dichotomy could imply that advertisement signaling occurs primarily during flight, when the upper wing surfaces are visible, and that resting butterflies seek concealment or at least display other types of signals.

Lepidoptera also display simpler patterns of specific hues on key areas of their wings, and it is these broad splashes of coloration that are more likely to function as sexual signals. Such patterns may vary in hue among species and sex, and in several cases they have been found to match spectral sensitivity peaks in conspecific receivers (Bernard and Remington 1991). In several species the male coloration, which may be accompanied by pheromones (Rutowski 1977a), is critical in courtship and territorial defense (Silberglied and Taylor 1978, Rutowski 1997, 1998), and it would be of interest to determine whether the optical patterns are graded and indicate signaler quality (see Wiernasz 1989, Brunton and Majerus 1995). As with courtship pheromones and acoustic and vibrational displays, the hue, shape, or size of an optical stimulus pattern could reflect a male's diet and potential for transfer of a valuable spermatophore, or his prowess in aggression and heritable viability.[15]

Signal Amplifiers. When an optical stimulus pattern is moved in a systematic manner, the movement may be acting as an "amplifier" (sensu Hasson 1990, 1991) of the stimulus pattern, or vice versa (cf. Section 3.1.3 on amplifiers in receiver mechanisms). Amplifiers would render a stimulus more conspicuous and thereby draw the receiver's attention to it. In arthropods that use compound eyes, movement amplifiers can be particularly important because poor spatial resolution may often prevent a receiver from detecting or fully evaluating a stationary stimulus pattern (cf. Taylor et al. 2000). Thus, optical stimulus patterns are often found on appendages, usually legs and wings, that can be waved or vibrated conspicuously. In butterflies, scale patterns on the wings may be amplified by fluttering movements of courting males in some species, and these patterns may be arranged in orientations that best enhance their apparent movement. Despite their extraordinary beauty, diversity, and conspicuous courtship, the movement stimuli of few butterfly species have been analyzed systematically (e.g., Magnus 1958; also see Rutowski 1977b, 1978, 1983, 2002). These stimuli may be difficult to simulate, and many butterflies court in flight and may not cooperate with laboratory testing protocols.

Technological Notes. As in acoustic communication, digital technology has played a crucial role in recent advances in our understanding of visual signaling and perception. Bioluminescent flashes are the easiest visual signals to simulate because many are simply point sources of a particular spectral range that vary in radiant flux over time (flash envelope). Thus, flash trains of nearly any design can be generated by the appropriately colored LED driven from a computer by digital : analogue (D : A) and voltage : frequency converters. This procedure has been used to test responses of female and male fireflies to features of conspecific signals in the laboratory and field (e.g., Moiseff and Copeland 1995). By varying a single temporal character at a time in flash playback trials, the importance of male flash rate—independent of other characters—was identified in the pair-formation of *P. consimilis* (Branham 1995).

Most reflected light signals are far more difficult to synthesize than bioluminescent flashes are, but through applications of video technology these displays too can be simulated digitally (Kunzler and Bakker 1998). By editing videorecorded or synthesized images and presenting them on a monitor, virtually any type of visual signal, no matter how complex in hue, spatial pattern, or movement, can be tested. Nonetheless, care must be taken to ensure that the equipment used can accurately display hues as perceived by the animals (Section 5.3.1) and rapid temporal patterns where they may represent signal characters (Section 5.3.1; D'Eath 1998, Fleishman et al. 1998). Additionally, some tests might be improved by developing and implementing an interactive playback of images, a technique that has proven useful in studying acoustic communication in birds (Dabelsteen 1992).

The crucial breakthrough in the video approach occurred during the late 1980s, when biologists first recognized that various animals would exhibit natural behavioral responses to such presentations (Evans and Marler 1991). Among arthropods, edited video images have been tested extensively with jumping and wolf spiders (Salticidae, Lycosidae). Here, playbacks of video images have identified banding patterns, hair tufts, and appendage movements that are critical for male courtship (Clark and Uetz 1990), demonstrated how females may evaluate a male's symmetry via his visual signals (Uetz and Smith 1999), and explored the function of optical stimulus patterns as amplifiers of movement signals (Hebets and Uetz 2000). Visual signals in other arthropod groups, including shore-dwelling crabs (see Aizawa 1998) and marine forms such as stomatopods, may prove to be equally amenable to testing by these video bioassays. But, even in arthropod species that assess mates and court on the wing, the video technique could be used for presenting optical stimuli in neuroethological tests on fixed or tethered subjects. Its primary value lies in allowing the investigator to extract and manipulate specific features of a complex optical background and determine whether such features are bona fide signal characters. Thus, we might learn which components of the pattern on an insect's wings are actually communicative, over what distances, and how a receiver's perception of the signal characters is influenced by the signaler's movement and its optical background.

5.3 Mechanisms of Visual Reception

Whereas most organisms possess some ability to detect and respond to the presence or absence of light, only three metazoan phyla, Mollusca (notably Cephalopoda), Arthropoda, and Chordata, have evolved eyes that can resolve complex optical images.[16] In general, the biochemistry of the visual pigments that absorb light is comparable in these visual groups, but optics of the receptor organs and neural processing vary greatly. Most arthropods see with a pair of compound eyes that contain anywhere from a mere eight (Collembola), or even fewer (Diptera: Nycterbiidae), to over 10 000 (Odonata) ommatidial units, each unit residing beneath an individual lens. Compound eyes may be situated at the distal ends of movable or fixed stalks projecting from the head (various Crustacea and Diptera), but in most insect groups they simply bulge on the anterior of the head, just beneath the vertex. Especially in active, aerial species, the compound eyes may cover a considerable proportion of the head surface, and their relative size and convex profile have led to our expression "bug-eyed." Indeed, some insects do appear as if their heads are all eye! Many insects also possess single-lens eyes, termed dorsal ocelli or stemmata, on the vertex or sides of the head. These organs may be very sensitive to light intensity but generally do not resolve complex images. However, in some chelicerates single-lens eyes that offer a high degree of acuity have evolved, albeit constrained by limited fields of view.

Previous chapters have repeatedly stressed the necessity of examining sensory perception in conjunction with signal transmission to understand communication. Nowhere does this dictum ring truer than in visual communication. Because of the human emphasis on vision, we are apt to succumb to relying on our own images and assume, subconsciously, that our perception of optical signals and phenomena coincides with the perception of the intended receivers of those signals. Thus, it is most critical that we attempt to see the world through an arthropod's eyes and come to appreciate the fundamentally different kinds of information that it extracts with them.

5.3.1 Compound Eyes: Multifaceted Vision

One way to understand how arthropod compound eyes function is by deconstructing them into their ommatidial units, determining the information that individual units provide, and then examining the manner in which the entire eye, and brain, assemble this information into an image. While this reductionist approach suffers because the full image does equal more than the sum of its ommatidial contributions, it is nonetheless an instructive sequence.

Ommatidial Structure and Function. Observation of an insect compound eye under the low magnification of a dissecting microscope reveals a honeycombed arrangement of hexagonal facets, the ommatidia, approximately 25 μm in diameter.[17] Each ommatidium is capped with a slightly convex lens that refracts incident light into the crystalline cone extending beneath (Figure 5.7). Light focused at the bottom of the crystalline cone is transmitted onward to sensory receptors tightly packed into a cylindrical structure, the rhabdom, that extends beneath the cone. The rhabdom is formed from the microvillar projections of the eight or nine retinula cells that surround it. These projections, termed rhabdomeres, are the sensory receptors containing the visual pigments, membrane-bound proteins,[18] and phototransduction takes place there. Light that has been transmitted through the lens, the crystalline cone, and into the rhabdom is absorbed in its rhabdomeres if the wavelength(s) matches or is similar to the absorption peaks of the pigments. Absorption stimulates a positive DC voltage shift, a generator potential, in a retinula cell(s), which then ascends to cells in the lamina, medulla, and lobula neuropils of the brain (Shaw 1984). Over a given dynamic range, the magnitudes of these generator potentials are commensurate with light intensity (radiance).

Color Perception. Three retinula cell categories, which are distinguished by the absorption spectra and peaks of their visual pigments, are normally found among the eight or nine cells in a given ommatidium (Figure 5.7c), and a phylogenetic analysis suggests that this arrangement is ancestral among insects (Briscoe and Chittka 2001). Functionally, the separation of these absorption spectra allows an insect to evaluate the spectrum of light incident at that ommatidium (cf. Figure 5.3). In the honeybee *A. mellifera*, the three categories of retinula cells or pigments have absorption peaks at 335, 435, and 540 nm (λ_{vac}); each has a sensitivity extending 50–100 nm above and below the peak (Menzel et al. 1986).[19] With this array of pigments segregated in different retinula cells, a honeybee can detect light ranging from UV to orange and also distinguish color within that spectral range. The honeybee's ability to distinguish color was originally recognized via experiments on associative learning in the context of foraging (von Frisch 1914), and it is now known to be most acute at approximately 400 and

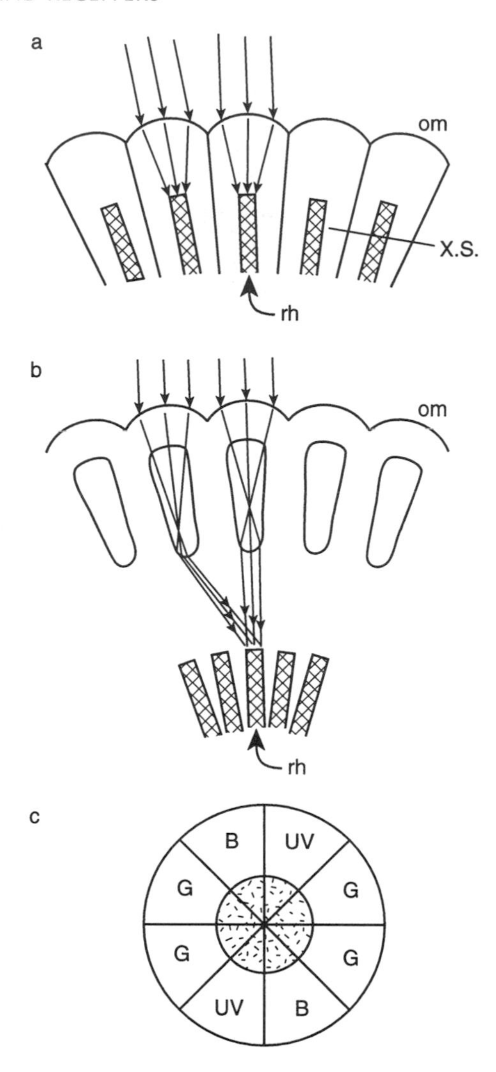

Figure 5.7 Compound eye structure and function. (a) Sagittal section of apposition eye, in which only light waves propagating along the axis of a given ommatidium (om) and entering through its lens are admitted to its rhabdom (rh). (b) Superposition eye, in which parallel light waves entering through the lenses of neighboring ommatidia reach the rhabdom of a given ommatidium; specific design for a refracting superposition eye is shown. (c) Cross-section of ommatidium (from a), showing eight retinula cells with photoactive central projections forming the rhabdom (stippled region). Letters indicate receptor (visual pigment) categories (UV: ultraviolet; B: blue; G: green); specific design for a closed rhabdom is shown (see Nilsson 1989).

500 nm. For light in these two regions, honeybees can discriminate hues differing as little as 5 nm in wavelength (von Helversen 1972).[20] Similar to mechanisms of color vision in vertebrates, perceived hue reflects the across-fiber pattern

yielded by the relative generator potentials evoked among an ommatidium's three categories of retinula cells; in Diptera, this pattern begins to be processed at the level of the medulla neuropil. Most spectra will evoke a specific set of relative generator potentials and hence yield a unique across-fiber pattern, but it is possible for two different, complex spectra to yield identical patterns that would be perceived as the same hue (see Figure 5.3).

Three categories of visual pigments or retinula cells, a trichromatic system, offer finer color discrimination than two would, and an ommatidium with a single pigment category would not be able to separate light intensity from wavelength (cf. Sections 3.1.3 and 4.3.5). However, additional retinula cell categories would offer even more acute discrimination, and we may ask why ommatidia are generally restricted to three? Possibly, the added neural connectivity required by four or more cell categories per ommatidium would impose enough of a burden on development and across-fiber neural processing that the benefit of heightened color discrimination would not offset its cost. Nonetheless, some terrestrial arthropods (e.g., papilionid butterflies; Arikawa et al. 1987) do have more than three categories of pigments and retinula cells, and all categories may co-occur within a single ommatidium. Such visual pigment diversity may offer a finer discrimination of hues that could be valuable for assessing both host plants on which to oviposit, and mates.

Among marine arthropods, stomatopods also use an expanded diversity of visual receptors. Several accounts suggest that as many as 16 different receptor categories, each with a different absorption peak, may be found in some species (Cronin and Marshall 1989, Marshall et al. 1991). Unlike the visual receptors in butterflies and other arthropods that have been closely examined, stomatopod receptors have sharply tuned absorption spectra (cf. Figure 2.3) owing to intraocular filtering, which may be developmentally adjusted for the general light environment (Cronin et al. 2001). Because these species inhabit shallow waters where the available spectrum changes markedly over a relatively short vertical distance, their sharp absorption spectra may improve "color constancy" (Osorio et al. 1997, Marshall et al. 1999b). That is, visual receptors tuned to the general background light typically fatigue, or adapt, in their response. Were broadly tuned receptors used, adaptation would cause a particular object to be perceived as one hue near the surface but as a different one several meters below. Such color inconstancy arises because a major portion of the eye's receptors are always desensitized, but the extent to which each of the several receptor categories is desensitized changes in different habitats. However, a stomatopod using an array of narrowly tuned visual receptors will suffer from this effect much less, albeit at the expense of diminished sensitivity to light intensity, because fewer of its many receptor categories are strongly adapted in any one habitat. Thus, an object that is spectrally dissimilar from the general backgrounds will appear relatively constant in hue. Stomatopods are known for their brilliant and elaborate coloration patterns, and color constancy could be critical for accurate perception of these features in sexual and social interactions. Given the polychromatic perception of color in stomatopods, the role of color patterns in their signaling and communication (see Hazlett 1979) deserves careful study in the future.

Sensitivity to Light Intensity. The sensitivity of an ommatidium to the intensity of incident light is a function of its diameter, the layering of visual pigment in the rhabdom, and optical isolation from neighboring ommatidia. An ommatidium with a wide lens can capture a high percentage of incident light and thereby evoke generator potentials whose signal : noise ratios are high. Consequently, the generator potentials stimulated by even relatively dim light will be significantly higher than the potentials resulting from spontaneous activity in the sensory receptor cells. More importantly, differences between the generator potentials of adjacent ommatidia will be easily detectable, which would improve both intensity and spectral contrasts in the full image assembled by the compound eye. Sensitivity to light afforded by large ommatidia may come at a high price, however, because resolution of the image that a compound eye assembles can be severely compromised.

Once incident light that is accepted by the lens enters the rhabdom, most of it is reflected internally owing to a high refractive index of the rhabdomeres. Thus, the rhabdom acts as a wave guide in the manner of a fiber-optic apparatus. Because little light escapes to surrounding cells, multiple layers of visual pigment in the rhabdom can increase the chance that a given photon is absorbed and contributes to a generator potential.

In a majority of diurnal insects, individual ommatidia are optically isolated from one another further by screening pigments. Thus, an opaque sleeve surrounds the crystalline cone, retinula cells, and rhabdom of each ommatidium and prevents light incident at neighboring ommatidia from entering. Compound eyes of this design are termed apposition eyes, and each of their cartridge-like ommatidia provides two basic items of information: spectral composition and intensity of light incident from the direction that ommatidium faces. Although the lens of an ommatidium does create an (inverted) image at the rhabdom, fusion of the rhabdomeres into a single light guide normally precludes resolving any aspect of this image.[21]

When sensitivity to light is of the utmost importance, as in nocturnal species, alternative optical designs that increase the effective aperture of the ommatidium are often found. Many insect species have various types of superposition eyes in which the optical isolation of ommatidia is incomplete (Nilsson 1989). Here, light entering the rhabdom of a given ommatidium includes not only those waves normal to the overlying lens but also parallel waves that enter through the lenses of neighboring ommatidia (Figure 5.7b). A variety of refractive and reflective devices gather light in this fashion, and they can increase by several orders of magnitude the amount of light available for absorption in any one ommatidium. For example, in *Ephestia* sp., a nocturnal moth using a superposition eye, the radiance needed to stimulate each photoreceptor with 1 photon·s^{-1} is approximately 10^{10} photons·m^{-2}·sr^{-1}·s^{-1}, comparable to human sensitivity and two orders of magnitude less than that needed in the honeybee *A. mellifera* (Kirschfeld 1974a, Land 1981). But, the increased sensitivity in superposition eyes may be achieved at the cost of image resolution, because the optical isolation of neighboring rhabdoms is greatly reduced. Additionally, specialized zones of acute vision (see below) do not occur.

Image Formation, Resolution, and Field of View. From the above perspective, admittedly a reductionist and simplistic one, formation of a full image by the compound eye retina (and lamina) is analogous to formation of an image by a computer screen. Because the compound eye surface is curved, the axis of each ommatidium projects in a slightly different direction than its neighbors. Thus, the ommatidia create a mosaic array of pixels, each representing the spectrum and intensity of light incident from a unique direction. This array is an erect (non-inverted) image of the eye's field of view, and its grain reflects the angular separation of adjacent ommatidial axes, $\Delta\Phi$ (Figure 5.8). Here, we may consider image grain to correspond with the compound eye's minimum spatial resolution, its ability to resolve two closely spaced objects that contrast with the visual background in hue or intensity. Minimum spatial resolution is calculated as $2\Delta\Phi$, a value representing the minimum angle that must occur between two such objects in order for the compound eye to recognize them as separate. Typical values for $\Delta\Phi$ in insect compound eyes are 1–2° of arc. Aeshnid dragonflies have the most acute resolution measured, with values as low as 20′, whereas the values in some insects and other arthropods exceed 3° (Land 1997). In contrast, $\Delta\Phi$ between adjacent receptors in the foveal region of the human eye is less than 30″ of arc, and the values in certain birds are even lower.

A fine-grained compound eye image, offering superior spatial resolution, may be achieved in an eye of given size by either decreasing ommatidium diameter or increasing the radius of curvature of the overall eye surface. These factors would decrease the angular separation of ommatidial axes, but both would limit other visual functions. Below, I analyze these trade-offs and arrive at the conclusion that any advantages in perception and overall fitness gained via finer grain and improved resolution may be more than offset by these other limitations.

When ommatidium diameter is decreased, sensitivity declines and diffraction may become strong enough that resolution is actually impaired rather than improved; this latter constraint on compound eye performance was first recognized over a century ago (Mallock 1894). Sensitivity is adversely affected because a smaller lens intercepts less light and may therefore focus too few photons into the rhabdom to evoke a significant generator potential or a potential significantly different from those evoked in neighboring ommatidia. The extent to which diffraction impairs the image focused by a lens is inversely related to D/λ, where D is the lens diameter, which represents an aperture in a solid object, and λ is wavelength of the incident light. Diffraction blurs the image by creating a pattern of concentrically spreading discs, termed the Airy pattern, where a single disc of light would otherwise be expected. Whereas the rhabdom of a given ommatidium does not normally resolve an image, diffraction affects overall resolution of the compound eye because the light reaching a rhabdom not only includes waves normal to the overlying lens but also waves

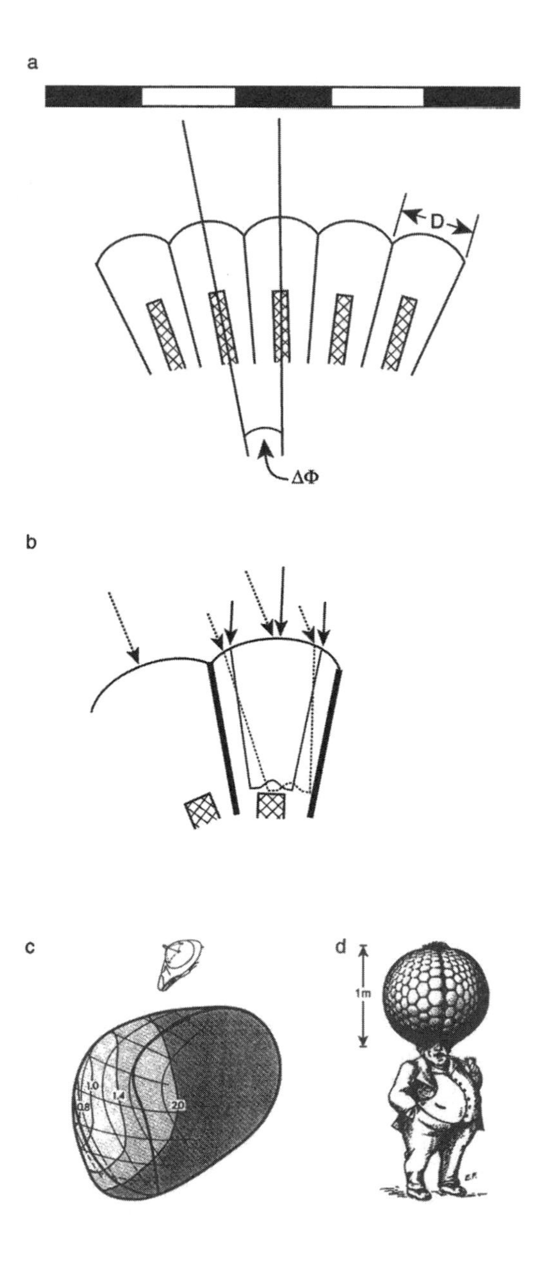

that are slightly off-axis (Figure 5.8b). Consequently, each ommatidium would provide information not only on the light incident from the direction it faces but from the directions faced by neighboring ommatidia as well. This effect may offset the more finely grained image gained from the reduced $\Delta\Phi$ associated with small ommatidia. The minimum diameters found among ommatidia are approximately 10 μm, and this limit is probably set by both sensitivity and diffraction (Land 1989). Because shorter wavelengths of light are diffracted less strongly than longer ones, the spectral shift toward UV sensitivity found among arthropods may represent an adaptation for maintaining image resolution.[22]

We may note that ommatidia in most compound eyes are considerably larger in diameter than the 10-μm limit, which suggests the importance of maintaining sensitivity over resolution. The incidence and diversity of superposition eyes among insects also indicate the priority of sensitivity, because these light-collecting devices do not allow for flattened zones affording acute vision (Land 1989; see below and Figure 5.8c). Additionally, it is critical to recognize that a compound eye's minimum spatial resolution only limits its ability to discern the detail present in a stimulus but not its ability to detect a stimulus. A stimulus that subtends an angle of arc much smaller than $\Delta\Phi$, the approximate angular diameter of an ommatidium, may be detected provided that its light, focused primarily within a single ommatidium, has sufficient contrast against the visual

---

Figure 5.8 Resolution of the compound eye. (a) The ability to distinguish parallel contrasts, as in a black and white grating, is inversely proportional to the interommatidial angle, $\Delta\Phi$ (see Figure 5.7). The maximum angular spatial frequency, repetition of black (or white) stripes per radian, that can be resolved $= 1/(2\Delta\Phi)$, but actual values in compound eyes are normally lower owing to blurring caused by diffraction (cf. Figure 5.2e). (b) Diffraction-induced blurring can arise at very small ommatidium diameters (D) because a greater proportion of light waves reaching the rhabdom of a given ommatidium are off-axis; i.e., light entering along the axis of an ommatidium (solid arrows and distribution curve above rhabdom) as well as some light parallel to the axes of neighboring ommatidia (dotted arrows and curve) will be focused on the rhabdom, especially when rhabdom diameter is large, as in dark-adapted superposition eyes. In general, diffraction of light entering an ommatidium is proportional to λ/D. (c) Compound eyes may bear flatter regions distinguished by smaller $\Delta\Phi$s (contour values on eye surface map) and higher resolving ability; a frontal zone of acute vision, featuring $\Delta\Phi\text{s} < 0.8°$, is indicated for the praying mantis *Tenodera australasiae*. (Reprinted from "Regional differences in photoreceptor performance in the eye of the praying mantis," *Journal of Comparative Physiology A*, vol. 131, p. 101, fig. 5, S. Rossel, copyright 1979 with permission by Springer-Verlag.) (d) Fly-man: a human fitted with compound eyes would require organs of grotesque proportion in order to match the resolution offered by our single-lens eyes. (Reprinted from "The resolution of lens and compound eyes," p. 365, fig. 7.4c, K. Kirschfeld, in *Neural Principles in Vision*, eds F. Zettler and R. Weiler, copyright 1976 with permission of Springer-Verlag.)

background focused in neighboring ommatidia. The detection of contrast is enhanced by sensitivity, and a sensitive compound eye may thereby detect the flash of a distant firefly or the silhouette of a small insect against the sky (e.g., Stavenga 1992). While no detail would be resolved in these single-ommatidium images, they may nonetheless represent valuable information on sexual advertisements of signaling insects.

When the radius of curvature of the eye surface is increased while overall eye size remains fixed, the eye becomes flattened and its field of view is reduced. For eyes of a given surface area and total number of ommatidia, a flatter eye would offer improved spatial resolution but only in a limited direction. This trade-off would likely restrict a flying insect's maneuverability and its capacity to detect predators or prey. Many active, aerial insects (e.g., Odonata, Diptera, Lepidoptera, and Hymenoptera) have compound eyes that either bulge conspicuously from the head or cover a significant portion of the head surface, and in some species the field of view may approach a solid angle of 4π sr: vision in all directions, with an absence of blind spots. Thus, it appears as if a wide field of view is a critical feature that is seldom relinquished. Where active, aerial insects have evolved specializations for enhanced resolution, they have relied on increasing the radius of curvature of only limited zones of the eye surface (Land 1989; Figure 5.8c). These flattened zones of acute vision are typically positioned anteriorly or dorsally on the eye and may improve the insect's ability to maneuver and to detect and assess potential mates (e.g., Stavenga 1992) or competitors. But, for every zone that is flattened, other regions compensate with a reduced radius of curvature and hence low resolution, and the eye's wide field of view is maintained (Figure 5.8c).

Given that high resolution cannot be achieved with exceedingly small ommatidia and that a flat compound eye limits the field of view markedly, do insects have other options for improving image detail? Yes, but the trade-off would again be severe. Ommatidia might be kept at their normal size, or even enlarged to increase sensitivity, while their number and the eye's overall dimensions are expanded. This option has been evaluated quantitatively, and the calculations indicate that a compound eye with spatial resolution equivalent to human vision would need a diameter of several meters (Mallock 1894, Kirschfeld 1976; Figure 5.8d). Because even modest improvements to resolution would require eyes far too unwieldy to be supported by a flying, or walking, insect, superior resolution is simply not a viable option as long as compound eyes are used and wide-field vision is important (but see Section 5.3.2). Thus, visual communication in insects must operate under this constraint, and the nature of optical signaling has been shaped by it (Section 5.4).

Perception of Polarized Light. Whereas poor visual resolution may severely limit an arthropod's ability to perceive the spatial details in optical stimuli, many species have markedly superior abilities in other aspects of vision. Detecting whether incident light is polarized and determining its plane of polarization are perhaps the most noteworthy of these other abilities. The mechanism responsible for arthropod polarization sensitivity resides entirely within the sensory receptors of the compound eyes, and it is based on the manner in which visual pigments are arranged in the microvillae of the rhabdomeres. Neither the lenses nor crystalline cones of the ommatidia contribute toward the mechanism by acting as polarizing filters (Shaw 1967).

Visual pigment molecules absorb maximal light energy when their long axes are arranged perpendicular to the path along which incoming waves are propagated and parallel to the plane in which the wave oscillates. The microvillae of an arthropod rhabdomere project perpendicular to incoming light, and pigment molecules tend to be arrayed parallel to the microvillar axes (Figure 5.9). Thus, when light is heavily polarized, some microvillae will absorb maximally because their axes are parallel to the plane of wave oscillation (polarization), whereas others will absorb minimally because they are perpendicular to the polarization

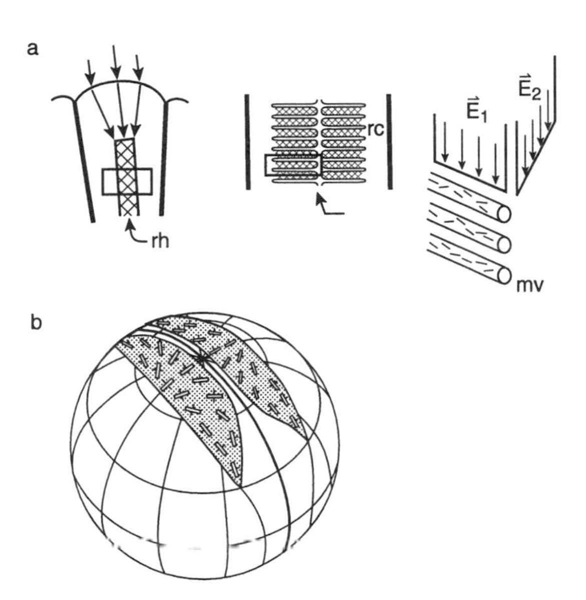

Figure 5.9 Evaluation of plane of polarization of light. (a) Microvillar projections (mv) of a retinula cell (rc) bear visual pigment molecules that are largely aligned with the microvillar axes. Dichroism: light waves entering the rhabdom (rh) and whose **E** vectors are parallel ($\mathbf{E}_1$) with the microvillar axes will generate maximum photoactivation of pigment, whereas waves whose **E** vectors are transverse ($\mathbf{E}_2$) to the microvillar axes will generate little (Goldsmith 1991). (b) Dorso-median zone of polarization sensitivity in the compound eyes of the honeybee *Apis mellifera*. Graph indicates spherical coordinates of the zone in each eye and the alignment of sensitive microvillar axes in different parts of the zone. (Redrawn from "Polarization sensitivity in compound eyes," p. 304, fig. 3b, S. Rossel, in *Facets of Vision*, eds D.G. Stavenga and R.C. Hardie, copyright 1989 with permission of Springer-Verlag.)

plane, a phenomenon termed dichroism. If all of the microvillae in the rhabdomere of a retinula cell project in the same direction, that cell will experience maximum excitation when oriented with its microvillar axes parallel to the polarization plane but minimum excitation when turned 90°. When fitted with specialized ommatidia whose rhabdomeres bear parallel microvillar arrays, an arthropod can determine the polarization plane by scanning the visual environment and adjusting its orientation until those receptors are maximally excited (Rossel and Wehner 1984).[23] This option is not available to vertebrates, whose retinal photoreceptors do not include parallel arrays of pigment molecules, and their sensitivity to polarized light is generally poor.

Neurophysiological investigations have identified polarization sensitivity operating according to the above principle in a diverse range of terrestrial as well as freshwater and marine arthropods. The sensitivity is often restricted to a limited region of the eye (Figure 5.9b) and associated with a single pigment or retinula cell category, normally that category sensitive to UV light (e.g., see von Helversen and Edrich 1974). This latter association is believed to allow for the separation of hue from polarization with minimum neural processing.[24]

Most behavioral investigations of polarization sensitivity have focused on its function in orientation and navigation, although its role in discerning suitable habitat and oviposition substrates, especially under dim light where intensity and spectral contrasts would be faint, has also been considered (Kelber 1999). But the light reflected by arthropods themselves will be polarized in the same manner that skylight and light reflected from vegetation are, and species that signal optically may generate characteristic spatial patterns and tempos of polarized light glinting off their reflective surfaces in specific planes. Because such spatial patterns and tempos might be readily detected and evaluated with a high degree of reliability by arthropod receivers, polarization sensitivity may be central to visual communication in certain species. This possibility, largely overlooked until now, should merit careful attention as the sophisticated polarization sensitivities of arthropod species are analyzed further (see Marshall et al. 1999a).

Perception of Tempo and Spatial Pattern. Superior perception of tempo, the ability to recognize and evaluate a rapidly flickering image as discrete pulses of light rather than a continuous blur, is another notable feature of arthropod vision. Unlike polarization sensitivity, though, high temporal sensitivity is primarily restricted to diurnal, aerial insects. This association probably occurs because tempo perception is a vital component of the visual processing necessary to fly through complex, obstacle-strewn environments. In order to complete such aerial maneuvers without collisions, an insect processes spatial information from the images that loom in its frontal visual field and that flow across its lateral fields (cf. Section 3.1.4). Owing to high angular velocity of this flow, processing may consist of nothing more than determining the tempo at which successive contrasts are perceived in a given ommatidial region and determining the spatial frequency of contrasts perceived in a snapshot of the visual field at a given instant; a flying insect's ability to determine spatial frequency is related to its tempo perception because a snapshot featuring sharp contrasts would have to be taken by integrating light absorbed over an extremely brief time interval. In both modes, the perception of contrast may be restricted to light absorbed by a single retinula cell category[25] and enhanced by lateral inhibition. A photoreceptor's generator potential is increased when its neighbors' potentials are low and decreased when its neighbors' potentials are high (e.g., Kirschfeld 1974b). Because sharp intensity or spectral contrasts normally signify the edges of objects, merely analyzing the tempos and spatial frequencies of contrasts may often be sufficient for avoiding obstacles, and possibly for recognizing perches and landing sites as well.

As in other visual abilities, superior tempo perception is maintained at a cost. Tempo perception normally requires a relatively high radiance of incident light, which implies that the sensitivity to light must be high. Should sensitivity be low, contrasts within a given ommatidial region over time or across the ommatidia at a given instant may not be sharp enough for tempo or spatial frequency to be perceived. Thus, a trade-off can arise between tempo perception and minimum spatial resolution, because the latter may entail reduced sensitivity by relying on small ommatidia.

Whereas the high flicker fusion frequencies characteristic of diurnal, aerial insects, which approach 300 Hz in some species (Autrum 1950),[26] and the ability to evaluate flicker frequencies below the fusion limit most likely evolved as part of flight control systems, insects might also rely on the ability in activities outside of navigation. Various experiments have shown that insects can recognize some spatial patterns, as may occur in association with valuable resources, by analyzing tempos or spatial frequencies of contrasts that they perceive from the patterns. Such analyses may also be central to the recognition of certain visual signals. Consider a male fly or butterfly bearing a series of transverse bands on his wings who advertises to females via characteristic wing movements. Females, as well as other males, might recognize the banding (e.g., Warzecha and Egelhauf 1995) and its movement by detecting a specific tempo of contrasts in a given ommatidial region combined with a specific spatial frequency of sharp contrasts in a snapshot image.

While there is no doubt that insects can and do evaluate the tempos and spatial frequencies of contrasts perceived from patterns, various experiments show that they also recognize elements in patterns that could not be determined by just evaluating these simple contrast characters. Findings, largely obtained in honeybees, demonstrate that more complex shapes and color patterns can be recognized, and one may infer that many other species equipped with similar perception can evaluate and recognize such spatial designs when encountered in a variety of contexts, including visual signals as outlined above.

What mechanisms might insects employ to evaluate spatial patterns? Perhaps influenced by an assumption of neural simplicity, some investigators claimed that patterns are recognized by a retinotopic matching process: components of an image are matched, pixel-by-pixel, with a template stored for the same retinal region in which the image is focused (Dill et al. 1993, Dill and Heisenberg 1995). Thus, retinotopic matching is not a position-invariant process, and an object or pattern such as a fly's visual advertisement signal would have to be viewed from a standard angle and at a standard range to be recognized. Although this process might afford superior precision and recognition of spatial detail, the viewing constraints would appear to relegate it to rather limited utility.

Recent experiments (e.g., Campbell 2001), including one (Ernst and Heisenberg 1999) performed by an early proponent of retinotopic matching, cast serious doubt on occurrence of the process. Rather, it is now argued that the visual system extracts and stores various abstract components of a spatial pattern and uses these components to recognize a novel pattern bearing one or more of them (Srinivasan et al. 1993, 1994, Srinivasan 1994, Ronacher and Duft 1996). For example, insects may evaluate the symmetry of objects or patterns and recognize the more symmetrical one even though its other features do not match those of any object or pattern previously encountered (Giurfa et al. 1996, Lehrer 1999). Such feature detection processes would be position-invariant (i.e., recognition would not be affected by translation of an object across the visual field). They may also be range- and angular or apparent size-invariant, because insects can determine absolute size of an object by combining its angular size with its estimated range (Horridge et al. 1992; see Schwind 1989 on mechanisms of range estimation by compound eyes). The experiments suggesting that feature detection processes operate in pattern recognition imply that insects and mammals may have converged on similar neural solutions to accomplish complex visual analyses. Most of the evidence for feature detection processes in insects is strictly behavioral, but one study (O'Carroll 1993) claims to demonstrate functional similarities between cells in the optic ganglia of the dragonfly brain and in the mammalian cortex.

At present, neuroethological investigations of pattern-recognition mechanisms in insects have been largely restricted to the honeybee *Apis mellifera* and the fruit fly *Drosophila melanogaster*, and specialized signals used in visual communication have not been studied. Nonetheless, findings from these two species clearly indicate that various advanced mechanisms exist with which insects can evaluate some spatial patterns, including the sorts of patterns that often represent critical characters of optical signals. Here, the creative application of video methodology to testing the perception of optical stimuli may reveal how such mechanisms might operate in signal recognition and evaluation.

Why Have Compound Eyes? By most accounts, compound eyes appear to represent a perceptual constraint for arthropods. Their optics restrict minimum spatial resolution to values at least 10 times greater than could be attained with a comparable single-lens eye, where, relatively unaffected by diffraction, individual photoreceptors may be as little as 1 μm in diameter and tightly packed. Moreover, any modifications that might improve upon the crude images that even the best compound eyes generate would entail unrewarding compromises. Reducing the inter-ommatidial angle $\Delta\Phi$ in all regions of the eye would sharpen acuity but yield a reduced field of view, and reducing $\Delta\Phi$ locally via flattening would offer sharp acuity in one direction but even poorer resolution elsewhere. Increasing the time interval over which absorbed light is integrated would improve sensitivity and hence image resolution for a stationary insect viewing a fixed object, but it would also impair tempo perception, an ability that aerial insects can ill afford to forgo.

The above considerations might lead one to conclude that once compound eyes originated in the arthropod lineage, these organs could not readily evolve toward a single-lens format offering better resolution. However, several variations on the single-lens theme have evolved among arthropods (Section 5.3.2), and at least one of these variations has undoubtedly descended from an ancestral compound-eye state. Thus, phylogeny and an unbridgable morphological chasm may not be responsible for the prevalence of the compound eye. Rather, certain optical features may render compound eyes advantageous, particularly for small, active organisms.

One optical feature that compound eyes offer more effectively than single-lens eyes can is an extremely wide field of view (Wehner and Srinivasan 1984). Were a fly fitted with two single-lens eyes in place of its compound ones, it would suffer from a much narrower field. Additionally, the retina in a large single-lens eye is located far behind the lens, and transparent eyeballs rather than visual neuropil and other brain sectors would fill much of the head. In order for a fly to use single-lens eyes with fields of view comparable to those offered by their compound eyes, it would have to support a set of apparatus far larger than it could ever bear. Because active, and particularly aerial, arthropods may value a panoramic view over sharp acuity, compound eyes prevail.

5.3.2 Single-lens Eyes

Eyes in which multiple photoreceptors are clustered beneath a single lens are found among arthropods in larval endopterygote insects, as supplements to compound eyes in both nymphs and adults of assorted exopterygote insects and adult endopterygotes, and in spiders (Araneae). In exopterygotes and adult endopterygotes, they are found on the vertex of the head and termed dorsal ocelli.[27] These organs generally do not resolve images, which are usually focused behind the retina, but may simply indicate the presence or absence of light. It is believed that dorsal ocelli can be used as horizon detectors for aerial species flying under dim light conditions (Wellington 1974). On the other hand, the single-lens eyes, termed stemmata, found laterally on the head of many endopterygote larvae can resolve images, albeit very crude ones in most cases. Because most larvae do not zoom through their habitat at high speeds, send or receive optical signals for sexual or social purposes, and were often oviposited on food resources by their mothers, vision beyond the coarsest of images may be a luxury that they can easily do without. However, the stemmata of adephagous Coleoptera larvae (e.g., tiger beetles), many of which are active predators, may offer reasonably acute resolution and other visual abilities (Gilbert 1994).

Segregation of Acuity and Panorama Functions in Spiders. Spiders, in contrast to most other arthropods, have evolved single-lens eyes that may be distinguished by highly developed resolution and magnification. Typically, a total of eight single-lens eyes, each with its own neuropil and pathways to the brain (Strausfeld and Barth 1993, Strausfeld et al. 1993), are arranged in two or three rows on the frontal carapace of the prosoma (cephalothorax; Figure 5.10a). The two anterior median, also termed principal, eyes have very narrow fields of view but superior resolution. These eyes may also feature telephoto vision (Williams and McIntyre 1980), and muscular activity can redirect them to focus on an object of interest. Minimum spatial resolutions as low as 2′24″ of arc are found in jumping spiders, which can recognize a conspecific by its visual information at a distance of 30 body-lengths (30 cm) (Jackson and Blest 1982). Wide-field vision in spiders is generally provided by the posterolateral eyes, whose acuity is quite poor. When the posterior lateral eyes detect rough outlines of objects or movements, the anterior median eyes receive an instruction to focus in the correct direction. This scheme for attaining both panoramic vision and sharp acuity by segregating the two functions within different, specialized eyes is probably effective for prey recognition in spiders, who employ sit-and-wait tactics. Because spider mating and social interactions may entail similar activity and movements, the segregation of visual functions might also be useful for recognizing and evaluating conspecifics and their optical signals (Forster 1982).

Chunk Sampling in Male Strepsiptera. An extraordinary variation on the spider visual scheme has apparently evolved in the Strepsiptera, twisted-wing parasites of Hymenoptera and several other insect orders. Superficially, strepsipteran males seem to have ordinary compound eyes, but a closer inspection of these small insects reveals that their eyes are comprised of a limited number (≈ 50) of unusually large facets (approximately 65 μm in diameter).[28] In the North American strepsipteran *Xenos peckii*, each facet's lens covers approximately 100 photoreceptors, and neuroanatomical investigation shows that the photoreceptors below each lens form a distinct, inverted image at the level of the retina

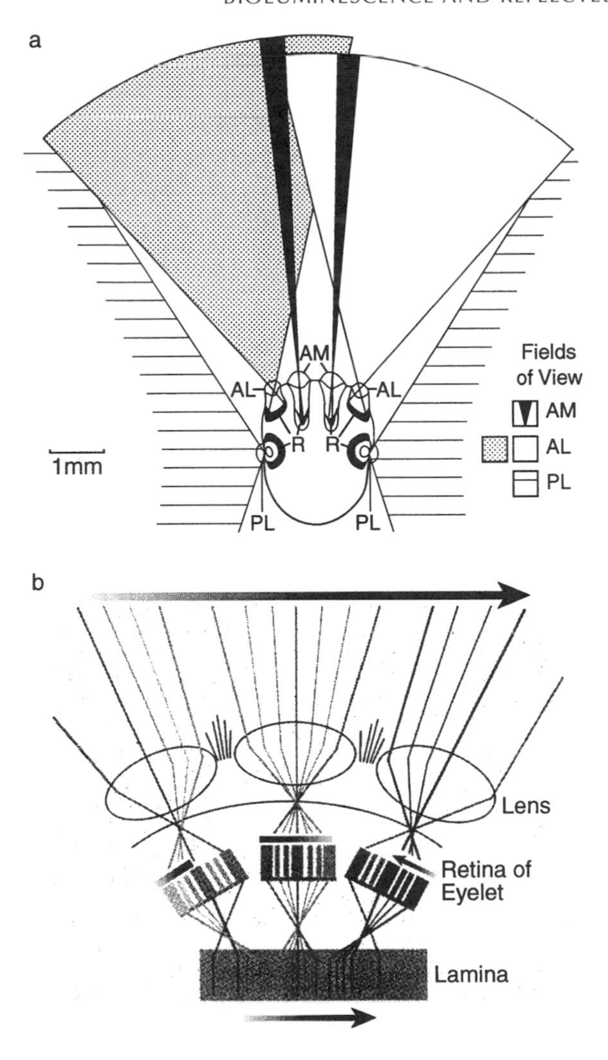

Figure 5.10 Optical mechanisms in arthropods for combining field of view and resolution. (a) The various single-lens eyes of jumping spiders (Salticidae) offer high resolution with a narrow field of view (AM: anterior-median eyes) or low resolution with a wide field of view (AL: anterior-lateral eyes; PL: posterior-lateral eyes). R indicates location of the retina (cf. Figure 5.7). (From L. Forster, "Visual communication in jumping spiders (Salticidae)," p. 161 in *Spider Communication*, eds P.N. Witt and J.S. Rovner. Copyright © 1982 by Princeton University Press. Reprinted by permission of Princeton University Press.) (b) The eye of the strepsipteran *Xenos peckii* is composed of approximately 50 single-lens "eyelets," each of which forms a high-resolution, inverted image at the level of the retina. Owing to chiasmata between the retina and lamina neuropils, it is inferred that the retinal images of the various eyelets are assembled into a single coherent, and erect, image at the level of the lamina. (Reprinted with permission from "Chunk versus point sampling: visual imaging in a small insect," E. Buschbeck, B. Ehmer, and R. Hoy, *Science*, vol. 286, p. 1179, copyright 1999 American Association for the Advancement of Science.)

(Buschbeck et al. 1999). These separate, inverted images are then assembled at the level of the lamina into a single image encompassing the eye's entire field of view (Figure 5.10b). Chiasmata between the retina and lamina reverse the separate, inverted images so that each is erect and hence combines with neighboring images in a coherent arrangement. Thus, each facet represents an "eyelet" that samples a chunk of the eye's field of view and processes the image of that chunk at high resolution.

Strepsipteran eyelets and the anteromedian eyes of spiders are both single-lens eyes in which photoreceptors are densely packed and hence offer sharp acuity (Buschbeck et al. 1999). Unlike the spider eyes, however, strepsipteran eyelets are multiple and wrap around much of the head, and their resolved images are combined at the level of the lamina to form a single, coherent panorama. Consequently, Strepsiptera enjoy both the surround vision typical of the compound eye and the sharp acuity of the single-lens eye at all times. Do the habits of strepsipterans offer any clues to selection pressures that may have led to evolution of their novel eyes? Male strepsipterans fly in search of females, who never leave their hymenopteran hosts, and this pursuit may represent a difficult perceptual task for small insects occurring at low population densities. Perhaps subtle cues from a parasitized host's behavior and appearance can reveal a female's presence to a male with acute vision, but equivalent factors certainly operate in many small, aerial insects. Presently, the evolution of the strepsipteran eye and of the order remain unclear; current hypotheses indicate affinities with either the Coleoptera or the Diptera (Kristensen 1999; cf. Figure 4.26). Until the phylogeny of this curious group is further resolved, attempts to infer the origins of its visual perception may be futile.

5.4 Perceptual Influences on Optical Signal Design

The diversity of color patterns among arthropods provides ample evidence that few limits on the creation of optical stimuli exist. Especially where scales cloak the wings and body and generate an array of pigmented and structural colors, a seemingly infinite variety of hues and patterns can be created. Nonetheless, stringent limitations on visual perception exist in most species, and relatively little of the detail in these patterns may represent actual signal characters that are recognized and assessed by conspecifics. Rather, the spatial patterns of coloration that we observe may have been largely shaped by the visual perception of vertebrates, particularly birds. Patterns of coloration can play a critical role in concealment from avian predators, or in confusing, distracting, startling, or aposematically warning them. Additionally, some components of patterns may have arisen as byproducts of the developmental programs responsible for distributing and arranging pigments and the various types of scales on the body and wing surfaces (see Nijhout 1991).

Because spatial resolution offered by the typical compound eye is at best two orders of magnitude poorer than that of human vision, it is most unlikely that insects and the majority of other arthropods can assess any but the cruder elements in conspecific coloration patterns. That is, even when aided by sophisticated pattern recognition mechanisms and separated by relatively short distances (<1–2 m; Rutowski 2002), arthropods may only discern bold elements such as relatively wide stripes and broad blotches of color. These restrictions would apply whether a receiver, or signaler, is moving or stationary, and even where the signal : noise ratio is apparently high. For example, in some lampyrid beetles (fireflies), two synthetic male flash trains that are each delivered at half the typical flash rate, 180° out of phase, and presented from opposite directions are as effective in stimulating females to reply as a single flash train delivered at the typical flash rate (Figure 5.11; Case 1984). Possibly, the need to boost visual sensitivity is so great that generator potentials evoked by absorbed light must be integrated over all ommatidia in both eyes. While this measure would improve the reliability with which females could detect dim male flashes and assess flash rate, it would preclude any assessment of flash movement (cf. Section 5.2.1). Moreover, reliable communication in dense populations might be difficult, if not impossible (Case 1984; cf. Section 4.4.1).

Unlike spatial pattern, though, arthropod sensitivity to color, polarization, tempo, and intensity of light are generally well developed by vertebrate standards, and each may have exerted strong influences on signal design (or vice-versa? See Chapter 6). These expectations are generally upheld where they have been checked. Arthropods are capable of "true color vision" (Kelber and Pfaff

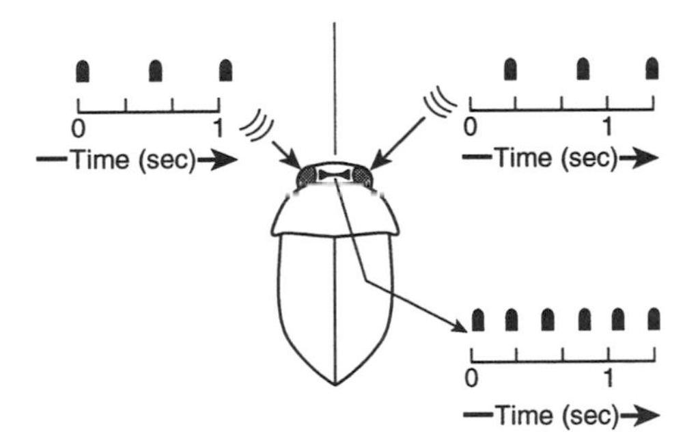

Figure 5.11 Experiment demonstrating visual summation in lampyrid beetles (fireflies). Each compound eye receives flash stimuli that are half the typical rate and presented 180° out of phase with respect to the contralateral stimuli; an opaque separator prevents an eye from directly perceiving the contralateral stimuli. Because the receiver responds as if the stimuli are presented at the typical rate, it is inferred that its brain does not treat input to each eye as a separate signal (from Case 1984).

1999), the ability to distinguish hue independent of intensity, and spectral peaks in signals match peaks in receiver sensitivity in various firefly (Case 1984) and butterfly species (Bernard and Remington 1991). Female fireflies may evaluate the rate (Branham and Greenfield 1996), intensity, and relative timing of male flashes (Vencl and Carlson 1998), thereby selecting for faster and brighter flashes and temporal signal interactions.

Ultimately, the major problems that arthropods face in communication originate from their small body size (see Bennet-Clark 1998a). These problems arise in all modalities, and they influence both the ability to localize signals and the nature of signals that can be generated and perceived. Witness the elaborate flight maneuvers that a moth must perform to track the filaments of an aerial pheromone plume (Section 3.1.4) and the inability of acoustic insects to produce, and hear, low-frequency far-field sounds (Sections 4.2.5 and 4.3). For visual communication, however, these problems are particularly insuperable. While developmental programs are capable of creating a wealth of optical stimuli, relatively little of the design in these stimuli is perceptually accessible at any but the closest range and hence most can not be interpreted as a signal character or represent information. This constraint may have led many arthropods to evolve visual communication along lines quite foreign to our own experience and imagination or to forsake long-range visual communication altogether in favor of more usable channels.

6

Sexual Selection and the Evolution of Signals

It is nearly axiomatic that animal communication signals do not arise de novo. Wherever historical or comparative information exists and phylogenetic inferences are possible, descent and modification from some pre-existing behavior or structure, either a signal functioning in another context or manner or a simple inadvertent cue, is usually indicated (e.g., Kessel 1955). Such origins and modifications may indicate that signal evolution is an economical process. They may also serve as a reminder that signals seldom arise in vacuo but tend to form in the presence of, and in response to, a perceptual background. It is assumed that perception of an earlier signal was already well developed, and when signals descend from cues, they may usually evolve either in response to pre-existing perception of the cues or coevolve alongside it.

For sexual advertisement and courtship signals, the principle that signals evolve in response to or coincident with perceptual ability is consistent with two of the main sexual selection (female choice) models receiving much current attention: (1) Male signals that females perceive and respond toward owing to pre-existing sensory abilities and preferences evolve (see Enquist and Arak 1993). Here, the novel signal is selected because it improves a male's fitness by increasing his opportunity to encounter and successfully court females. Moreover, the female receiver may also enjoy improvements in the processes of pair-formation and mate selection owing to her response to the male signal. This model is known as (exploitation of) receiver bias, and it may involve development of novel male signals from previous cues, modification of male signals previously used in other sexual or non-sexual contexts, or development of a novel male signal in conjunction with a shift in function of the receiver bias from a previous non-sexual context to a sexual one.[1] Importantly, the actual perception and preference trait of the female receiver do not evolve during the process. (2) A male signal trait coevolves with a female's preference trait. Here, the preferred male signal may reliably indicate a male's genetic quality, in which case a discriminating female benefits by producing offspring distinguished by high viability as well as aesthetic appeal; this mechanism is known as the viability indicator or good-genes model. Otherwise, when the preferred male signal is merely attractive, a discriminating female benefits by producing aesthetically appealing sons, who in turn will be sought after as mates in the following generation; this mechanism is known as the Fisherian, or arbitrary, model. In this chapter we examine the evidence for these models among arthropod advertisement and courtship signals and analyze the specific ways in which they may operate.

6.1 Exploitation of Receiver Bias

Formal demonstration of the receiver bias model for sexual selection of a male trait demands four lines of evidence: (1) females both perceive and prefer that trait, as expressed by conspecific males, over its alternatives or absence; (2) absence of the male trait is the ancestral character state, not simply a secondary loss; (3) females of the ancestral group must prefer the male trait even though it is not present in conspecifics; (4) a peripheral or central nervous feature is responsible for the perception and preference (Basolo 1995). Complete evidence is difficult to obtain, and indisputable support for the receiver bias model is actually quite rare. Suitable phylogenetic information is often lacking, and even when phylogeny has been fully resolved, uneven evolutionary rates may have intervened and obscured evidence for receiver bias (see Endler and Basolo 1998). For example (Figure 6.1a), a novel female preference may arise that is

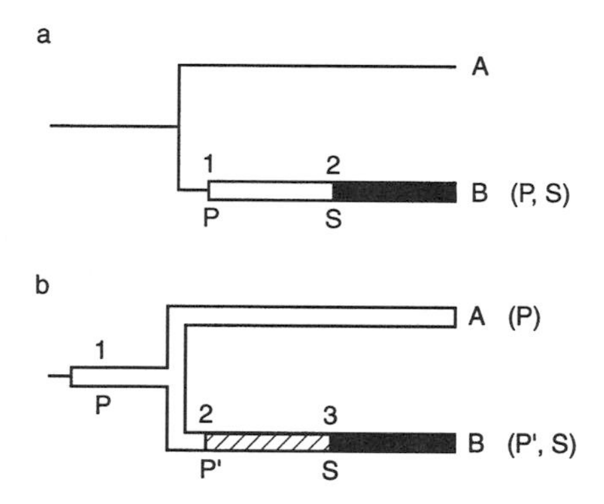

Figure 6.1 Problems in evaluating receiver bias as a mechanism of signal evolution. (a) A novel preference, P, arises at juncture 1 and is later exploited by a novel signal, S, at juncture 2. But, S arose prior to the occurrence of a speciation event, and phylogenetic information would not reveal involvement of receiver bias in evolution of the signal, S, found in species B; i.e., species A does not prefer S (from Endler and Basolo 1998). (b) A novel preference, P, arises at juncture 1 and is then replaced by another novel preference, P′, at juncture 2 in the ancestors of species B. The novel preference P′ is later exploited by a novel signal, S, at juncture 3. Again, phylogenetic information would not reveal involvement of receiver bias in evolution of the signal, S, found in species B; i.e., species A does not prefer S.

later exploited by a novel male signal, but the latter evolves rapidly and does so before any speciation takes place. Whereas signal evolution via exploitation of receiver bias has occurred, an examination of the phylogeny would indicate that the signal and preference coevolved, implying operation of Fisherian or possibly good-genes sexual selection—or simply that different, but effective, advertisements evolved in the various species. Or, the female trait in the ancestral (plesiomorphic) group may have drifted subsequent to speciation such that females no longer prefer the same male signal as females in the derived (apomorphic) group do (Figure 6.1b). In this case too, evolution via exploitation of receiver bias has occurred but cannot be confirmed.

Among arthropods —and animals in general—Proctor's (1991) study of male vibratory courtship in the water mite *Neumania papillator* (Acari: Parasitengona) remains one of the most unambiguous demonstrations of signal evolution via exploitation of receiver bias. Male *N. papillator* court females by trembling their first and second pairs of legs such that vibrations closely mimicking those made inadvertently by copepod prey are generated (Figure 6.2). These vibrations elicit a positive turning response in females, who posture themselves such that males can more easily transfer their spermatophore; the positive female response, termed a net stance, resembles her initial reaction toward a swimming copepod. Thus, it appears likely that a novel male signal has developed that exploited the female receiver's bias toward a particular category of vibrations and that this

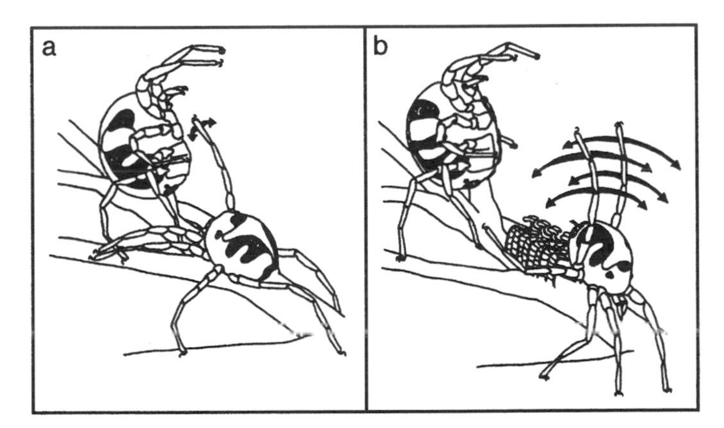

Figure 6.2 Receiver bias in the evolution of male courtship signaling in the water mite *Neumania papillator*. (a) Leg trembling behavior by male (lower right); female is in net stance posture. (b) Male (lower right) continues to tremble while female turns toward him; male fans his forelegs over the spermatophores deposited in front of the female. (Reprinted from *Animal Behaviour*, vol. 42, H.C. Proctor, "Courtship in the water mite *Neumonia papillator*: males capitalise on female adaptations for predation," pp. 589–598, 1991, by permission of the publisher Academic Press, London.) Trembling mimics the vibratory cues generated by copepod prey and elicits receptive responses in female *N. papillator*.

bias was co-opted to function in a sexual context.[2] In the congeneric species *N. distincta*, females also respond positively to synthetic vibrations mimicking those of male *N. papillator* (and copepod prey), but male *N. distincta* do not produce such signals (Proctor 1992; cf. Figure 6.1). Because phylogenetic analysis indicates that the character states found in *N. distincta* are ancestral, and the female preference results from sensitivity and response to a particular category of substrate vibrations, all four stipulations listed above for evolution of a male trait via exploitation of receiver bias are met.

In part, the evolutionary mechanism outlined above is an appealing explanation for signal evolution in various situations because the novel traits may be favored by selection acting directly on both the male signaler and the female receiver. The male would enjoy an increased encounter rate with females because his signal is more easily perceived and localized or elicits a postural response that is more conducive to successful courtship and copulation. These advantages would translate to higher fitness provided they outweigh risks that might result from being treated as the very item on which the preference ordinarily focused. Obviously, these risks may be potentially serious in those cases where the co-opted preference is involved in prey capture. Thus, we might expect courtship evolution often to entail two sequential or simultaneous elements: (1) The novel courtship signal mimics a cue eliciting a female response, such as her cessation of movement or positive taxis, that increases the male's female encounter rate or his likelihood of successful courtship and copulation. (2) The courtship behavior concomitantly becomes slightly differentiated from the mimicked cue by modification or the addition of new elements that lessen the chance of being attacked, and consumed, by the female.[3] For example, many male orb-weaving spiders (Araneae: Araneidae) pluck a specialized mating thread and thereby transmit vibrations toward the female in the center of the web. The female may respond to these vibrations as if they are the twitching of prey caught in the web, and the male on the mating thread may be able to escape quickly should he find himself in a dangerous predicament (Section 4.2.4). While such features may simply allow the male to contact the female in relative safety, they might also be perceived by the female as distinct from prey cues and lead eventually to a coevolutionary elaboration of her receiver bias.[4] Analogous events and conflicting pressures may have led to the evolution of courtship pheromones derived from host-plant substances. Male scents similar to host-plant odors would have exploited a female receiver bias for such odors, but shifts, possibly coevolutionary, toward odors that are chemically conspicuous against the background of host-plant volatiles may also be expected (Section 3.2).

The female, for her part in evolution of the novel signal, may enjoy the opportunity to pair with a mate in a manner that reduces time, energy, and risk. Because the novel signal exploits a receiver bias, females can perceive it readily and may be expected to localize it easily if they are the sex that searches and orients during pair-formation. This opportunity would represent a direct, material benefit for the female.

In theory, any novel signal that a male can produce that would increase his fitness, and possibly that of the female, by making pairing easier while not generating significant disadvantages in a natural selection context (e.g. vulnerability to predators) could evolve by exploiting a receiver bias. Signals evolving as such may often represent modifications or exaggerations of cues associated with other behaviors. In the water mite and spider examples above, male vibratory courtship might have descended from unspecialized vibrations that inadvertently resulted from ordinary walking and moving. Alternatively, they might be alterations of previous, possibly less effective, signals.

6.1.1 Exploitation of Anti-predator Receiver Bias?

It is usually assumed that when a signal evolves by exploiting a receiver bias co-opted from another activity, the bias is one that elicits a positive motor activity such as attraction or turning toward the stimulus. Thus, receiver biases that focus on food or habitat would appear most likely to be co-opted and exploited, and various purported cases of signal evolution via exploitation of receiver bias do involve such perception (e.g., Backwell et al. 1995, Sakaluk 2000). Nonetheless, a high proportion of an animal's perception is normally devoted to detecting predators and other natural enemies and then responding in a negative fashion. Can these receiver biases and the negative motor responses associated with them be exploited as well? What special conditions or mechanisms might enable signal evolution via this route?

Earlier, I noted that the pattern of widespread hearing common to both sexes and infrequent sound production occurring only in males, as found in various insect orders and families, is consistent with signal evolution via exploitation of receiver bias (Section 4.5). Because these patterns are not based on completed phylogenies, wherein nodes represent identified speciation events, and females of the ancestral groups perceive sound but do not respond positively toward it as if orienting to a male, the receiver bias model cannot be confirmed—or refuted. That is, we have no means of determining whether these non-homologous acoustic signaling devices, occurring sporadically within families that otherwise exhibit only hearing, originated via exploitation of receiver bias and were subsequently modified via a coevolutionary sexual selection mechanism or both originated and were modified solely by the latter (Figure 6.1). But, we might nevertheless speculate on whether and how a novel male acoustic signal could originate by exploiting an anti-predator hearing bias. In more general terms, we shall ask how sexual advertisement signals can evolve when ancestral receiver biases for that basic stimulus category were negative.

Ultrasonic signaling in the galleriine pyralid moth *Achroia grisella* (Section 4.2.1) poses the general dilemma outlined above. Like other pyralid moths, both male and female *A. grisella* exhibit striking negative reactions to ultrasound pulsed as in bat echo-locations: Flying moths drop immediately to the ground, and moths moving on a surface may simply stop (Greenfield and Weber 2000). These responses would provide effective defense against both aerial-hawking and surface-gleaning bats, which respectively capture their prey by echo-location and by localizing sounds made during movement or other activities. Thus, a male ultrasonic advertisement signal would have been perceived by females at its inception, but it would not have attracted them. How, then, did this signal ever evolve to its present state of development, and was exploitation of receiver bias involved in the process?

One way to understand how a stimulus or cue that is ostensibly repellent or inhibitory can evolve as an attractive advertisement is to recognize that the receiver bias model only requires that a novel signal trait improve a male's encounter rate with females or his chances for successful courtship and copulation. In *A. grisella*, and in general, stimuli that resemble cues from a predator may fulfill this basic requirement by causing a female to land, if in flight, and cease moving. As long as a female responds as such, a male who has also been advertising via another signaling modality may benefit from increased pairing opportunities. For example, a male who wing-fans and thereby advertises with olfactory, vibratory, or visual signals might increase the efficacy of these advertisements if the wing-fanning process adds an associated acoustic device whose output arrests a female's movement within range of the olfactory, vibratory, or visual display. Because the arrested female would perceive both the attractive olfactory, vibratory, or visual display and the sound, her inhibitory responses to the latter may be suppressed.[5] Presumably, the male's added acoustic emission is not so conspicuous to bats or other predators as to outweigh his improved female encounter rate or the female's ease and safety in pairing.

Most acoustic signals are not simply locomotion arresters, though, and mechanisms other than exploitation of receiver bias must have influenced their subsequent evolution as attractive advertisements. Here, chance and coevolutionary sexual selection mechanisms may have played leading roles. Male *A. grisella* pulse their ultrasonic advertisement signals as echo-locating bats do, but their pulse rates (>70 pulses$\cdot s^{-1}$)[6] are significantly faster than those of all bat echo-locations (<40 pulses$\cdot s^{-1}$) except the ones emitted during the final 250–500 ms prior to capturing a flying insect. These high pulse rates in male *A. grisella* advertisement signals may reflect two evolutionary mechanisms. First, the moths have probably been influenced by a motor constraint (see Podos 1996) originating in their indirect flight (wing movement) mechanism, which cannot operate efficiently at low rates of muscle contraction (Section 4.2.7). This motor constraint would fortuitously ensure high wingbeat frequencies and pulse rates that automatically distinguish a male signal from a predatory bat's echo-locations. Thus, the ancestral receiver bias would have been free to evolve, either by drift or selection, from a general inhibitory response to a specialized one that included positive phonotaxis toward high-pulse rate (male) signals. Second, once the female receiver bias was modified to include a specialized positive response, the male signal would have likely coevolved with that response via Fisherian or good-genes sexual selection and assumed the exaggerated signal characters (e.g., high pulse rates) presently observed.

6.1.2 *Male Receiver Bias and Female Advertisement?*

Can the fundamental receiver bias model also account for the evolution of female advertisements? Probably not, insofar as the stipulations listed above are adhered to. A parsimonious explanation for many female advertisements is that they originated as inadvertent cues leaked or emitted to the environment and perceived by highly sensitive males (Section 3.1.5). Thus, the first step in the evolution of female signals from previous cues would have been a fine-tuning of the pre-existing male receiver bias. For example, where males detected and localized females by odors that were simple byproducts of secondary metabolism, any refinement of olfactory receptor neurons (ORNs), binding sites, and binding proteins (PBPs) that increased male sensitivity would have been selected. Whether the changes in female odor that characterize advertisement pheromone evolution occurred at this juncture or later, some initial evolutionary shift in the male receiver bias—which violates the model's central assumption—would seem to be unavoidable. That is, coevolution between male perception and female cues may define the evolution of female advertisement signals.

The evolution of oviposition pheromones (Section 3.3.2) and of many social signals in the Hymenoptera and Isoptera (Sections 3.4–3.7) may parallel that of female advertisements. For example, when worker ants begin to respond toward autocommunication signals or inadvertent cues (e.g., recognition or alarm pheromones) left by foragers returning to the nest and follow these odors back to the resource visited by the foragers (Section 3.7), a modification, or development, of the worker's receiver bias for these stimuli is indicated. If the odors also become modified such that they serve as a specialized recruitment signal, coevolution between the worker's receiver bias and the initial forager's signal may be inferred. What sets this process apart from the receiver bias model sensu stricto is that the receiver is under at least as much selection pressure as the signaler is. Consequently, receiver features are not just passively exploited by evolving signals but rather participate interactively in evolution of the novel signal by coevolving alongside it from the start. It is such participation that may be responsible for the phenomenon of receiver extravagance (see Section 3.1.4).

6.2 Coevolutionary Mechanisms

Male advertisement signals that originate via the receiver bias model may be selected because female receivers can obtain a subtle, yet direct, material (somatic) benefit: a quicker and easier, and hence safer, means of pairing with a mate. Thus, males broadcasting the preferred signals, those exploiting pre-existing receiver biases, may enjoy a higher rate of encountering females and mating because the females attracted suffer lower mortality en route to the pairing location or simply find a male faster, and after less expenditure of energy, than females attracted to other signals do. Consequently, these males may enjoy a greater chance of passing genes associated with their signal traits to the next generation than males broadcasting other signals will. In contrast, sexual selection models based on coevolutionary (indirect, genetic) mechanisms explain how male signals may originate and be maintained in the complete absence of direct, material benefits accrued by the female.

Coevolutionary sexual selection models are based on the assumption that substantial additive genetic variance exists for both male features (e.g. advertisement signal characters) and female preferences for particular male features. When both male signals and female preferences are heritable traits, a genetic covariance may arise in which males who transmit the preferred signals will carry alleles for preference for such signals, which would be expressed in the next generation by their daughters. Likewise, females who strongly prefer certain male signals may carry alleles for those signals, which would be expressed by their sons. These two processes mutually reinforce each other, and populations in which a majority of males express exaggerated values of a signal character will eventually include a preponderance of females who strongly prefer signals with those extreme values, and vice versa (e.g., Gilburn et al. 1993; also see Bakker 1993). The genetic covariance, or linkage disequilibrium, between the male feature and the female preference traits is a third assumption of the co-evolutionary models.

6.2.1 *Fisherian Selection*

When Fisher (1915, 1958) first conceived the basic mechanism underlying the coevolutionary models, he suggested that the process might be initiated if males displaying a particular feature (signal) enjoy an advantage in the context of natural selection. However, formal population genetic models (Lande 1981, Kirkpatrick 1982) show that the coevolutionary process can be initiated by chance alone; that is, random drift may lead to accumulation, within a given subpopulation, of alleles for strong preference for an arbitrary male feature. When that arbitrary, preferred feature continues to remain independent of a natural selection advantage, Fisherian (arbitrary) selection is said to occur.

Here, males displaying the preferred feature may actually suffer reduced survivorship owing to energy expenditure or conspicuousness. Because females do not receive any direct, material benefit by mating with males displaying preferred features and these males do not hold any natural selection advantages, discriminating females would not be expected to produce offspring of higher viability. Rather, the sole benefit accrued by these females would be aesthetically pleasing sons, who bear their father's heritable, preferred trait. Nonetheless, these sons on average would be more attractive than the offspring of females who mate indiscriminately, and this advantage within the sexual selection arena would offset any disadvantage suffered in survivorship (e.g., Gray and Cade 1999b) or development. If the sexual selection advantage confers exact compensation for the male's natural selection disadvantage, the population may assume evolutionary equilibrium: the male feature and the intensity of female preference for it remain unchanged (Figure 6.3).

6.2.2 Good-genes Selection

When the genetic benefits that a discriminating female accrues by mating with a preferred male include viability genes that may be inherited by her offspring, either male or female, good-genes selection is said to occur. Several variations of the good-genes process have been described via formal models (Pomiankowski 1987, 1988, Grafen 1990, Maynard Smith 1991b), but all assume that a preferred male signal serves as a reliable indicator (cf. Zahavi 1977) of heritable viability. Signals can be reliable if males of low viability are unable to display

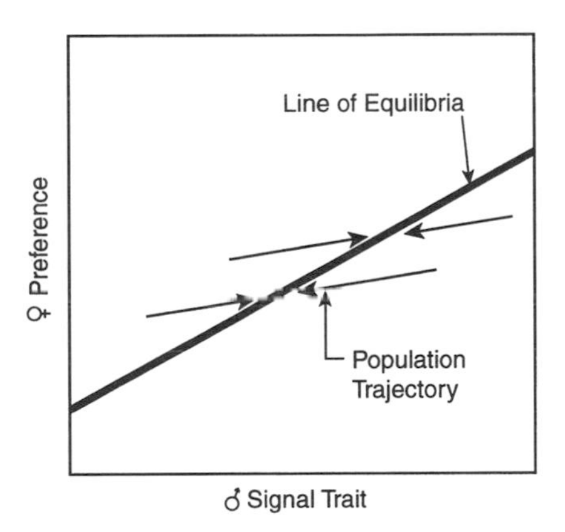

Figure 6.3 Sexual selection via the Fisherian process. Populations in which the mean level of the male trait and the mean level of the female preference fall along the thick line are in an equilibrial state wherein natural and sexual selection are balanced. If perturbed away from the line of equilibria, the population returns by following a trajectory to another equilibrium point (from Lande 1981, Kirkpatrick 1982).

them fully or if a male's deficiencies in viability are normally revealed by virtue of his display. Thus, signals that are associated with morphological symmetry or whose intensity reflects energy expenditure might be reliable indicators (e.g., Kotiaho et al. 1996, 1999).

In general, signals that are not reliable are not expected to persist in a population (Zahavi 1977), as there may be little selection pressure favoring them. Rather, receiver responses to unreliable signals may be selected against because they can often lead to deception and a fitness-reducing outcome (e.g., pairing with a non-viable mate). Because signal production usually incurs risk and energy expenditure, selection acting on receivers would ultimately eliminate most potentially deceptive signaling unless it is infrequent and its consequences for receivers are not overly harsh (Johnstone 1994, 1995b). The principle of reliability may be expected to apply equally to intrasexual signaling, in the context of male–male competition (Section 4.4.2, Acoustic Duels), as to advertisements that attract mates and influence their receptivity.

Recent laboratory and field studies provide some empirical support for the good-genes process in various vertebrates and arthropods (Moller and Alatalo 1999). In several fish (e.g., Reynolds and Gross 1992), amphibians (e.g., Welch et al. 1998), and birds (e.g., Petrie 1994), the offspring of females who mate with preferred males have been found to enjoy higher survivorship or more rapid maturation than the offspring of indiscriminate females or those mated with average or non-preferred males. And, among arthropods, females mating with preferred males may produce offspring who exhibit faster rates of development (Moore 1994, Jia and Greenfield 1997), a factor that would lead to higher survivorship should food be limiting, or more favorable offspring sex ratios (Wilkinson et al. 1998).[7]

6.2.3 Caveats and General Issues

Good-genes selection is a mechanism of mate choice as opposed to intrasexual selection, and any attempt to demonstrate it must ensure that males who sire offspring of higher viability mated because females preferred them and not because they competitively excluded rival males. Cases in which these sires mated owing to their success in male–male competition cannot be construed as supporting the good-genes mechanism.[8] Thus, an experiment must generally be performed in which one eliminates the potential for intrasexual selection to determine the male(s) with whom a female mates: Under natural conditions, male mating success will nearly always be influenced by some level of intrasexual selection.[9] For example, male advertisement signals are often used by male receivers to assess the fighting prowess of their neighbors (Section 4.4.2). Males who assess themselves as lower in such prowess may then retreat to inferior signaling locations at which females are less likely to arrive or be encountered. This scenario describes the nature of sexual selection found in some lekking formations, including the substrate-based mating aggregations of various arthropods (see Shelly and Whittier 1997).

Once the potential for intrasexual competition is eliminated, one must confront the possibility that hidden material benefits remain. Some of these benefits are automatically removed by the controlled conditions under which an experiment is normally conducted and are not cause for concern. Whereas females in natural populations might gain avoidance of a transmissible disease, risk, and energy expenditure by localizing and mating with males who signal vibrantly, such benefits are unlikely to come into play in a controlled experiment and influence fecundity or offspring survivorship. However, other hidden material benefits may be more difficult to remove. Particularly among arthropods, females may obtain nutritional or defensive benefits from extragametic materials transferred at copulation (Sections 3.2 and 4.4.1). In some cases these materials might be factored out if their quantity is proportional to the benefit obtained and can be measured.

Maternal (cytoplasmic) influences on offspring viability must also be eliminated in order to demonstrate the good-genes mechanism. These influences are generally dealt with experimentally by establishing a breeding design in which characteristics of the offspring of full sisters, half-sisters, and unrelated females may be contrasted via statistical techniques (Figure 6.4). But even this quantitative genetic measure cannot deal effectively with the possibility that a female might control and influence her offspring's characteristics in accordance with her mate's features. For example, a female's hormonal contributions to her developing eggs could reflect the male's dominance status (cf. Gil et al. 1999).

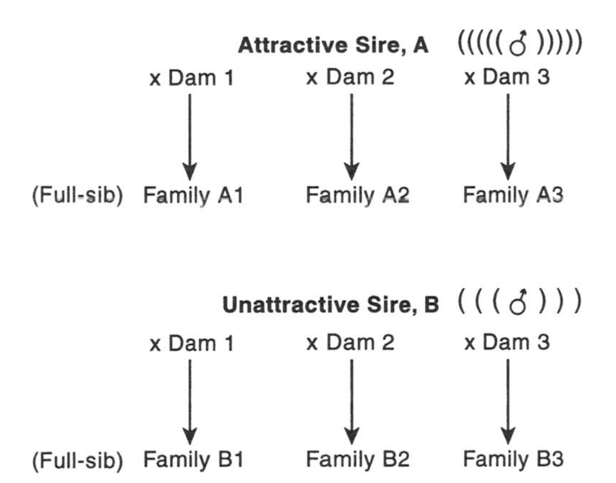

Figure 6.4 Breeding design for investigating good-genes effects in sexual selection. Attractive and unattractive sires are each mated with multiple dams to establish full-sib (offspring) families nested within half-sib families. ANOVA is used to determine the sire's contribution to attractiveness in male offspring.

Such factors may have influenced various of the findings reported above as good-genes effects.

The need to rule out material benefits and maternal influences should not be construed to imply that these factors cannot co-occur with a good-genes process (Section 4.4.1). On the contrary, a male's ability to provide material benefits such as spermatophore nutrients may reflect his overall viability and represent a heritable trait, which could readily coevolve with female preference for a male signal reliably indicating that benefit. Additive genetic variance and heritability are as likely to exist for sexually selected characters as for characters in general (Bakker and Pomiankowski 1995), and signal evolution can be expected to proceed via a coevolutionary mechanism whether or not females regularly obtain direct benefits (see Jennions and Petrie 2000). However, the influences of direct, material benefits and coevolutionary mechanisms on signal evolution would be confounded should both occur. Thus, empirical demonstration of the good-genes process is usually most definitive in cases where direct benefits are not a potential influence or can be removed experimentally.

Demonstration of the Fisherian process would require the same elements as the good-genes process does but with the stipulation that the offspring sired by preferred males do not enjoy higher viability (e.g., Jones et al. 1998). This stipulation may be very difficult to confirm, because viability can be a phenotypically plastic trait whose expression depends on environmental conditions experienced during development. While offspring of preferred and unpreferred sires may perform equally well under benign laboratory conditions, a marked difference might arise in the field should environmental factors such as temperature or food supply be less than favorable (e.g., Jia and Greenfield 1997). Thus, failure to detect enhanced viability in the offspring of preferred sires under experimental laboratory conditions does not necessarily indicate the Fisherian and invalidate the good-genes process. In the following section we examine how such trait plasticity may have a profound influence on the general way in which coevolutionary sexual selection mechanisms operate in natural populations.

6.2.4 Phenotypic Plasticity of Signal and Preference Traits

Finding that sexually selected male features such as signal characters normally have substantial amounts of additive genetic variance and are heritable confirms one of the main assumptions underlying the coevolutionary mechanisms of the good-genes and Fisherian processes. But just as the heritability of signal characters satisfies one basic element, it raises further questions. Because strong selection, as imposed by mate choice, is expected to diminish genetic variance, heritability for male sexual characters may not persist over time and there should exist little potential benefit for females to continue exercising mate choice among males. Nonetheless, both heritability of signal characters and female discrimination among males do persist in many species. How can this apparent paradox be resolved?

In theory, genetic variation within a population may be maintained by any of the following basic factors: mutation-selection balance, heterozygote advantage, frequency-dependent selection, antagonistic pleiotropy, and environmental heterogeneity in the presence of genotype × environment interaction (Roff 1997). Additionally, genetic variation in a sexually selected trait may be maintained if fitness increases exponentially with exaggeration of the trait (Pomiankowski and Møller 1995) or because the expression of such traits is generally condition-dependent and there exists abundant genetic variance in condition (Rowe and Houle 1996).[10] An alternative view would contend that intersexual selection in natural populations is actually a rather weak force (see Alatalo et al. 1998), because females may mate indiscriminately with any male whose signal characters exceed an absolute threshold set at a relatively low value. But open-ended female preferences and active comparison, as opposed to passive filtering, of available mates (Section 4.4.1) suggests that this possibility does not necessarily, or even regularly, occur.

In practice, the factors responsible for maintaining genetic variation of traits—sexually selected or otherwise—within populations have seldom been tested.[10] But, many traits, including sexually selected ones, are known to exhibit substantial phenotypic plasticity in which their expression is heavily influenced by environmental conditions experienced by embryos and immature individuals (e.g., Olvido and Mousseau 1995). When a trait exhibits phenotypic plasticity, it is possible for the various genotypes to interact with the environment in different ways such that no single variant performs maximally in every environment. Genotype × environment interaction (GEI) of this nature can maintain genetic variation within a population provided there exists sufficient overlap of generations (Ellner and Hairston 1994) or gene flow across habitats (Slatkin 1978). For example, a male signal character expressed by a particular genotype (or genetic variant) may be exaggerated when developed under certain environmental conditions but rather stunted when developed under others, whereas another variant may express a reversed pattern of exaggeration (Figure 6.5). If all genetic variants respond with these fundamental patterns, genetic variation for the male signal trait would be maintained (Roff 1997) and coevolutionary mechanisms of sexual selection should escape the paradox and continue to operate and influence signal evolution.

Or would they? Signals that evolve under coevolutionary mechanisms of sexual selection are expected to be reliable indicators of either heritable viability (good-genes process) or aesthetic appeal (Fisherian process). However, a signal whose expression is subject to strong phenotypic plasticity and GEI could be decidedly unreliable, and we may ask how a preference for it can ever persist (Jia et al. 2000). When females prefer exaggerated values of a male signal

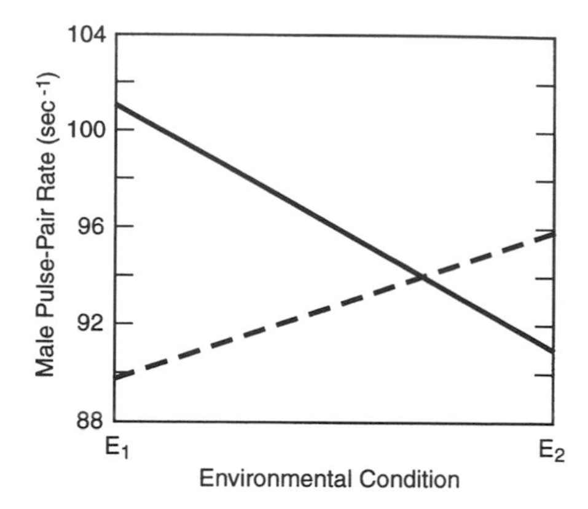

Figure 6.5 Phenotypic plasticity and genotype × environment interaction (GEI) in signal expression. Male lesser wax moths (*Achroia grisella*) who have been reared under favorable diet and temperature conditions (E_1) signal at different pulse-pair rates (cf. Figures 4.13 and 4.14) from males reared under less favorable conditions (E_2). Moths that had been artificially selected for high pulse-pair rates under E_1 (solid line) showed a 10% decrease in that signal character when reared under E_2, whereas moths artificially selected for low pulse-pair rates under E_1 (dashed line) showed a 7% increase when reared under E_2 (from Jia et al. 2000). Thus, neither genotype (pulse-pair rate line) performs better—signals at a higher, more attractive rate—in both conditions. Male pulse-pair rate is a heritable trait (cf. Table 4.2).

character but exaggerated signals are expressed by different genetic variants in different years or in different habitats, a choosy female will not necessarily mate with a particular variant. How then can she be assured of obtaining genes that may confer high viability to her offspring? Likewise, if the environment changes markedly before the following season, a choosy female may end up producing male offspring whose signals are unexaggerated and lack aesthetic appeal. In general, whenever environmental conditions change with any frequency and GEI influences signal expression, the linkage disequilibrium between male signal traits and female preference traits that drives the very evolution of these features would be repeatedly disrupted.

The undermining of reliability and the disruption of linkage disequilibrium might appear to imply that coevolutionary mechanisms of sexual selection can only operate under rarified conditions wherein environments never change or that signal traits are seldom influenced by GEI. Because neither of these implications is likely to be true, the points raised above indicate that our treatment of signal evolution has been unrealistically simplified. Many traits in addition to male signal features exhibit phenotypic plasticity and are influenced by GEI, and some may be closely linked, possibly via pleiotropic action, with signal traits (Jia et al. 2000). Where signal and viability traits are associated and both respond in parallel fashion to environmental conditions, a female choosing an attractive male may be reasonably assured that she is mating with a viable one (e.g., see

Lesna and Sabelis 1999). Conversely, this assurance may imply that such suites of associated traits are a prerequisite for operation of the good-genes process in signal evolution.

In presenting the situation above, I made the tacit assumption that female preference is a fixed trait and relied on this assumption to demonstrate that females who choose attractive males could be assured of mating with viable ones. Females always choose a male signal in which a particular exaggerated character is expressed, and while this consistency leads to mating with different genetic variants under different environmental conditions, these variants would be the most viable ones under the respective conditions. However, there is no a priori reason to expect preference to be invariant. On the contrary, additive genetic variance and heritability of female preference are basic elements of the population genetic models on which the coevolutionary mechanisms of sexual selection are founded, and several studies now report both phenotypic and genetic variance for female preference functions (e.g., Ritchie 2000). In the lesser wax moth, *Achroia grisella* (Lepidoptera: Pyralidae), for example, females vary repeatably in the relative weight attached to different signal characters while assessing males, and this variation is heritable (Jang and Greenfield 2000). Phenotypic plasticity and GEI associated with female preference traits are unknown at this point, and it is not clear how they might influence signal evolution. But any realistic treatment of a coevolutionary sexual selection process will have to acknowledge these factors and incorporate them in modeling efforts.

7

Signal Evolution: Modification and Diversification

Diversity is undoubtedly one of the hallmarks of arthropod biology. Interspecific variation characterizes many features of arthropod behavior, development, morphology, and physiology, and it is particularly evident among sexually selected traits such as mate-recognition signals and preferences. Here, though, its nature and occurrence pose several major problems in evolutionary biology. Whereas most traits can reveal phylogeny because the (ancestral) features of closely related species resemble each other, a species' advertisement and courtship signals and mate discriminations usually differ markedly from those of its closest relatives (Henry et al. 1999). This variation in mate-recognition signals and discriminations may represent the very factor that allowed speciation to occur, and in some cases interpopulation variation may have been instrumental in accelerating speciation by providing effective pre-mating isolation at an incipient stage in the process (Hollocher et al. 1997a,b, Jiggins et al. 2001; also see note 34 in Chapter 3). Nonetheless, this factor does not explain how signal traits diversify in the first place and create the astounding variety of odors, sounds, vibrations, and color patterns described in the preceding chapters. More to the point, given that aberrant signals and preferences would lack matched receivers and signalers, respectively, how do they ever arise and yield evolutionary changes in natural populations? The stabilizing selection that acts on the "static" characters (sensu Gerhardt 1991)[1] of mate-recognition signals could not be expected to tolerate such aberrations.

7.1 Genetic Coupling?

The conundrum outlined above has drawn many an evolutionary biologist, and some intriguing solutions have been offered. One early proposal that continued to attract attention for many years was Alexander's (1962) novel hypothesis that male signal traits are genetically coupled with female preference traits. The genetic coupling hypothesis made both neuroethological and genetic assumptions. First, elements of the central nervous system that regulate male signal features such as tempo and other motor patterns are also expressed as a sensory template in conspecific females, who compare perceived signals with their template and respond favorably when the two match. Second, a given gene (or genes) must pleiotropically influence both the male signal and the female preference by controlling the critical neural element. Thus, a male bearing a mutant allele would generate a novel signal, and a female bearing that allele would exhibit a preference for the novel signal.

Initially, hybridization studies were used to evaluate the genetic coupling hypothesis, and some findings among acoustic insects were interpreted as supportive. Hoy et al. (1977) bred *Teleogryllus oceanicus* × *T. commodus* (Orthoptera: Gryllidae) hybrids and found that the calling songs of F_1 males were intermediate in various characters between the songs of the parental species; similarly, the F_1 females preferred these intermediate, hybrid calls.[2] Further analyses revealed that the female F_1 hybrids preferred calls produced by male hybrids from the same cross over calls from the reciprocal one, indicating that the critical gene is sex-linked and that maternal effects are present.

As compelling as these findings may seem, however, they could just as easily reflect an alternative explanation that the match between male signals and female preferences in F_1 hybrids resulted from a coevolutionary sexual selection process, which had established a linkage disequilibrium between signal and preference genes in the parental species. That is, F_1 hybrids simply had a blend of the parental species' genes for both signaling and preference. Bauer and von Helversen (1987) set out to test this alternative by investigating the temperature coupling of signal production and signal preference in *Chorthippus* spp. grasshoppers (Orthoptera: Acrididae). As with various acoustic insects, both chirp rates in male songs and female preferences for those rates are modified by temperature in *Chorthippus*; at a given temperature females generally prefer the average chirp rates that males produce (von Helversen and von Helversen 1981; cf. Section 4.2.7 and Figure 4.8). Interestingly, Bauer and von Helversen found that male signaling and female call preference were influenced by thoracic and head temperature, respectively (Figure 7.1).[3] Because two separate influences were found, a single neural element controlled by a critical gene(s) was not indicated and the genetic coupling hypothesis appeared unlikely—but not impossible. As Butlin and Ritchie (1989) later pointed out, a given gene could pleiotropically influence two separate neural elements, and the *Chorthippus* findings do not necessarily support the coevolutionary hypothesis.

To distinguish between the genetic coupling and coevolutionary hypotheses, F_2 or higher order hybrids or backcrosses would have to be examined for evidence that signaling and preference are linked traits influenced by different genes (Butlin and Ritchie 1989, Boake 1991). Such findings would confirm the coevolutionary hypothesis, but in their absence one could merely conclude that genetic coupling remains a possibility. Linkage between the genes influencing

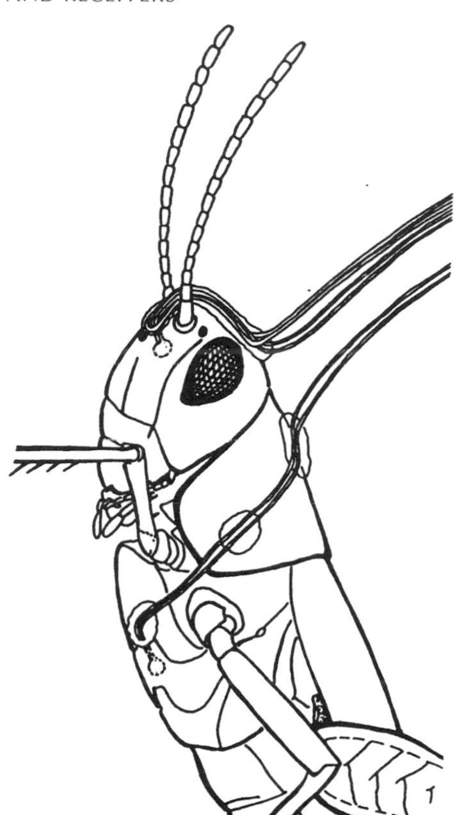

Figure 7.1 Experiment investigating genetic coupling in *Chorthippus* grasshoppers. Thermocouples inserted into the head and thorax heated the brain and stridulation muscles differentially. (Reprinted from "Separate localization of sound recognizing and sound producing neural mechanisms in a grasshopper," *Journal of Comparative Physiology* A, vol. 161, p. 96, fig. 1, M. Bauer and O. von Helversen, copyright 1987 with permission of Springer-Verlag.) Findings indicated that male stridulation rate is influenced by thoracic temperature, whereas female preference for stridulation rate is influenced by head temperature.

signaling and preference might be close, and further crossing would be required to reveal them. To date, no study has confirmed the existence of genetic coupling in any signaling system. No evidence indicates that females rely on templates that are sensory expressions of the motor patterns in male signals in order to recognize those signals, and studies of *Teleogryllus* actually suggest that these cricket species do not (Pollack and Hoy 1979, 1981). Additionally, findings indicating that signal and preference traits map to different chromosomal locations unambiguously reject the genetic coupling hypotheses in several species of *Drosophila* and moths (e.g., Löfstedt et al. 1989a).

Genetic coupling is at odds with the coevolutionary processes of sexual selection, which assume that signal and preference traits are influenced by separate genes associated by linkage disequilibrium only (Section 6.2). Because of mounting evidence that coevolutionary processes operate widely, it seems rather improbable that genetic coupling could be an important factor in signal evolution. But, mate-recognition signals do evolve and diversify, sometimes rapidly (Gleason and Ritchie 1998), and we remain challenged to explain how these seemingly improbable changes can take place, regardless of the mode of speciation.

7.2 Aberrant Signals and Preferences: Conducive Factors and Toleration

One way of addressing this challenge is to identify the conditions under which modifications in signals and preferences might be selected against less strongly. Instead of conceiving specialized genetic and neuronal architectures that may occur but rarely, we might focus on situations in which mutational changes in mate-recognition systems could remain in a population should they arise. Ritchie and Phillips (1998) adopted this approach and noted several conditions wherein bearers of modified signals and preferences may not be severely disadvantaged. In particular, they proposed that mutant elements in mate-recognition systems are not selected against strongly when they arise in females and in signals as opposed to males and preferences. Because most females in a population are expected to mate, those who possess a slightly aberrant preference or signal may not suffer unduly (DeJong and Sabelis 1991). Similarly, mate-recognition signals are often less variable among individuals than preferences are, and an individual bearing a modified signal may still be quite acceptable to the vast majority of the population. For example, in many species of moths, a slightly modified ratio of components in the female sex pheromone will still attract males, albeit less effectively than the ratio normally transmitted by female signalers (Roelofs 1978).

Under either condition proposed above, a population that acquires a significant proportion of an aberrant signal or preference, whether due to a founder event, disruptive selection, or environmental changes and natural selection favoring the aberration, might retain it for long enough that a coordinated preference or signal, respectively, coevolve. Once such coordination arises, the initial aberration should stabilize.

Ritchie and Phillips (1998) also proposed that another condition, sex linkage, might favor the retention of aberrant, or mutant, signals and preferences. This condition is associated with fundamental properties of sexually selected traits that are now coming to light, and it merits separate attention. Because most mutations are recessive, a mutant signal or preference should have a higher chance of being retained until coordination might arise if it is sex-linked and thereby expressed in the heterogametic sex. Thus, a mutant trait is more likely to be expressed before it is potentially lost due to stochastic factors. A recent survey (Reinhold 1998) shows that a higher than expected proportion of sexually selected traits are sex-linked, lending some support to this proposal. That mate-recognition signals and preferences tend to be sex-linked may simply reflect the likelihood that such traits are often sex-limited in expression (Reinhold 1999).

Clearly, we will need far more information than is presently available on the inheritance and genetic architecture of signals and preferences to evaluate whether any of these proposed conditions actually allow signal evolution to proceed. In several cases, a single gene has been found to exert a major influence on a signal or preference, but these may represent exceptions, which are largely restricted to pheromonal communication (e.g., Haynes and Hunt 1990). There, modifications may involve the addition or deletion of a particular step in the sequence of reactions leading from a precursor to a pheromone component; alternatively, a gene may influence chemical properties of the binding sites on olfactory receptor neurons. But, most signal and preference traits are probably influenced polygenically (e.g., Pugh and Ritchie 1996, Shaw 1996, 2000), and we currently have little understanding of linkage, heritability, and the numbers of genes involved in controlling these quantitative characters.

7.3 Are Complex and Multi-modal Signals Favored?

A somewhat different approach to the challenge of accounting for evolutionary change in signaling systems was taken by von Helversen and von Helversen (1994), who considered the sorts of changes in signal characters that might be compatible with Fisherian sexual selection. Recognizing that coevolutionary mechanisms such as Fisherian selection are likely to be widespread but should select against most novel mate-recognition signals, they reasoned that modified signals having entirely novel elements rather than simply altered values of pr-existing elements might fare much better. Their reasoning was based on both observation and theoretical inference. That females of some Orthoptera prefer complex male advertisement calls over simple ones (e.g., Stumpner and von Helversen 1992) indicated that preferences for complexity per se could exist and favor modified signals bearing novel elements.[4] They then proposed the hypothetical case where some (mutant) males in a population add a novel sound at the end of a normal call. Were this novelty merely a change in a continuous character such as pulse rate, carrier frequency, or call length, it would likely be removed by stabilizing selection: mutant females who favor the novel call would tend to produce sons who only attract mutant females. However, the proposed novelty does not entail any change to the original call, and mutant females favoring this call should therefore produce sons who potentially attract all females, not only mutants. Thus, both signal and preference mutants may remain in the population and could ultimately contribute to evolutionary change in the species' signaling system and speciation.

If valid, the von Helversens' (1994) proposal may imply that the mate-recognition signals of closely related species should generally differ in their fundamental elements rather than simply bearing different values of a continuously varying character. For example, we might expect to find congeneric acoustic species with calls that are trills, trills + chirps, and trills + pulse trains. Similarly, the pheromones of congeneric olfactory species might include compound A, A + B, and A + C (e.g., Mustaparta 1997). Moreover, mate-recognition signals should frequently acquire elements in two or more different modalities (e.g., sound plus pheromone). Such multi-modal signaling may be quite common among animals (see Johnstone 1995a, 1996, Hölldobler 1999, Partan and Marler 1999), and many examples of multi-modal sexual advertisement could have originated by default. These are the kinds of signaling novelties that are tolerated by coevolutionary sexual selection mechanisms.

Certainly, many arthropod taxa do exhibit interspecific variation in fundamental signal elements, but an equivalent number of cases probably occur in which closely related acoustic species mainly differ in pulse rate (e.g., Alexander 1956, Walker 1964) and related olfactory species differ in component ratios of the same two compounds (Greenfield 1983). Additionally, signal evolution may often proceed by deletion, as opposed to addition, of elements (e.g., Otte 1992), and it is not clear how these changes could occur. In this situation, a mutant female favoring a (mutant) call in which an element has been deleted would produce sons attractive to mutant females only. Consequently, selection favoring complex and multi-modal signals may be responsible for some portion of signal evolution and diversification, but, as with the other proposals presented here, it is unlikely to represent a magical solution to this profound problem.

Rather, what appear to be needed for advancing our understanding of signal evolution are more thorough investigations of the underlying genetic and neural architectures and how they change during diversification. Here, the mapping of quantitative trait loci (QTLs) influencing signal generation and reception might prove fruitful, and efforts could be made to apply this technique to populations undergoing diversification. Modeling approaches, such as neural net simulations (see Arak and Enquist 1993), also hold some promise as ways for exploring what sorts of changes in signals and receivers are possible and how likely they are to occur in reasonable time spans. It is to be hoped that the appropriate blend of such empirical and theoretical inquiries will yield some long-awaited answers.

Notes

Chapter 1

1. Lest the reader construe this point as a modern revelation, I note that several Medieval and Renaissance scholars, including Albert Magnus, Cardano, and Aldrovandi, arrived at similar conclusions concerning size constraints on the senses of insects (see Réaumur 1926).

2. See Hasson (1997) for an expanded framework of signaling and communication.

Chapter 2

1. Equation 2.2 is applicable where signals exhibit discrete character states. See Beecher (1989) for methods of determining information and transmission capacity of channels where signal characters vary in continuous fashion along one or more dimensions.

Chapter 3

1. Several nineteenth century observers (Fabre 1879 (6th edn, 1912), Forel 1908) described the orientation of male Lepidoptera to females and suggested that female scents attracted the males. But a strong undercurrent of incredulity was present in these reports, probably due to the extreme sensitivity of lepidopteran receivers and to human unfamiliarity with the chemical channel. For example, Fabre (sér. 7, p. 393), describing his experiments and observations of moths and sexual odors, wrote:

> Mais de la femelle Bombyx ou Grand-Paon, que se dégage-t-il matériellement? Rien d'après odorat. Et ce rien, lorsque les mâles accourent, devrait saturer de ses molécules un orbe immense, de quelques kilomètres de rayon! Ce que ne peut faire l'atroce puanteur du serpentaire, l'inodore maintenant le ferait! Si divisible que soit la matière, l'esprit se refuse à telles conclusions. Ce serait rougir un lac avec un grain de carmin, combler l'immense avec zéro.
>
> But in the silkmoth or emperor moth [*Pavonia pavonia*, a large Eurasian saturniid], what substance is the female emitting? Nothing, according to our sense of smell. And this apparent nothingness, when the males are arriving, ought to

> be saturating an immense space, several kilometers in radius, with odor particles! That which the unbearably fetid odor of the serpentaire [*Arum dracunculus*; Araceae; the dragon arum of southern Europe] cannot accomplish, this odorless substance now will! However dispersible this substance may be, the mind refuses such conclusions. It would be like dying an entire lake red with a speck of carmine, filling the whole volume with nothing [my translation].

We may appreciate the depth of this unfamiliarity by noting that olfaction is our only major sense whose impairment is not defined by a word from everyday vocabulary: Anosmias remain relatively obscure and unrecognized in humans despite evidence that most of us suffer at least one type of these impairments (Amoore 1967, 1974, 1977).

2. For vertebrate comparisons, threshold concentrations for olfaction in humans and dogs are approximately 10^9 and 10^6 molecules·cm^{-3}, respectively.

3. In their treatment of female odors in moths and their attractiveness to males, both Fabre (1879 (6th edn, 1912)) and Forel (1908) described incidents of what appeared to be communication across several kilometers. Similar claims were made by several contemporary North American observers (reports in Cardé and Charlton 1984).

4. An indication of the impact of the 1959 silkworm moth report is the coining of the term "pheromone" by Karlson and Lüscher (1959) later that same year.

5. In many species, female advertisement pheromones are indicated by male responses to various blends of components of the pheromones known from closely related species. Here, the species' actual pheromone has not been isolated and identified, but existence and approximate composition of the chemical signal may be inferred from the receiver's behavior.

6. The female advertisement pheromones of sympatric species may be different geometric or optical (e.g., Hansen et al. 1983) isomers of the same compound or different blend ratios of the same isomers (Cardé 1986). Alternatively, sympatric species may use identical chemicals but exhibit sexual activity at different times of the day (e.g., Haynes and Birch 1986) or season (Greenfield and Karandinos 1979). Allopatric species, however, may use identical chemicals as advertisement pheromones (e.g., Leal et al. 1994); see Henry et al. (1999) for an example of allopatric species displaying equivalent vibration signals.

7. Tang et al. (1992), Kamimura and Tatsuki (1993), Spurgeon et al. (1995), and Gemeno and Haynes (2000) all indicate age-related increases in pheromone flux rate and length of the daily advertisement period in female Lepidoptera.

8. Schal et al. (1998) report an exception in which hemolymph lipophorin transports female sex pheromone to the site of emission.

9. Tang et al. (1991) report an exception in the cabbage looper (*Trichoplusia ni*), in which pheromone biosynthesis is regulated by ecdysteroid titer rather than a neuropeptide (PBAN).

10. Both Greenfield and Coffelt (1983) and Wagner and Rosovsky (1991) report moth species in which wing movement greatly increases the rate at which odor molecules evaporate from the body.

11. Males and females of some arboreal cockroaches of tropical forests are vertically stratified (Schal 1982). Are the advertisement pheromones of these species transmitted vertically?

12. An earlier notion that the antennae of insect receivers function as dielectric devices which detect and evaluate pheromones based on the electromagnetic (infrared, IR) emission spectra of these compounds (Callahan 1991) has received no confirmation.

13. Female Lepidoptera too may exhibit subtle, yet potentially vital, responses to female advertisement pheromones, but these effects have rarely been investigated. Ljungberg et al. (1993) found that female cotton leafworm moths (*Spodoptera littoralis*; Noctuidae) can detect conspecific pheromone, and Schneider et al. (1998a) reported EAG (electroantennogram) responses in female and male *Panaxia quadripunctaria* (Arctiidae) to both female and male pheromone. The antennae of *P. quadripunctaria* are equally developed in both sexes, and it is possible that females occasionally locate host plants or habitat by cuing on the advertisement pheromones of conspecific females (see Sections 3.3 and 4.4; Stamps 1987). Similar female responses to conspecific female pheromone have been found in another arctiid moth, *Utetheisa ornatrix* (R.J. O'Connell, personal communication).

14. Actual molecular identification of olfactory receptors has been accomplished only recently in arthropods (see Clyne et al. 1999, 2000 and Vosshall et al. 1999 for findings on olfactory receptor genes in *Drosophila melanogaster*).

15. Nonetheless, pheromone analogs that elicit behavioral responses are known for many arthropod species (e.g., Berg et al. 1995). In general, analogs are compounds whose structure mimics the actual pheromone such that they are able to bind, to varying degrees, with PBPs and at ORN receptor sites.

16. Hansson (1995) and Mustaparta (1997) report that some receptors on lepidopteran ORNs may bind with two different pheromone compounds; see Imaizumi and Pollack 1999 (Section 4.3.5) and Kitamoto et al. 1998 (Section 5.3.1) for acoustic and visual examples of this phenomenon. Hansson (1995) indicates that such binding reflects tolerance for the two compounds at a given receptor site on an ORN rather than a given ORN having two different receptor categories, each binding selectively to a different compound.

17. This concentration difference would represent an increase of only 1.59%, which is unlikely to be detectable by any arthropod species, and particularly not by an aerial one, where the difference would have to be discerned rapidly. For vertebrate comparisons, humans can only discriminate odor concentrations that differ by 50% or more, and the threshold differences in fish are 10–30% (Dusenbery 1992).

18. Host-plant volatiles heighten male response to female sex pheromone in many Lepidoptera (Dickens 1997, Landolt and Phillips 1997).

19. Male silkworm moths (*Bombyx mori*; Bombycidae), which are flightless and track female pheromone while running, turn in the direction of a stimulated antenna (Kanzaki et al. 1992) and may simultaneously compare input received by the left and right antennae during orientation. *B. mori* also exhibit a zigzagging locomotory pattern that may represent a vestige remaining from their aerial ancestors. In contrast to the localization of sound stimuli by acous-

tic insects, which normally requires binaural input (Section 4.3.5), most species of moths can efficiently track aerial pheromone plumes with a single intact antenna.

20. Basil and Atema (1994) have developed analogous in vivo recording techniques for lobsters orienting within aqueous odor plumes.

21. Male copepods track the pheromone trails that swimming females leave behind. Both tropo- and klino-tactic orientation mechanisms may be used to localize females up to 30 mm (=15–25 bodylengths) distant (Yen et al. 1998).

22. Some critics (e.g., Witzgall 1997) have pointed out that the continuation of casting may reflect responses to either residual wind remaining in the tunnel after its fan is turned off or visual cues. These criticisms would suggest that counterturning is not necessarily a self-directed process.

23. Perhaps, minor components are assessed during the initial wing-fanning response, when sensitivity can be elevated (see Loudon and Koehl 2000). Thus, pheromone identity may be determined before orientation is initiated. Todd and Baker (1999) suggest that minor components are evaluated in a tonic fashion, which could imply that perceived blend ratios change as a receiver flies through a plume filament.

24. Sand scorpions rely on a mechanism of this nature to range the distance to prey whose movements generate vibrations in the substrate. By detecting two different types of mechanical waves, compressional (P) and Rayleigh (R) (see Section 4.1.2; Aicher and Tautz 1990), which travel at different velocities, a scorpion may determine its distance from the vibration source by measuring the difference between arrival times of the two wave types (Brownell 1977, 1984).

25. Tiger moths in the genus *Holomelina* (Arctiidae) use female advertisement pheromones whose components include both 16-carbon and 19-carbon compounds, which should differ in vapor pressure and hence generate variable blend ratios at different ranges. Nonetheless, component blends in these pheromones are stable ratios, which might reflect the way in which the material is emitted (Schal et al. 1987); that is, these pheromones may be atomized (see Krasnoff and Roelofs 1988) rather than vaporized. On the other hand, Liu and Haynes (1994) report that pheromone blend ratios in the cabbage looper (*Trichoplusia ni*; Noctuidae) do change with temperature as a result of the various components having different vapor pressures. Either male responses change commensurately with temperature (see Sections 4.2.7, 5.2.1 and Chapter 7 for acoustic and optical examples of temperature coupling), or the males are oblivious to these alterations in blend ratios of the female pheromone.

26. George C. Williams, winner of the Crafoord Prize in 1999 (Maynard Smith et al. 1999).

27. A relatively small number of enzymatic steps account for much of the diversity found among female sex pheromones in the Lepidoptera. Many of the differences between pheromones arise from the sequence in which these steps act during biosynthesis (Baker 1989).

28. An early report (Hendry et al. 1975) claiming major dietary influences on composition of the female sex pheromone in the oak leaf roller moth (*Archips semi-*

feranus; Tortricidae) was subsequently found to be incorrect (Miller et al. 1976; see also Hendry 1976).

29. Contrary to some popular notions, butterflies are a recently evolved clade of the Lepidoptera and comprise only three of 29 superfamilies (Hesperioidea, Hedyloidea, Papilionoidea) of the Ditrysia (Scoble 1992).

30. Similarly, female sex pheromones in marine crustaceans and in arachnids may have originated as cues released at the time of molting (Kittredge and Takahashi 1972, Miyashita and Hayashi 1996).

31. Exocrine glands functioning in pheromone emission have apparently been lost secondarily in females of acoustic moths such as *Achroia grisella* (Pyralidae: Galleriinae). Thus, a cost of maintaining these glands may be inferred (see Section 1.1).

32. Williams's (1992) point, termed the "female pheromone fallacy," might be accurately applied to some female cuticular chemicals in arthropods that elicit sexual responses from conspecific males upon contact. Such "contact pheromones," which are present at all times, may not have been modified by selection imposed by male responses and could be categorized as cues (Section 1.1).

33. Males in some Lepidoptera may contribute limited amounts of parental investment in the form of extragametic materials transferred to the female along with the spermatophore (Boggs 1990). However, among species in which males respond and orient to female advertisement pheromones, male discrimination among receptive females (i.e., male choice) has been reported in only one species, the winter moth *Operophtera brumata* (Geometridae; van Dongen et al. 1998).

34. Some population genetic models (e.g., Liou and Price 1994, Kelly and Noor 1996, Noor 1999) demonstrate that processes reinforcing pre-mating isolation can result in corresponding post-mating isolation—and speciation—but the conditions required by these models, nearly complete hybrid inviability from the start, are restrictive and may seldom occur (Butlin 1998). One situation that can satisfy these requirements is a "mimetic shift," a change in the model species being mimicked, that is accompanied by a commensurate shift in mating signals and preferences (Jiggins et al. 2001).

35. *Paranthrene simulans* (Sesiidae), which emerges and is sexually active early in the season when few related species are present, uses the Z,Z geometric isomer of 3,13 octadecadien-1-ol-acetate as its advertisement pheromone and is not inhibited by the E,Z isomer of this compound. On the other hand, sesiid species that fly later in the season and use this advertisement pheromone are strongly inhibited by other isomers. These species fly in the presence of others that are using blends of the Z,Z and E,Z isomers, and inhibition by the E,Z antagonist would be advantageous.

36. An exception may occur in certain species of fireflies (Coleoptera: Lampyridae) in which flightless females emit continuous bioluminescent glows near their burrow entrances and thereby attract roving males. The range over which these advertisements function is not known (Section 5.2.1; Lloyd 1997).

37. Aldrich (1995) lists various examples of male sex pheromones in the Hemiptera that attract natural enemies. See Zuk and Kolluru (1998) for a general review of the exploitation of sexual signals by predators and parasitoids.

38. Several genera of araneid spiders in addition to *Mastophora* use chemical aggressive mimicry to trap male moths, but they do not swing a sticky bolas to ensnare their prey and the attractive chemicals have not been identified (Yeargan 1994, Stowe et al. 1995). Other, non-lethal, cases of male olfactory deception among arthropods involve attraction to floral scents that mimic the pheromones of conspecific females. In the andrenid bee *Andrena nigroaenea*, males are duped by and attracted to the floral scent of the orchid *Ophrys sphegodes* and serve as its pollinator. Here, the floral scent is chemically identical to female *A. nigroaenea* odor (Schiestl et al. 1999).

39. In several moth species (*Estigmene acrea*, Arctiidae; *Trichoplusia ni*, Noctuidae), male lekking and attraction of females occur at dusk but are followed later during the same night by male searching for pheromone-emitting females (Willis and Birch 1982, Landolt and Heath 1990). Is lekking a feasible alternative because suitable host plants, and hence female encounter sites, may be clustered?

40. In tephritid fruit flies, the male's advertisement is a combination of visual and olfactory signals, but his odor alone will attract females (Landolt et al. 1992).

41. Male pheromonal advertisement at resource locations occurs in the greater wax moth, *Galleria mellonella* (Pyralidae: Galleriinae), in which males remain and display near honeybee colonies (Röller et al. 1968, Leyrer and Monroe 1973). *G. mellonella* larvae feed on the wax and stored food in the colonies. Similarly, male carrion beetles, *Nicrophorus* spp. (Staphylinidae: Silphinae), advertise pheromonally at the vertebrate carcasses on which their larvae develop (Eggert and Müller 1989a,b).

42. The specific function of male pheromones in many Hemiptera is not clear, and their effective range is generally unknown. That many of these pheromones are volatile compounds suggests, however, that they serve in courtship rather than advertisement.

43. Other than scent-marking, the only means by which an arthropod can generate a continuous beacon is architectural construction. The turrets constructed by males in some shore-dwelling crabs may function in this capacity (e.g., Christy 1988a,b; see Section 6.1).

44. Females may be able to detect conspecific female pheromone,[13] and dispersion of pair-forming activities away from an agroecosystem is a conceivable evolutionary response.

45. Studies on a variety of moth species (e.g., Collins and Cardé 1989, Collins et al. 1990, Jurenka et al. 1994) suggest that pheromone titer and component ratios, as well as receiver responses, could be easily modified by selection. Both signaler and receiver traits may each be controlled by a single gene (Haynes and Hunt 1990), and these genes may be unlinked (Lofstedt et al. 1989a; see Section 6.2 and Chapter 7).

46. The *Utetheisa* model does not apply to all Lepidoptera, or even all Arctiidae. Krasnoff and Roelofs (1990) report that PA titer is not related to male courtship success in several arctiid moths (*Phragmatobia fuliginosa*, *Pyrrharctia isabella*) and suggest that their pheromones represent vestigial traits.

47. See Leal et al. (1998) for chemical characterization of the male sex pheromone emitted by the stink bug *Piezodorus hybneri* (Pentatomidae).

48. Like the advertisement pheromones of female Lepidoptera, bark beetle advertisement pheromones contain blends of components, and receiver responses are

influenced by antagonist compounds. But, unlike Lepidoptera, intraspecific variation in blend ratios in bark beetle pheromones is high, and species specificity is correspondingly low. Cues from the host tree, particularly volatile chemical emissions, are critical in the mutual attraction of conspecific bark beetles (Schlyter and Birgersson 1989).

49. Could host trees evolve a low titer of these precursors to thwart beetle attack? Possibly not, if these substances are needed for defense against other natural enemies (Byers and Birgersson 1990) or for other vital functions. Additionally, our current observations may reflect a particular episode in the coevolution between bark beetles and their host trees that is not representative of their overall history.

50. Other examples of conspecific cuing among arthropods include ovipositing females responding to feces of feeding larvae and female locusts (*Schistocerca* spp.; Orthoptera: Acrididae) responding to odors in the froth caps of egg masses that were previously deposited in the ground (Saini et al. 1995).

51. Guaiacol (2-methoxyphenol), a key aggregation pheromone component, is produced by the action of gut bacteria on gut contents (Dillon et al. 2000). Thus, the digestion of plant material is required for production of locust aggregation pheromone.

52. The massive choruses of periodical cicadas, *Magicicada* spp. (Homoptera: Cicadidae), in central and eastern North America are an example of mixed-species aggregations of sexually advertising adults (see Section 4.4). Choruses may include up to three cicada species, which are differentiated by their songs (Alexander and Moore 1962).

53. The entire queen pheromone may include several unidentified compounds, possibly produced in the head, in addition to the five-component mandibular gland pheromone (Winston and Slessor 1998).

54. The Cape honeybee (*Apis mellifera capensis*) differs from other subspecies in that its workers regularly lay eggs.

55. Various cuticular hydrocarbons have been identified as nestmate recognition cues in the desert ant *Cataglyphis niger* (Lahav et al. 1999).

56. Johnstone (1997) has analyzed this problem in a more general way and claims that under many circumstances it is evolutionarily stable for a signaler to reveal only partial identity to receivers.

57. Lockwood and Rentz (1996) describe a leaf-rolling grasshopper, *Hyalogryllacris* sp. (Orthoptera: Gryllacrididae), that construct silken webs as nesting sites and can return to their own webs via chemical cues (autocommunication signals?) in the silk.

58. Gamboa et al. (1986) proposed the identification of offspring by parents as a potential origin of kin recognition, but they also noted the lack of evidence for such identification in the Hymenoptera. In this regard, it is worth noting that females in the subsocial earwig *Labidura riparia* (Dermaptera: Labiduridae) offer care to their brood and avoid errors by accurately recognizing specific odors that they themselves impart to the offspring via the nest; that is, odors intrinsic to the offspring are not involved (Radl and Linsenmair 1991). Could this form of autocommunication evolve toward a phenotype-matching mechanism of kin recognition?.

59. The suggestion has been made that hissing sounds made by Madagascan cockroaches (*Gromphadorina portentosa*; Blattodea: Blaberidae) might serve as individual signatures in dominance interactions (Clark et al. 1995), but this function has not been confirmed experimentally.

60. Both snails (Stirling and Hamilton 1986) and snakes (Ford and Low 1984) may be able to determine odor trail polarity by textural cues deposited by the trail-layer.

61. Wellington (1957) suggested that tent caterpillars exhibited a division of labor between "leaders," who explored resources and laid trails, and "followers," who were recruited by leaders. Subsequent studies of *Malacosoma* have not confirmed regular polyethism, however.

Chapter 4

1. Orthoptera, Homoptera, Hemiptera, Coleoptera, Diptera, Lepidoptera, and Hymenoptera.

2. Blattodea, Plecoptera, Trichoptera, and Neuroptera.

3. Equivalent in intensity, as measured by equations 4.2 and 4.3.

4. Extremely high intensity sound piles up as shock waves and will travel faster than sound of lower intensity, but this phenomenon is not relevant to biological signals.

5. An rms value is used because its square is proportional to average power of the waves. Thus, SPLs of different waveforms, no matter how complex, can be accurately compared.

6. The particle displacements (A) in sound waves are extremely small. For a 94-dB sound measured in the far field, A is 5.5×10^{-7} m; for a typical cricket call at 1 m, A may be approximately 5×10^{-12} m.

7. The relationship between relative humidity, sound frequency, and attenuation via atmospheric absorption is a complex one. The relative humidity at which attenuation is maximum ranges from 9 to 20% as sound frequency rises from 1500 to 10 000 Hz; for these frequencies, attenuation declines markedly at higher relative humidities (Kinsler and Frey 1962).

8. At low sound frequencies, the radiation pattern of song in the grass cicada *Tympanistalna gastrica* may have equivalent intensity in all directions and approach the monopole model (Michelsen and Fonseca 2000). This pattern probably reflects the manner in which the cicadas radiate sound from their abdomens.

9. Poisson ratio = transverse strain / longitudinal strain of a medium subject to tension, strain being the proportional change in length.

10. Masters (1984a) reports that values of Young's modulus of elasticity range from 3 to $20 \times 10^9\ \text{N}\cdot\text{m}^{-2}$ in radial silk strands and are $\cong 0.05 \times 10^9\ \text{N}\cdot\text{m}^{-2}$ in spiral strands of the orb-weaving spider *Nuctenea sclopetaria*. Young's modulus of elasticity (E) measures the longitudinal force necessary to generate a given proportional change in length of a particular material; vibration attenuation is inversely proportional to $\sqrt{(E/\rho_l)}$, where ρ_l = mass per unit length (linear density). For comparison, E of wood is approximately $1 \times 10^{10}\ \text{N}\cdot\text{m}^{-2}$.

11. The male calling songs of some cicada species do include frequency modulations, and Fonseca et al. (2000) propose that sharply tuned neurons in the cicada ear

process these spectral features. Cicada hearing may be the most sophisticated among the acoustic insects, as each ear can include 1000 or more scolopidial units, many of which exhibit unusually steep "roll-offs" in sensitivity (increases in threshold of 30–50 dB per octave) above and below the best frequency. Frequency modulations also occur in the vibratory courtship signals of *Chrysoperla* lacewings (Chrysopidae). Wells and Henry (1992) report downward frequency sweeps, from 77 to 40 Hz within 600-ms signals, and that females prefer signals containing the full range of frequencies.

12. In tree crickets (Gryllidae: Oecanthinae), which have thin, flimsy wings, the entire tegmen, not just the harp, serves as the sound radiator.

13. See Monge-Najera et al. (1998) for description of an alternative wing percussion mechanism found in *Hamadryas* butterflies (Nymphalidae).

14. Additional examples of specialized devices used for generating substrate-borne vibration are found in the Hemiptera (Cydnidae, Pentatomidae, Reduviidae). Cydnids generate vibration via both stridulatory and tymbal devices (Gogala 1970, Gogala et al. 1974, Čokl 1983), whereas pentatomids (Harris et al. 1982, Čokl 1983, Čokl et al. 2000) and reduviids (Gogala and Čokl 1983, Roces and Manrique 1996) rely on stridulation. Reduviids stridulate with a proboscis/prosternum device. Incidental airborne sound may be generated by various of these stridulatory and tymbal devices, but its amplitude is low (35 dB at 1 cm in cydnid bugs) and receivers do not exhibit specialized responses to it (Gogala et al. 1974).

15. This principle applies to the first mode of vibration, in which complex surface warping does not arise (Fletcher 1992). The sound-radiating membranes of crickets and katydids may primarily vibrate in this mode.

16. The escapement mechanism can still yield a nearly pure tone, the resonant frequency, when teeth are not evenly spaced on the file.

17. Tonal purity of song is often measured by its Q-value, the carrier frequency at which SPL is maximum (peak frequency) divided by bandwidth at −3 dB SPL. Some insect songs that appear to be produced by excitation of simple resonators do not have the high Q-values expected. Low tonal purity may result when the effective mass or elasticity of the sound apparatus changes over the course of singing and alters its vibration frequency (Bennet-Clark 1999).

18. This mechanism effectively replicates the box of a loudspeaker, which prevents acoustic short-circuiting around the edge of the loudspeaker diaphragm.

19. Acoustic impedance as used here differs from specific acoustic impedance, Z (equation 4.6, Section 4.1.1), and refers to the impedance of a given area. It is measured in units of $Pa{\cdot}s{\cdot}m^{-3}$ and may be considered as the amount of pressure that must be applied to achieve a given fluid velocity (mass flow) over that area.

20. This phenomenon follows from the reciprocity theorem in physical acoustics (see Kinsler and Frey 1962, Fletcher 1992).

21. North American tree crickets in the genus *Oecanthus* produce male calling songs that consist of continuous trains of pulses (trills), continuous pulse trains interrupted by silent intervals (broken trills), or brief groups of pulses (chirps) which are rhythmically repeated (Walker 1957). Pulse rate is critical to female receivers in species that produce trills or broken trills, but females of broken-trill species will respond to continuous trills delivered at the correct pulse rate. Similarly, females in chirping species evaluate chirp rate but are indifferent to

pulse rate within the chirps; that is, pulseless chirps, comprised of continuous sound, are as attractive as normal ones.

22. A 50% duty cycle assumes that wing closing and opening are performed at the same speed, which is very rarely the case.

23. Thus, pulse rates exceeding 500 Hz, twice the myogenic muscle contraction rate, can be produced. In cicada species where individual tymbal ribs buckle, pulse rates of several kilohertz are possible.

24. It is generally unknown whether open-ended preferences, as often found in assessment of mate quality, also shift toward higher pulse rates with increasing temperature, but there is no a priori reason to expect them not to.

25. Shuvalov (1999) reported that the density at which immature crickets are reared can influence the number of pulses per chirp in adult calls, however there is no indication of reafference, auditory feedback, or learning in this response.

26. Stephen and Hartley (1995) have reported that acoustic feedback influences the position at which crickets hold their wings during stridulation and hence their carrier frequency. However, Prestwich et al. (2000) were unable to confirm these findings by comparing crickets calling in air and in a low-density gas (heliox/air mixture, wherein carrier frequency should have increased ≈50% according to the Stephen and Hartley hypothesis) and dispute the interpretations.

27. The acoustic efficiencies of most insects compare unfavorably with those of loudspeakers, which typically exceed 5%.

28. The vertebrate inner ear may be traced back to various modifications of the labyrinthine organs, beginning in fish (Popper et al. 1992).

29. Displacement thresholds in crustacean sensilla range from 0.1 to 0.6 μm; in contrast, maximum particle displacement (A in equation 4.3) in a 60-dB (0 dB re 1 μPa), 100-Hz sound underwater ≅ 0.001 μm.

30. From equations 4.2 and 4.3, maximum particle displacement in sound waves can be determined as $A = \Delta p_{\max}/(2\rho_0 \cdot c \cdot \pi \cdot f)$. Because the product $\rho_0 \cdot c$ is approximately 3500 × greater in water than in air, particle displacements of an underwater sound wave arriving at a small air bubble (radius $\ll \lambda$) will be amplified approximately 3500 × in the fluid at the bubble, provided that $f \cong$ the bubble's natural frequency and the bubble is readily deformed.

31. The reported ability of some lobsters to perceive underwater sound (Meyer-Rochow et al. 1982) might also result from dual sensitivity of vibration detectors, as insulated pressure-sensitive ears have not been identified.

32. The lowest threshold reported, 12 dB (at the best frequency), is found in the South African pneumorid grasshopper *Bullacris membracioides* (van Staaden and Römer 1997); this insect does not use an ordinary tympanal organ (Section 4.3.1). Among insects using tympanal ears, the most sensitive hearing indicated, a threshold ≅ 20 dB, is in the parasitoid tachinid fly *Ormia ochracea*. In female acoustic insects, hearing thresholds may be subject to hormonal influence: Allatectomy may raise the threshold, and juvenile hormone application (and aging) may reduce it (Stout et al. 1991), thereby ensuring that maximal sensitivity occurs at reproductive maturity.

33. Female mosquitoes also appear to have extremely sensitive hearing, with thresholds only slightly higher than males (Göpfert and Robert 2000).

34. Multiple sensitivity peaks have also been found within individual auditory fibers of the Polynesian cricket *Teleogryllus oceanicus* (Imaizumi and Pollack 1999). For a visual analogy, see Kitamoto et al. (1998) and endnote 19 in Sect 5.3.1. It is suggested that this neuronal feature represents a mechanism for expanding the range of spectral sensitivity.

35. Hartley and Stephen (1992) indicate that tettigoniids who produce extremely brief, high-frequency calls may be prevented from performing much conspecific frequency analysis regardless of the differential tuning of their receptors. That is, frequency analysis would require detection and processing of a minimum number of tympanal vibrations.

36. In contrast, humans blessed with superior musical talent can distinguish tones differing as little as 1 Hz in carrier frequency, but our ability to distinguish tempo is poor. Human acoustic fusion frequency is approximately 15–20 Hz, above which all pulse rates are perceived as continuous sound.

37. In the phaneropterine tettigoniid *Ancistrura nigrovittata* both temporal and spectral features of male songs are critical to females (Dobler et al. 1994a, 1994b).

38. Maximal rates as high as 1000 Hz may exist in some arthropods.

39. Female *Gryllus bimaculatus* assess males based on a weighted summation of several song characters, including pulse rate within chirps (Doherty 1985a,b).

40. The intensity difference between the ipsilateral and contralateral ears may be as great as 10 dB.

41. Comparably small pyralid moths (see Sections 4.4.1, 4.5 and 6.1) are also able to localize sound sources, albeit of high ultrasonic frequencies. In *Achroia grisella* the ears are situated ventrally on the 1st abdominal segment and are separated by < 2 mm. The localization mechanism(s) that this species uses is currently unknown.

42. Pollack (1998) describes this algorithm as "turn toward the side on which the calling song's temporal pattern is represented more strongly."

43. Crickets can distinguish sounds separated by as little as 11° in azimuth. This spatial acuity or minimum audible angle broadens when the sound sources are lateral rather than in front of the cricket. Sounds separated in altitude can also be distinguished, but with considerably less precision (Wyttenbach and Hoy 1997).

44. Bailey and Stephen (1984) report angular acuity of ±5° in the sagine katydid *Pachysagella australis*, and Mason et al. (2001) indicate even greater accuracy in the parasitoid tachinid fly *Ormia ochracea*. In the latter case, the accuracy is achieved by pooling the neural responses of a great number of individual receptor units, a phenomenon termed "temporal hyperacuity."

45. See Cocroft et al. (2000) for additional mechanisms by which very small arthropods may discern the direction from which a bending wave approaches.

46. Cicada ears are also protected during calling by changes in the corrugation of the tympanal membrane, which desensitize them by approximately 20 dB (Hennig et al. 1994).

47. Viability as used here refers to a factor operating within the context of natural selection. Examples would include survivorship and developmental rate.

48. Tuckerman et al. (1993) report that female *Scudderia curvicauda* (Orthoptera: Tettigoniidae) preferentially mate with larger males and do so by evaluating song

characters; however, spermatophore size is not correlated with male body size in this species.

49. Also, see Enquist and Arak (1994), who suggest that visual preference for symmetry may reflect a receiver bias arising from ancestral elements in the nervous system (Section 6.1; Dumont and Robertson 1986).

50. The hump-winged cricket *Cyphoderris strepitans* represents a notable exception. Females chew off a segment of a male's wings at the time of mating, and a male's mating success may be reliably estimated by the missing portion of his wings (Snedden 1996).

51. Studies of female choice in orthopteran choruses (Snedden and Greenfield 1998) suggest that receivers may respond to an individual calling with particular temporal characteristics even when its perceived SPL is up to 6 dB lower than that of neighbors calling with different characteristics. At greater SPL differences, the receiver may respond only to the louder callers regardless of their other acoustic features.

52. Visual signals may also draw the attention of predators, and appropriate modifications in signaler, and receiver, behavior can be expected. Koga et al. (1998) report that male fiddler crabs, *Uca beebei*, display less leg waving and turret construction and that females sample fewer males when in the presence of predators (cf. Hedrick and Dill 1993).

53. Differential bat predation among habitats may also influence receiver traits. Rydell (1998) suggests that earless ghost moths (*Hepialus humuli*; Hepialidae) that restrict their flight and sexual activity to areas cluttered with tall grass suffer low rates of predation by bats. Despite their lack of hearing, these moths may remain relatively safe because few bats can hunt in these areas.

54. Male–female duets also occur in cicadas (Gwynne 1987), in ultrasonic arctiid moths (Conner 1999), in aquatic Hemiptera communicating with acoustic signals (e.g., *Corixa dentipes*; Jansson 1973, Theiss 1983), and in the vibratory communication of lacewings (Neuroptera: Chrysopidae; Wells and Henry 1992), planthoppers (Homoptera: Fulgoroidea; DeWinter and Rollenhagen 1990), and wandering spiders (Araneae: Ctenidae; Schüch and Barth 1985).

55. For an example in visual signaling, see Oliveira and Custodio (1998) on the dual function of claw waving in fiddler crabs, *Uca tangeri*.

56. Field and Rind (1992) report similar acoustic dueling by rival males in a New Zealand weta, *Hemideina crassidens* (Orthoptera: Stenopelmatoidea).

57. Grafe (1999) reports that females in the West African running frog *Kassina fusca* respond preferentially to the last call in a series produced by spatially separated males.

58. A group of three or more callers may generate an alternating chorus if each individual responds to only a limited subset of its neighbors. Thus, a given individual may alternate with its nearest neighbor while synchronizing with more distant callers.

59. Ryan et al. (1981) report that males in the Central American frog *Smilisca sila* evade attacks from phonotactic bats by synchronizing their calls. However, *S. sila* synchrony is a sporadic event, and these frogs are not rhythmic callers.

60. See Neda et al. (2000) for a description of synchronized applause among culturally homogeneous concertgoers in Eastern Europe.

61. The North American periodical cicada *Magicicada cassini* would be an appropriate species to investigate. *M. cassini* males form massive choruses in which individuals synchronize their calls and do not waver from a 5-s call period throughout lengthy bouts (Figure 4.23).

62. One report of vibratory chorusing comes from a shore-dwelling ghost crab, *Ocypode jousseaumei*, of the Arabian Peninsula (David Clayton, personal communication). *O. jousseaumei* males advertise by rapping the sand with the major chela at a rhythm $\cong$ 0.7–1.0 raps$\cdot$s^{-1}, and neighbors' raps are imperfectly synchronized (cf. Figure 4.19). But the raps are generated by a conspicuous movement, and it is not clear whether visual signals play a role in chorusing (cf. Aizawa 1998, Backwell et al. 1998).

63. A simpler maneuver, the "round dance," is performed by *A. mellifera* foragers returning from nearby food sources. Waddington and Kirchner (1992) offer several speculations on signal coding in these maneuvers.

64. Spangler (1991) suggests that the carrier frequency of honeybee dance sounds, which equals the rate of wing vibration, may convey distance information to recruits (cf. Towne 1995).

65. *Apis florea*, which is considered an ancestral form according to various phylogenies (Engel and Schultz 1997), dances on a horizontal comb in the open and does not generate sounds in the process.

66. *Apis mellifera* too will dance on a horizontal surface, its waggle pointing toward the food source, if forced to do so by reorienting the comb (von Frisch 1967).

67. Recent studies of other *Melipona* spp. and *Trigona carbonaria* have not revealed recruitment via a dance language (Hrncir et al. 2000, Jarau et al. 2000; Nieh et al. 1999).

68. For example, the snapping sounds made by pistol shrimp may be used to incapacitate mollusc prey.

69. In particular, chordotonal organs with associated air sacs and functioning in proprioception or in monitoring walking and flight movements may have been the precursors of tympanal hearing organs (see Boyan 1993, Fullard and Yack 1993).

70. Rydell and Lancaster (2000) report that among the Scandinavian moth fauna, earless species tend to be large, have high wing-loading (ratio of body weight: wing area), and sustain elevated body temperatures prior to and during flight. As a result, they might be able to outfly and avoid bats without hearing and detecting them at a distance. Thus, selection to evolve ears may not have been strong, but it is also possible that ears would represent a trade-off in which flight ability is necessarily reduced. Nonetheless, phylogeny cannot be ignored, as eared species are concentrated in three superfamilies, Pyraloidea, Geometroidea, Noctuoidea (Figure 4.27).

71. Moreover, effective far-field hearing and fully functional tympana are restricted to adult insects (see Boyan 1998 on hearing development; cf. Yager 1996b). This restriction may reflect the difficulty that insects would experience in molting once a thin tympanum is formed.

72. Even among the parasitoid sarcophagid and tachinid flies in which females localize their acoustic insect hosts (cicadas and ensiferan orthopterans, respectively) by tracking the male calling songs, males as well as females have tympanal ears (Robert et al. 1994). However, male perception is less sensitive at lower frequencies

and only approximates female sensitivity at frequencies above 10 kHz. Male hearing might function in courtship or, more likely, the detection of natural enemies such as insectivorous bats.

73. Several examples of secondary loss of sound production may be found among crickets (Gryllidae). Most genera of most subfamilies of crickets include tympanate species in which males call via stridulation of the forewings (Otte and Alexander 1983). But certain species distributed among several subfamilies do not call, and their muteness may have evolved in response to high attack rates by phonotactic parasitoids or other natural enemies (cf. Walker 1974, 1977). Among Tettigonioidea, flightlessness and extreme brachyptery have evolved secondarily in many groups, but males usually retain forewing remnants that are sufficient for stridulation. Thus, secondary muteness may not necessarily occur as a correlated response to selection for other traits. See Weller et al. (1999) concerning the secondary loss of ultrasound-producing tymbals in various species of arctiid moths.

74. But see Davis (1998) for the suggestion that sound perception in the Tineoidea preceded the origin of bat echo-location by ≈30 million years and arose under other selection pressures.

75. But see Rust et al. (1999), who suggest that the ability of tettigoniids to hear low ultrasound was present at least 55 Mya, coincident with or prior to the appearance of bats.

76. See Bailey and Römer (1991) and Mason and Bailey (1998) for other examples of mismatch between receivers and signalers in the Tettigoniidae. Some tettigoniids may be able to control the opening and closing of their acoustic spiracles, which amplify high-frequency sounds. Thus, best frequencies may be altered such that high-frequency hearing dominates when the spiracles are open whereas low-frequency hearing dominates when they are closed (Bailey 1993). A female with apparently mismatched ears could then favor detection of males under some circumstances but of (terrestrial) natural enemies under others.

77. Similarly, the ultrasonic courtship signals found in various arctiid moths probably originated as defensive signals used to deter insectivorous bats (Weller et al. 1999).

Chapter 5

1. Exceptions include "invisible" (transparent) marine species, subterranean forms, and nocturnal species.

2. Arthropod photoreceptors may be located on other regions of the body, but these organs are not image-forming devices. A notable example are the genitalic photoreceptors in male papilionid butterflies, which presumably facilitate copulatory movements out of the viewing range of the compound eyes (Arikawa et al. 1980, Arikawa and Aoki 1982). The importance of retaining image-forming eyes close to the brain is illustrated well by the stalk-eyed flies (Diptera: Diopsidae), where the major visual neuropils are removed from the central sector of the brain and kept at the end of the stalk, adjacent to the retina (Buschbeck and Hoy 1998). This arrangement can be considered as a division of the brain into separate sectors, with the visual ones adjacent to the eyes.

3. The sensitivity of the human eye drops to 1% of its peak value at these lower and upper wavelength cutoffs. However, we can perceive light outside of these boundaries if it is sufficiently intense. Strictly speaking, our visual pigments can absorb light of wavelengths much shorter than 430 nm but are prevented from doing so by the corneal lens, which filters out these wavelengths; only 0.1% of 365-nm light is transmitted through the lens.

4. The advantage of favoring light at the ultraviolet end of the spectrum would be more pronounced in terrestrial arthropods. For freshwater and marine species, all wavelengths of light are 26% shorter owing to transmission through water, and diffraction, which is proportional to λ/D, would pose less of a limitation on visual perception than when light is transmitted through air (Section 5.3.1). Many vertebrates have some sensitivity to UV light, but they do not have sensitivity peaks for wavelengths <350 nm (cf. Section 5.3.1).

5. Snell's Law also applies to sound waves, in which case n_1/n_2 is replaced by c_1/c_2, the speeds of sound in the two media (Section 4.1.1).

6. Known luciferases vary in molecular weight from 21 000 to 420 000 (Hastings and Morin 1991).

7. In *Photinus consimilis*, females do not reply differentially to flashes with peak λ values ranging from 555 to 605 nm (Branham 1995). However, at threshold light intensities, as would occur in the field when a male first comes within a female's range of vision, a flash whose spectrum matches the peak sensitivity of the female receiver may be more likely to elicit her reply than a flash whose spectrum is shifted.

8. Visual signals relying on reflected light may also be temperature-dependent, particularly if they involve amplification by appendage movement (Section 5.2.2); for example, see the analysis by Doherty (1982) of male leg-waving in the fiddler crabs *Uca minax* and *U. pugnax*.

9. See Backwell et al. (1998, 1999) for an analogous event involving reflected light signals, the synchronous leg-waving displays generated by advertising males in fiddler crabs.

10. Case et al. (1994) suggest that bioluminescence is rare in terrestrial and freshwater habitats because such environments typically include many places in which an organism may conceal itself. Thus, it may remain hidden at most times and only emerge and signal, possibly revealing itself optically, when necessary. In marine habitats, however, hiding places are extremely uncommon, so an organism might remain deep in the water column during the day and only rise to the surface at night, when it may safely signal with brief flashes of bioluminescence (Case et al. 1994). But this hypothesis does not explain why so few of the many terrestrial arthropods that emerge into open areas at night use bioluminescence for signaling.

11. Other exploiters of firefly flashes include competing males who may interlope in the dialogues of their neighbors (see Lloyd 1997; see Section 4.4.2 for interloping in acoustic duets).

12. See Prum et al. (1998, 1999) for examples of avian hues generated via coherent scattering.

13. Many beetles feature iridescence generated by cuticular microstructures on the elytra (Srinivasarao 1999).

14. See Nijhout (1991) for a thorough treatment of butterfly scale patterns. In most cases the patterns are mosaics of scales that each generate a specific color, but in some species individual scales may generate more than one color and create particular spectral effects, depending on the observer's perception (e.g., Vukusic et al. 2000).

15. See Hill (1990, 1991) for studies on sexual selection in avian plumage, which demonstrate that male coloration may indicate superior viability to a discriminating female. In butterflies, Silberglied (1984) suggested that within the context of sexual selection male coloration generally functions in male–male competition rather than female choice, as had been proposed by Darwin (1871 (2nd edn, 1874)). However, this claim was made prior to several critical findings reported here (Chapter 5).

16. Simple optical images may be formed via the lens eyes found among polychaete worms (Annelida; see Land 1981, Goldsmith 1991).

17. Arthropod ommatidia generally range from 10 (lower limit) to approximately 60 μm in diameter, but some Xiphosura have much larger ones; ommatidium diameters as large as 300 μm are reported for *Limulus*.

18. See Schwemer (1989) for a review of the photochemistry of arthropod vision.

19. Generally, a retinula cell contains only one category of pigment (molecule), but some butterflies may include two pigment categories within a given retinula cell (Kitamoto et al. 1998). Presumably, this arrangement broadens spectral sensitivity. See Imaizumi and Pollack (1999) for the analogous neuronal feature in individual acoustic fibers (Sect 4.3.5; endnote 34).

20. For a human comparison, note that within our most sensitive region of the spectrum, 500–600 nm, we can typically discriminate colors differing by as little as 1–2 nm in wavelength. But the finer spectral discrimination reported in human vision could reflect the more sensitive assays that are possible. The three human visual pigments have absorption peaks at 420 (violet), 534 (green), and 564 nm (yellow-green).

21. Some arthropods (e.g., brachyceran Diptera) have open, or unfused, rhabdoms in which the individual rhabdomeres do not overlap when viewed axially (Nilsson 1989). However, this feature serves to enhance sensitivity, and images are not resolved at the level of the ommatidium.

22. In similar fashion, a UV source may be used for illumination in light microscopy to improve resolution.

23. In theory, an alternative method for determining the plane of polarization would be to use several photoreceptors, each with parallel microvillar arrays oriented in different directions, and process the across-photoreceptor pattern of excitation within an ommatidium. This procedure would require a minimum of three different photoreceptor orientations per ommatidium, but only two (orthogonal) orientations are normally seen. Thus, it does not appear to be a likely mechanism for aerial insects who must determine the polarization plane in the sky (Rossel 1989), but it may suffice for some surface-dwelling species (see Schwind 1984).

24. Kelber (1999) reports finding polarization sensitivity in all retinula cell categories in the Australian butterfly *Papilio aegeus*. While this retinal feature would lead to the perception of "false colors," it may enhance the ability to detect subtle differ-

ences in reflectance patterns from potential oviposition substrates, and possibly mates. Wehner and Bernard (1993) noted that the "twisted rhabdom" feature found in the ommatidia of honeybees (*Apis mellifera*) and many other insects prevents perception of the false colors that would otherwise arise when polarized light is confounded with reflectance spectra.

25. In the honeybee *A. mellifera*, only the green receptor is used for orientation analysis (Giger and Srinivasan 1996), and for perception of spatial pattern and motion as well.

26. Three hundred hertz is extremely high by the standards of human vision, in which flicker fusion frequencies are approximately 45 Hz.

27. Three dorsal ocelli represent the primitive (plesiomorphic) character state for insects (Pterygota + Archaeognatha and Thysanura; cf. Figure 4.26); in many species, one or all three of these organs have been secondarily lost.

28. A similar organization of the eye is apparent in fossils of trilobites (Fordyce and Cronin 1993).

Chapter 6

1. See Endler and Basolo (1998) for a review of variations of the receiver bias model; see Ryan et al. 1990 and Christy (1995) for specific variations and see Shaw (1995) for treatment of phylogenetic issues pertaining to the model.

2. This variant of the receiver bias model is termed a "sensory trap" (Christy 1995).

3. Andrade (1996) points out that under such circumstances a male may contribute parental investment to his offspring by allowing himself to be cannibalized by his mate. Thus, in some species the second element of courtship evolution, differentiation from the mimicked cue, might not be prominent.

4. Coevolutionary elaboration of a receiver bias and the exploiting signal might also arise via a "chase-away" process (Holland and Rice 1998) in which females evolve "resistance" to the male signal and males, in response to such resistance, evolve greater signal exaggeration.

5. Both Acharya and McNeil (1998) and Farris et al. (1998) indicate that moths may ignore mating signals when bat echo-location signals are broadcast at high intensities. Conversely, is it possible that a moth might ignore bat echo-location signals if the mating signal is sufficiently intense or includes a non-acoustic element (e.g., a chemical signal) that could distinguish it from a predator call?

6. *A. grisella* beat their wings ≈45 strokes·s^{-1} at 25°C, which generates 90 ultrasound pulse pairs·s^{-1} (Section 4.2.1; Figure 4.13).

7. See Gilburn and Day (1994a,b) for additional reports of good-genes effects in arthropods.

8. A classic illustration of this problem is the study by Partridge (1980), which reported that female *Drosophila melanogaster* who were encaged with an equivalent population of males produced offspring that developed more competitively than the offspring of females who were assigned individual males as mates. As pointed out by Kingett et al. (1981), female choice did not necessarily occur in the population cage.

9. In a broader context, male reproductive success under natural conditions is generally subject to intrasexual selection in the form of sperm competition (e.g., see Section 3.2.2).

10. Recently, several studies have begun to address the problem of maintaining additive genetic variation in signal characters subject to (directional) sexual selection. For example, Jia et al. (2001) showed that genetic variation in pulse rate in the male advertisement signal of the lesser wax moth (*A. grisella*) is potentially maintained by genotype × environment interaction. Genetic variants for pulse rate exist and can be separated by artificial selection, but the variants that produce high and low pulse rates under favorable temperature and dietary conditions exhibit a reversed pattern when they undergo immature development under unfavorable conditions. Thus, neither variant would display superior attractiveness in all environments.

Negative genetic correlations (genetic tradeoffs, or antagonistic pleiotropy) between signal and life history traits may also maintain additive genetic variance, provided that dominance variance is relatively high and trait expression is not sex-limited. In lesser wax moths, a genetic tradeoff between signal attractiveness and immature development rate may occur (L. Brandt, unpublished data). To the extent that these two traits influence overall fitness, no one signal variant would be favored over another.

There exists remarkably little evidence that (male) sexual attractiveness and adult longevity, or general survivorship, are negatively correlated. Rather, a recent meta-analysis of phenotypic data available from the literature (Jennions et al. 2001) indicates that sexually attractive individuals actually enjoy higher survivorship than their less attractive counterparts, perhaps owing to superior "condition." Because additive genetic variance for condition (i.e., mass, mass : length ratio, or residuals of a mass : length regression) may be high in many natural populations, additive genetic variance for sexually selected features (e.g., male advertisement signals) that reflect condition is retained (Rowe and Houle 1996). The ultimate source of additive genetic variance for condition may be a genetic tradeoff between life history traits, such as between condition (size, which can influence sexual attractiveness as well as competitive ability) and development rate.

Chapter 7

1. Gerhardt (1991) distinguishes between static signal characters, which have relatively low coefficients of variation and function in species recognition, and dynamic signal characters, which have higher coefficients of variation and function in mate choice. Static signal characters tend to match corresponding preferences and are regulated by stabilizing selection, whereas a discrepancy normally exists between dynamic signal characters and corresponding preferences (Section 4.4.1).

2. Doherty and Gerhardt (1983, 1984) have reported similar coupling of male signals and female preferences in anuran hybrids (*Hyla chrysoscelis* × *Hyla femoralis*).

3. Pires and Hoy (1992b) have tested temperature coupling and its control in the field cricket *Gryllus firmus*. Like *Chorthippus*, male song in *G. firmus* is influenced by thoracic temperature, but female preference is influenced by both head and thoracic temperature.

4. Similarly, complex song repertoires are favored in various bird species.

References

Acharya, L. and McNeil, J.N. 1998. Predation risk and mating behavior: the responses of moths to bat-like ultrasound. Behavioral Ecology 9: 552–558.

Adam, L.-J. 1983. Hind femora influence directionality of the locust ear: a possible physical solution for front/rear ambiguity. Journal of Comparative Physiology A 152: 509–516.

Adams, E.S. 1991. Nestmate recognition based on heritable odors in the termite *Microcerotermes arboreus*. Proceedings of the National Academy of Sciences, U.S.A. 88: 2031–2034.

Adams, E.S. 1994. Territory defense by the ant *Azteca trigona*: maintenance of an arboreal ant mosaic. Oecologia 97: 202–208.

Adler, J. 1975. Chemotaxis in bacteria. Annual Review of Biochemistry 44: 341–356.

Aicher, B. and Tautz, J. 1990. Vibrational communication in the fiddler crab, *Uca pugilator*. 1. Signal transmission through the substratum. Journal of Comparative Physiology A 166: 345–353.

Aiken, R.B. 1985. Sound production by aquatic insects. Biological Reviews of the Cambridge Philosophical Society 60: 163–211.

Aizawa, N. 1998. Synchronous waving in an ocypodid crab, *Ilyoplax pusilla*: analyses of response patterns to video and real crabs. Marine Biology 131: 523–532.

Alatalo, R.V., Kotiaho, J., Mappes, J. and Parri, S. 1998. Mate choice for offspring performance: major benefits or minor costs? Proceedings of the Royal Society of London B 265: 2297–2301.

Alcock, J. 1982. Natural selection and communication among bark beetles. Florida Entomologist 65: 17–32.

Alcock, J. 1998. Animal Behavior, 6th edn. Sinauer Associates, Sunderland, MA.

Alcock, J. and Bailey, W.J. 1995. Acoustical communication and the mating system of the Australian whistling moth *Hecatesia exultans* (Noctuidae: Agaristinae). Journal of Zoology 237: 337–352.

Aldrich, J.R. 1995. Chemical communication in the true bugs and parasitoid exploitation. In Chemical Ecology of Insects, Vol. 2 (eds R.T. Cardé and W.J. Bell), pp. 318–363. Chapman and Hall, New York.

Alexander, R.D. 1956. A comparative study of sound production in insects, with special reference to the singing Orthoptera and Cicadidae of the Eastern United States. PhD dissertation. Ohio State University, Columbus.

Alexander, R.D. 1960. Communicative mandible snapping in Acrididae (Orthoptera). Science 132: 152–153.

Alexander, R.D. 1962. Evolutionary change in cricket acoustical communication. Evolution 16: 443–467.

Alexander, R.D. 1975. Natural selection and specialized chorusing behavior in acoustical insects. In Insects, Science, and Society (ed. D. Pimentel), pp. 35–77. Academic Press, New York.

Alexander, R.D. and Moore, T.E. 1962. The evolutionary relationships of 17-year and 13-year cicadas, and three new species (Homoptera, Cicadidae, *Magicicada*). Miscellaneous Publications of the Museum of Zoology, University of Michigan, Ann Arbor. No. 121: 1–59.

Almaas, T.J. and Mustaparta, H. 1991. *Heliothis virescens*: response characteristics of receptor neurons in sensilla trichodea type 1 and type 2. Journal of Chemical Ecology 17: 953–972.

Amoore, J.E. 1967. Specific anosmia: a clue to the olfactory code. Nature 214: 1095–1098.

Amoore, J.E. 1974. Evidence for chemical olfactory code in man. Annals of the New York Academy of Sciences 237: 137–143.

Amoore, J.E. 1977. Specific anosmia and concept of primary odors. Chemical Senses and Flavour 2: 267–281.

Anderbrant, O. 1993. Pheromone biology of sawflies. In Sawfly Life History Adaptations to Woody Plants (eds M.R. Wagner and K.F. Raffa), pp. 119–154. Academic Press, San Diego, CA.

Andersson, M. 1994. Sexual Selection. Princeton University Press. Princeton, NJ.

Andrade, M.C.B. 1996. Sexual selection for male sacrifice in the Australian redback spider. Science 271: 70–72.

Anton, S. and Hansson, B.S. 1995. Sex pheromone and plant associated odor processing in antennal lobe interneurons of male *Spodoptera littoralis* (Lepidoptera: Noctuidae). Journal of Comparative Physiology A 176: 773–789.

Applebaum, S.W. and Heifetz, Y. 1999. Density-dependent physiological phase in insects. Annual Review of Entomology 44: 317–341.

Arak, A. 1988. Callers and satellites in the natterjack toad: evolutionarily stable decision rules. Animal Behaviour 36: 416–432.

Arak, A. and Eiríksson, T. 1992. Choice of singing sites by male bush crickets (*Tettigonia viridissima*) in relation to signal propagation. Behavioral Ecology and Sociobiology 30: 365–372.

Arak, A. and Enquist, M. 1993. Hidden preferences and the evolution of signals. Philosophical Transactions of the Royal Society of London B 340: 207–213.

Arak, A. and Enquist, M. 1995. Conflict, receiver bias and the evolution of signal form. Philosophical Transactions of the Royal Society of London B 349: 337–344.

Arak, A., Eiríksson, T. and Radesäter, T. 1990. The adaptive significance of acoustic spacing in male bush crickets *Tettigonia viridissima*: a perturbation experiment. Behavioral Ecology and Sociobiology 26: 1–7.

Arbas, E.A. 1997. Neuroethological study of pheromone-modulated responses. In Insect Pheromone Research: New Directions (eds R.T. Cardé and A.K. Minks), pp. 320–329. Chapman and Hall, New York.

Arikawa, K. and Aoki, K. 1982. Response characteristics and occurrence of extraocular photoreceptors on lepidopteran genitalia. Journal of Comparative Physiology A 148: 483–489.

Arikawa, K., Eguchi, E., Yoshida, A. and Aoki, K. 1980. Multiple extra-ocular photoreceptive areas on genitalia of butterfly *Papilio xuthus*. Nature 288: 700–702.

Arikawa, K., Inokuma, K. and Eguchi, E. 1987. Pentachromatic visual system in a butterfly. Naturwissenschaften 74: 297–298.

Arnold, G.P. 1974. Rheotropism in fishes. Biological Reviews of the Cambridge Philosophical Society 49: 515–576.

Atema, J. 1995. Chemical signals in the marine environment: dispersal, detection, and temporal signal analysis. Proceedings of the National Academy of Sciences, U.S.A. 92: 62–66.

Atema, J. 1996. Eddy chemotaxis and odor landscapes: exploration of nature with animal sensors. Biological Bulletin 191: 129–138.

Atema, J. 1998. Tracking turbulence: processing the bimodal signals that define an odor plume. Biological Bulletin 195: 179–180.

Autrum, H.-J. 1940. Über Lautausserungen und Schallwahrnehmung bei Arthropoden. II. Das Richtungshören von Locusta und Versuch einen Hörheorie für Tympanalorgane vom Locustidentyp. Zeitschrift für Vergleichende Physiologie 28: 326–352.

Autrum, H.-J. 1950. Die Belichtungspotentiale und das Sehen der Insekten (Untersuchungen an *Calliphora* und *Dixippus*). Zeitschrift für Vergleichende Physiologie 32: 176–227.

Backwell, P.R.Y., Jennions, M.D., Christy, J.H. and Schober, U. 1995. Pillar building in the fiddler crab *Uca beebei*: evidence for a condition dependent ornament. Behavioral Ecology and Sociobiology 36: 185–192.

Backwell, P., Jennions, M., Passmore, N. and Christy, J. 1998. Synchronized courtship in fiddler crabs. Nature 391: 31–32.

Backwell, P.R.Y., Jennions, M.D., Christy, J.H. and Passmore, N.I. 1999. Female choice in the synchronously waving fiddler crab *Uca annulipes*. Ethology 105: 415–421.

Backwell, P.R.Y., Christy, J.H., Telford, S.R., Jennions, M.D. and Passmore, N.I. 2000. Dishonest signalling in a fiddler crab. Proceedings of the Royal Society of London B 267: 719–724.

Bailey, W.J. 1978. Resonant wing systems in Australian whistling moth *Hecatesia* (Agarasidae: Lepidoptera). Nature 272: 444–446.

Bailey, W.J. 1991. Acoustic Behaviour of Arthropods. Chapman and Hall, London.

Bailey, W.J. 1993. The tettigoniid (Orthoptera, Tettigoniidae) ear: multiple functions and structural diversity. International Journal of Insect Morphology and Embryology 22: 185–205.

Bailey, W.J. 1998. Do large bushcrickets have more sensitive ears? Natural variation in hearing thresholds within populations of the bushcricket *Requena verticalis* (Listroscelidinae: Tettigoniidae). Physiological Entomology 23: 105–112.

Bailey, W.J. and Field, G. 2000. Acoustic satellite behaviour in the Australian bushcricket *Elephantodeta nobilis* (Phaneropterinae: Tettigoniidae: Orthoptera). Animal Behaviour 59: 361–369.

Bailey, W.J. and Haythornthwaite, S. 1998. Risks of calling by the field cricket *Teleogryllus oceanicus*; potential predation by Australian long-eared bats. Journal of Zoology 244: 505–513.

Bailey, W.J. and Römer, H. 1991. Sexual differences in auditory sensitivity: mismatch of hearing threshold and call frequency in a tettigoniid (Orthoptera: Tettigoniidae: Zaprochilinae). Journal of Comparative Physiology A 169: 349–353.

Bailey, W.J. and Stephen, R.O. 1978. Directionality and auditory slit function: theory of hearing in bushcrickets. Science 201: 633–634.

Bailey, W.J. and Stephen, R.O. 1984. Auditory acuity in the orientation behaviour of the bushcricket *Pachysagella australis* (Orthoptera: Tettigoniidae: Saginae). Animal Behaviour 32: 816–829.

Bailey, W.J. and Yeoh, P.B. 1988. Female phonotaxis and frequency discrimination in the bushcricket *Requena verticalis*. Physiological Entomology 13: 363–372.

Bailey, W.J., Cunningham, R.J. and Lebel, l. 1990. Song power, spectral distribution and female phonotaxis in the bushcricket *Requena verticalis* (Tettigoniidae, Orthoptera): Active female choice or passive attraction? Animal Behaviour 40: 33–42.

Bailey, W.J., Greenfield, M.D. and Shelly, T.E. 1993a. Transmission and perception of acoustic-signals in the desert clicker, *Ligurotettix coquilletti* (Orthoptera: Acrididae). Journal of Insect Behavior 6: 141–154.

Bailey, W.J., Withers, P.C., Endersby, M. and Gaull, K. 1993b. The energetic costs of calling in the bush cricket *Requena verticalis* (Orthoptera: Tettigoniidae: Listroscelidinae). Journal of Experimental Biology 178: 21–37.

Baker, T.C. 1986. Pheromone-modulated movements of flying moths. In Mechanisms in Insect Olfaction (eds T.L. Payne, M.C. Birch and C.E.J. Kennedy), pp. 39–48. Clarendon Press, Oxford.

Baker, T.C. 1989. Sex pheromone communication in the Lepidoptera: new research progress. Experientia 45: 248–262.

Baker, T.C. and Cardé, R.T. 1979. Courtship behavior of the Oriental fruit moth (*Grapholita molesta*): experimental analysis and consideration of the role of sexual selection in the evolution of courtship pheromones in the Lepidoptera. Annals of the Entomological Society of America 72: 173–188.

Baker, T.C. and Haynes, K.F. 1989. Field and laboratory electroantennographic measurements of pheromone plume structure correlated with oriental fruit moth behavior. Physiological Entomology 14: 1–12.

Baker, T.C. and Haynes, K.F. 1996. Pheromone mediated optomotor anemotaxis and altitude control exhibited by male oriental fruit moths in the field. Physiological Entomology 21: 20–32.

Baker, T.C. and Linn, C.E. Jr. 1984. Wind tunnels in pheromone research. In Techniques in Pheromone Research (eds H.E. Hummel and T.A. Miller), pp. 75–110. Springer-Verlag, New York.

Baker, T.C. and Vickers, N.J. 1997. Pheromone-mediated flight in moths. In Insect Pheromone Research: New Directions (eds R.T. Cardé and A.K. Minks), pp. 248–264. Chapman and Hall, New York.

Baker, T.C., Willis, M.A. and Phelan, P.L. 1984. Optomotor anemotaxis polarizes self-steered zigzagging in flying moths. Physiological Entomology 9: 365–376.

Baker, T.C., Fadamiro, H.Y. and Cosse, A.A. 1998. Moth uses fine tuning for odour resolution. Nature 393: 530.

Bakker, T.C.M. 1993. Positive genetic correlation between female preference and preferred male ornament in sticklebacks. Nature 363: 255–257.

Bakker, T.C.M. 1999. The study of intersexual selection using quantitative genetics. Behaviour 136: 1237–1266.

Bakker, T.C.M. and Pomiankowski, A. 1995. The genetic basis of female mate preferences. Journal of Evolutionary Biology 8: 129–171.

Balakrishnan, R. and Pollack, G.S. 1996. Recognition of courtship song in the field cricket, *Teleogryllus oceanicus*. Animal Behaviour 51: 353–366.

Ballard, J.W.O., Olsen, G.J., Faith, D.P., Odgers, W.A., Rowell, D.M. and Atkinson, P.W. 1992. Evidence from 12s ribosomal-RNA sequences that onychophorans are modified arthropods. Science 258: 1345–1348.

Barth, F.G. 1982. Spiders and vibratory signals: sensory reception and behavioral significance. In Spider Communication: Mechanisms and Ecological Significance (eds P.N. Witt and J.S. Rovner), pp. 67–122. Princeton University Press, Princeton, NJ.

Barth, F.G. 1997. Vibratory communication in spiders: adaptation and compromise at many levels. In Orientation and Communication in Arthropods (ed. M. Lehrer), pp. 247–272. Birkhäuser Verlag, Basel.

Barth, F.G. 1998. The vibrational sense of spiders. In Springer Handbook of Auditory Research, Vol. 10; Comparative Hearing: Insects (eds R.R. Hoy, A.N. Popper and R.R. Fay), pp. 228–278. Springer-Verlag, New York.

Barth, F.G. 2000. How to catch the wind: spider hairs specialized for sensing the movement of air. Naturwissenschaften 87: 51–58.

Barth, F.G. and Holler, A. 1999. Dynamics of arthropod filiform hairs. V. the response of spider trichobothria to natural stimuli. Philosophical Transactions of the Royal Society of London B 354: 183–192.

Basil, J. and Atema, J. 1994. Lobster orientation in turbulent odor plumes: simultaneous measurement of tracking behavior and temporal odor patterns. Biological Bulletin 187: 272–273.

Basolo, A.L. 1995. Phylogenetic evidence for the role of a preexisting bias in sexual selection. Proceedings of the Royal Society of London B 259: 307–311.

Bauer, M. and von Helversen, O. 1987. Separate localization of sound recognizing and sound producing neural mechanisms in a grasshopper. Journal of Comparative Physiology A 161: 95–101.

Beckers, R., Deneubourg, J.L. and Goss, S. 1993. Modulation of trail laying in the ant *Lasius niger* (Hymenoptera: Formicidae) and its role in the collective selection of a food source. Journal of Insect Behavior 6: 751–759.

Beecher, M.D. 1989. Signaling systems for individual recognition: an information theory approach. Animal Behaviour 38: 248–261.

Belanger, J.H. and Arbas, E.A. 1998. Behavioral strategies underlying pheromone-modulated flight in moths: lessons from simulation studies. Journal of Comparative Physiology A 183: 345–360.

Belanger, J.H. and Willis, M.A. 1996. Adaptive control of odor guided locomotion: behavioral flexibility as an antidote to environmental unpredictability. Adaptive Behavior 4: 217–253.

Bell, W.J. 1984. Chemo-orientation in walking insects. In Chemical Ecology of Insects (eds W.J. Bell and R.T. Cardé), pp. 93–109. Chapman and Hall, London.

Bell, W.J. and Tobin, T.R. 1981. Orientation to sex pheromone in the American cockroach: analysis of orientation mechanisms. Journal of Insect Physiology 27: 501–508.

Bell, W.J., Kipp, L.R. and Collins, R.D. 1995. The role of chemo-orientation in search behavior. In Chemical Ecology of Insects, Vol. 2 (eds R.T. Cardé and W.J. Bell), pp. 105–152. Chapman and Hall, New York.

Belwood, J.J. 1990. Anti-predator defences and ecology of neotropical forest katydids, especially the Pseudophyllinae. In The Tettigoniidae: Biology, Systematics, and Evolution (eds W.J. Bailey and D.C.F. Rentz), pp. 8–26. Springer-Verlag, Berlin.

Belwood, J.J. and Morris, G.K. 1987. Bat predation and its influence on calling behavior in neotropical katydids. Science 238: 64–67.

Bennet-Clark, H.C. 1970. Mechanism and efficiency of sound production in mole crickets. Journal of Experimental Biology 52: 619–652.

Bennet-Clark, H.C. 1971. Acoustics of insect song. Nature 234: 255–259.

Bennet-Clark, H.C. 1984. A particle velocity microphone for the song of small insects and other acoustic measurements. Journal of Experimental Biology 108: 459–463.

Bennet-Clark, H.C. 1987. The tuned singing burrow of mole crickets. Journal of Experimental Biology 128: 383–409.

Bennet-Clark, H.C. 1989. Songs and the physics of sound production. In Cricket Behavior and Neurobiology (eds F. Huber, T.E. Moore, and W. Loher), pp. 227–261. Cornell University Press, Ithaca, NY.

Bennet-Clark, H.C. 1997. Tymbal mechanics and the control of song frequency in the cicada *Cyclochila australasiae*. Journal of Experimental Biology 200: 1681–1694.

Bennet-Clark, H.C. 1998a. Size and scale effects as constraints in insect sound communication. Philosophical Transactions of the Royal Society of London B 353: 407–419.

Bennet-Clark, H.C. 1998b. How cicadas make their noise. Scientific American 278(5): 58–61.

Bennet-Clark, H.C. 1999. Resonators in insect sound production: how insects produce loud pure-tone songs. Journal of Experimental Biology 202: 3347–3357.

Bennet-Clark, H.C. and Daws, AG. 1999. Transduction of mechanical energy into sound energy in the cicada *Cyclochila australasiae*. Journal of Experimental Biology 202: 1803–1817.

Bennet-Clark, H.C., Leroy, Y. and Tsacas, L. 1980. Species and sex-specific songs and courtship behavior in the genus *Zaprionus* (Diptera: Drosophilidae). Animal Behaviour 28: 230–255.

Berg, B.G., Tumlinson, J.H. and Mustaparta, H. 1995. Chemical communication in heliothine moths. 4. Receptor neuron responses to pheromone compounds and formate analogs in the male tobacco budworm moth *Heliothis virescens*. Journal of Comparative Physiology A 177: 527–534.

Berg, H.C. and Brown, D.A. 1972. Chemotaxis in *Escherichia coli* analyzed by three-dimensional tracking. Nature 239: 500–504.

Bergman, P. and Bergstrom, G. 1997. Scent marking, scent origin, and species specificity in male premating behavior of two Scandinavian bumblebees. Journal of Chemical Ecology 23: 1235–1251.

Berkey, C. and Atema, J. 1999. Individual recognition and memory in *Homarus americanus* male-female interactions. Biological Bulletin 197: 253–254.

Bernard, G.D. and Remington, C.L. 1991. Color vision in *Lycaena* butterflies: spectral tuning of receptor arrays in relation to behavioral ecology. Proceedings of the National Academy of Sciences, U.S.A. 88: 2783–2787.

Berryman, A.A., Dennis, B., Raffa, K.F. and Stenseth, N.C. 1985. Evolution of optimal group attack, with particular reference to bark beetles (Coleoptera: Scolytidae). Ecology 66: 898–903.

Beugnon, G. and Dejean, A. 1992. Adaptative properties of the chemical trail system of the African weaver ant *Oecophylla longinoda* Latreille (Hymenoptera: Formicidae: Formicinae). Insectes Sociaux 39: 341–346.

Billen, J. and Morgan, E.D. 1998. Pheromone communication in social insects: sources and secretions. In Pheromone Communication in Social Insects: Ants, Wasps, Bees and Termites (eds R.K. Vander Meer, M.D. Breed, K.E. Espelie and M.L. Winston), pp. 3–33. Westview Press, Boulder, CO.

Birch, M.C., Poppy, G.M. and Baker, T.C. 1990. Scents and eversible scent structures of male moths. Annual Review of Entomology 35: 25–58.

Bjostad, L.B. 1988. Insect electroantennogram responses to semiochemicals recorded with an inexpensive personal computer. Physiological Entomology 13: 139–145.

Bjostad, L.B. and Roelofs, W.L. 1980. An inexpensive electronic device for measuring electroantennogram responses to sex pheromone components with a voltmeter. Physiological Entomology 5: 309–314.

Bjostad, L.B., Gaston, L.K. and Shorey, H.H. 1980. Temporal pattern of sex pheromone release by female *Trichoplusia ni*. Journal of Insect Physiology 26: 493–498.

Blest, A.D. 1957. The function of eyespot patterns in the Lepidoptera. Behaviour 11: 206–256.

Blest, A.D., Pye, J.D. and Collett, T.S. 1963. Generation of ultrasonic signals by a New World arctiid moth. Proceedings of the Royal Society of London B 158: 196–207.

Blum, M.S. 1974. Pheromonal bases of social manifestations in insects. In Pheromones (ed. M.C. Birch), pp. 190–199. North-Holland, Amsterdam.

Boake, C.R.B. 1991. Coevolution of senders and receivers of sexual signals: genetic coupling and genetic correlations. Trends in Ecology and Evolution 6: 225–227.

Boake, C.R.B. and Capranica, R.R. 1982. Aggressive signal in "courtship" chirps of a gregarious cricket. Science 218: 580–582.

Boeckh, J. 1984. Neurophysiological aspects of insect olfaction. In Insect Communication, pp. 83–104. Academic Press, London.

Boggs, C.L. 1990. A general model of the role of male donated nutrients in female insects reproduction. American Naturalist 136: 598–617.

Böhm, H., Schildberger, K. and Huber, F. 1991. Visual and acoustic course control in the cricket *Gryllus bimaculatus*. Journal of Experimental Biology 159: 235–248.

Boland, W. 1995. The chemistry of gamete attraction: chemical structures, biosynthesis, and (a)biotic degradation of algal pheromones. Proceedings of the National Academy of Sciences, U.S.A. 92: 37–43.

Bonabeau, E., Theraulaz, G., Deneubourg, J.L., Aron, S. and Camazine, S. 1997. Self organization in social insects. Trends in Ecology and Evolution 12: 188–193.

Boore, J.L., Collins, T.M., Stanton, D., Daehler, L.L. and Brown, W.M. 1995. Deducing the pattern of arthropod phylogeny from mitochondrial DNA rearrangements. Nature 376: 163–165.

Boppré, M. 1978. Chemical communication, plant relationships, and mimicry in the evolution of danaid butterflies. Entomologia Experimentalis et Applicata 24: 264–277.

Boppré, M. 1984. Chemically mediated interactions between butterflies. Symposia of the Royal Entomological Society of London 11: 259–275.

Boppré, M. 1986. Insects pharmacophagously utilizing defensive plant chemicals (pyrrolizidine alkaloids). Naturwissenschaften 73: 17–26.

Boppré, M. 1990. Lepidoptera and pyrrolizidine alkaloids: exemplification of complexity in chemical ecology. Journal of Chemical Ecology 16: 165–185.

Boppré, M. and Schneider, D. 1985. Pyrrolizidine alkaloids quantitatively regulate both scent organ morphogenesis and pheromone biosynthesis in male *Creatonotos* moths (Lepidoptera: Arctiidae). Journal of Comparative Physiology A 157: 569–577.

Boppré, M. and Vane-Wright, R.I. 1989. Androconial systems in Danainae (Lepidoptera): functional morphology of *Amauris*, *Danaus*, *Tirumala* and *Euploea*. Zoological Journal of the Linnean Society 97: 101–133.

Bossert, W.H. 1968. Temporal patterning in olfactory communication. Journal of Theoretical Biology 18: 157–170.

Bossert, W.H. and Wilson, E.O. 1963. Analysis of olfactory communication among animals. Journal of Theoretical Biology 5: 443–469.

Bowden, R.M., Willamson, S. and Breed, M.D. 1998. Floral oils: their effect on nestmate recognition in the honeybee, *Apis mellifera*. Insectes Sociaux 45: 209–214.

Boyan, G.S. 1993. Another look at insect audition: the tympanic receptors as an evolutionary specialization of the chordotonal system. Journal of Insect Physiology 39: 187–200.

Boyan, G.S. 1998. Development of the insect auditory system. In Handbook of Auditory Research, Vol. 10; Comparative Hearing: Insects (eds R.R. Hoy, A.N. Popper and R.R. Fay), pp. 97–138. Springer-Verlag, New York.

Bradbury, J.W. and Vehrencamp, S.L. 1998. Principles of Animal Communication. Sinauer Associates, Sunderland, MA.

Brand, J.M., Bracke, J.W., Markovetz, A.J., Wood, D.L. and Browne, L..E. 1975. Production of verbenol pheromone by a bacterium isolated from bark beetles. Nature 254: 136–137.

Branham, M.A. 1995. Sexual selection in bioluminescent signaling: female choice for high male flash rates in the firefly *Photinus consimilis* (Coleoptera: Lampyridae). MA thesis, University of Kansas, Lawrence.

Branham, M.A. and Greenfield, M.D. 1996. Flashing males win mate success. Nature 381: 745–746.

Breed, M.D. 1983. Nestmate recognition in honey bees. Animal Behaviour 31: 86–91.

Breed, M.D. 1998. Chemical cues in kin recognition: criteria for identification, experimental approaches, and the honey bee as an example. In Pheromone Communication in Social Insects: Ants, Wasps, Bees and Termites (eds R.K. Vander Meer, M.D. Breed, K.E. Espelie and M.L. Winston), pp. 57–78. Westview Press, Boulder, CO.

Breer, H. 1997. Molecular mechanisms of pheromone reception in insect antennae. In Insect Pheromone Research: New Directions (eds R.T. Cardé and A.K. Minks), pp. 115–130. Chapman and Hall, New York.

Briscoe, A.D. and Chittka, L. 2001. The evolution of color vision in insects. Annual Review of Entomology 46: 471–510.

Brockmann, A., Bruckner, D. and Crewe, R.M. 1998. The EAG response spectra of workers and drones to queen honeybee mandibular gland components: the evolution of a social signal. Naturwissenschaften 85: 283–285.

Brower, L.P., Brower, J.V.Z. and Cranston, F.P. 1965. Courtship behavior of the queen butterfly, *Danaus gilippus berenice* (Cramer). Zoologica (NY) 50, 1: 1–39.

Brown, J.L. 1982. Optimal group size in territorial animals. Journal of Theoretical Biology 95: 793–810.

Brown, W.D., Wideman, J., Andrade, M.C.B., Mason, A.C. and Gwynne, D.T. 1996. Female choice for an indicator of male size in the song of the black-horned tree cricket *Oecanthus nigricornis* (Orthoptera: Gryllidae: Oecanthinae). Evolution 50: 2400–2411.

Brownell, P.H. 1977. Compressional and surface waves in sand: used by desert scorpions to locate prey. Science 197: 479–482.

Brownell, P.H. 1984. Prey detection by the sand scorpion. Scientific American 251(6): 86–97.

Brunton, C.F.A. and Majerus, M.E.N. 1995. Ultraviolet colours in butterflies: intra- or inter-specific communication? Proceedings of the Royal Society of London B 260: 199–204.

Brusca, R.C. 2000. Unraveling the history of arthropod biodiversification. Annals of the Missouri Botanical Garden 87: 13–25.

Brush, J.S., Gian, V.G. and Greenfield, M.D. 1985. Phonotaxis and aggression in the coneheaded katydid *Neoconocephalus affinis*. Physiological Entomology 10: 23–32.

Buck, J.B. 1938. Synchronous rhythmic flashing of fireflies. Quarterly Review of Biology 13: 301–314.

Buck, J. 1988. Synchronous rhythmic flashing in fireflies. II. Quarterly Review of Biology. 63: 265–289.

Buck, J. and Buck, E. 1976. Synchronous fireflies. Scientific American 234: 74–85.

Buck, J. and Buck, E. 1978. Toward a functional interpretation of synchronous flashing by fireflies. American Naturalist 112: 471–492.

Buck, J., Buck, E., Case, J.F. and Hanson, F.E.. 1981a. Control of flashing in fireflies. V. Pacemaker synchronization in *Pteroptyx cribellata*. Journal of Comparative Physiology A 144: 287–298.

Buck, J., Buck, E., Hanson, F.E., Case, J.F., Mets, L. and Atta, G.J. 1981b. Control of flashing in fireflies. IV. Free run pacemaking in synchronic *Pteroptyx*. Journal of Comparative Physiology A 144: 277–286.

Buckle, G.R. and Greenberg, L. 1981. Nestmate recognition in sweat bees (*Lasioglossum zephyrum*): does an individual recognize its own odor or only odors of its nestmates? Animal Behaviour 29: 802–809.

Burgess, J.W. 1976. Social spiders. Scientific American 234: 100–106.

Burgess, J.W. 1979. Web-signal processing for tolerance and group predation in the social spider *Mallos gregalis* Simon. Animal Behaviour 27: 157–164.

Burk, T.E. 1982. Evolutionary significance of predation on sexually signalling males. Florida Entomologist 65: 90–104.

Buschbeck, E.K. and Hoy, R.R. 1998. Visual system of the stalk-eyed fly, *Cyrtodiopsis quinqueguttata* (Diopsidae: Diptera): an anatomical investigation of unusual eyes. Journal of Neurobiology 37: 449–468.

Buschbeck, E., Ehmer, B. and Hoy, R. 1999. Chunk versus point sampling: visual imaging in a small insect. Science 286: 1178–1180.

Bushmann, P.J. and Atema, J. 2000. Chemically mediated mate location and evaluation in the lobster, *Homarus americanus*. Journal of Chemical Ecology 26: 883–899.

Butenandt, A. 1963. Bombycol, sex attractive substance of silkworm, *Bombyx mori*. Journal of Endocrinology 27: ix–xvi.

Butenandt, A. and Karlson, P. 1954. Uber die Isolierung eines Metamorphose-Hormons der Insekten in kristallisierter Form. Zeitschrift für Naturforschung 9b: 389–391.

Butenandt, A., Beckmann, R., Stamm, D. and Hecker, E. 1959. Uber den Sexuallockstoff des Seidenspinners *Bombyx mori*: Reindarstellung und Konstitution. Zeitschrift für Naturforschung B 14: 283–284.

Butenandt, A., Zayed, S.M.A. and Hecker, E. 1963. Uber den Sexuallockstoff des Seidenspinners. 3. Ungesattigte Fettsauren aus den Hinterleibsdrusen (sacculi laterales) des Seidenspinnerweibchens (*Bombyx mori* L.). Hoppe-Seyler's Zeitschrift für Physiologische Chemie 333: 114–126.

Butlin, R. 1998. What do hybrid zones in general, and the *Chorthippus parallelus* zone in particular, tell us about speciation? In Endless Forms: Species and Speciation (eds D.J. Howard and S.H. Berlocher), pp. 367–378. Oxford University Press, Oxford.

Butlin, R.K. and Ritchie, M.G. 1989. Genetic coupling in mate recognition systems: What is the evidence? Biological Journal of the Linnean Society 37: 237–246.

Butlin, R.K. and Trickett, A.J. 1997. Can population genetic simulations help to interpret pheromone evolution? In Insect Pheromone Research: New Directions (eds R.T. Cardé and A.K. Minks), pp. 548–562. Chapman and Hall, New York.

Byers, J.A. 1989a. Chemical ecology of bark beetles. Experientia 45: 271–283.

Byers, J.A. 1989b. Behavioral mechanisms involved in reducing competition in bark beetles. Holarctic Ecology 12: 466–476.

Byers, J.A. 1995. Host-tree chemistry affecting colonization of bark beetles. In Chemical Ecology of Insects, Vol. 2 (eds R.T. Cardé and W.J. Bell), pp. 154–213. Chapman and Hall, New York.

Byers, J.A. and Birgersson, G. 1990. Pheromone production in a bark beetle independent of myrcene precursor in host pine species. Naturwissenschaften 77: 385–387.

Byers, J.A. and Wood, D.L. 1981. Antibiotic-induced inhibition of pheromone synthesis in a bark beetle. Science 213: 763–764.

Cade, W. 1975. Acoustically orienting parasitoids: fly phonotaxis to cricket song. Science 190: 1312–1313.

Cade, W.H. 1979. The evolution of alternative male reproductive strategies in field crickets. In Sexual Selection and Reproductive Competition in Insects (eds M.S. Blum and N.A. Blum), pp. 343–379. Academic Press, New York.

Cade, W.H. 1981. Alternative male strategies: genetic differences in crickets. Science 212: 563–564.

Cade, W.H. 1984. Effects of fly parasitoids on nightly calling duration in field crickets. Canadian Journal of Zoology 62: 226–228.

Cade, W.H. and Wyatt, D.R. 1984. Factors affecting calling behaviour in field crickets, *Teleogryllus* and *Gryllus* (age, weight, density, and parasites). Behaviour 88: 61–75.

Caldwell, R.L. 1985. A test of individual recognition in the stomatopod *Gonodactylus festae*. Animal Behaviour 33: 101–106.

Caldwell, R.L. 1992. Recognition, signaling and reduced aggression between former mates in a stomatopod. Animal Behaviour 44: 11–19.

Callahan, P.S. 1991. Dielectric waveguide (open resonator) models of the corn earworm sensilla-sensilla relationship to infrared coherent molecular scatter emissions from semiochemicals (Lepidoptera: Noctuidae). Annals of the Entomological Society of America 84: 361–368.

Callahan, P.S. and Carlysle, T.C. 1971. Function of epiphysis on foreleg of corn earworm moth, *Heliothis zea* (Lepidoptera: Noctuidae). Annals of the Entomological Society of America 64: 309–311.

Campbell, H.R. 2001. Orientation discrimination independent of retinal matching by blowflies. Journal of Experimental Biology 204: 15–23.

Cardé, R.T. 1983. Flight periodicity in insects and the influence of atmospheric hydrocarbons. American Naturalist 121: 746–748.

Cardé, R.T. 1986. The role of pheromones in reproductive isolation and speciation of insects. In Evolutionary Genetics of Invertebrate Behavior: Progress and Prospects (eds R.T. Cardé and M.D. Huettel), pp. 303–317. Plenum Press, New York.

Cardé, R.T. and Charlton, R.E. 1984. Olfactory sexual communication in Lepidoptera: strategy, sensitivity and selectivity. In Insect Communication (ed. T. Lewis), pp. 241–265. Academic Press, London.

Cardé, R.T. and Mafra-Neto, A. 1997. Mechanisms of flight of male moths to pheromone. In Insect Pheromone Research: New Directions (eds R.T. Cardé and A.K. Minks), pp. 275–290. Chapman and Hall, New York.

Cardé, R.T. and Minks, A.K. 1997. Insect Pheromone Research: New Directions. Chapman and Hall, New York.

Carlson, A.D. and Copeland, J. 1985. Communication in insects. 1. Flash communication in fireflies. Quarterly Review of Biology 60: 415–436.

Carlson, A.D., Copeland, J., Raderman, R. and Bulloch, A.G.M. 1976. Role of interflash intervals in a firefly courtship (*Photinus macdermotti*). Animal Behaviour 24: 786–792.

Carson, R. 1962. Silent Spring. Houghton-Mifflin Co., Boston, MA.

Case, J.F. 1984. Vision in mating behavior of fireflies. In Insect Communication (ed. T. Lewis), pp. 195–222. Academic Press, London.

Case, J.F. and Strause, L.G. 1978. Neurally controlled luminescent systems. In Bioluminescence in Action (ed. P.J. Herring), pp. 331–336. Academic Press, London, U.K.

Case, J.F., Haddock, S.H.D. and Harper, R.D. 1994. The ecology of bioluminescence. In Bioluminescence and Chemiluminescence: Fundamentals and Applied Aspects (eds A.K. Campbell, L.J. Kricka and P.E. Stanley), pp. 115–122. John Wiley and Sons, New York.

Charalambous, M., Butlin, R.K. and Hewitt, G.M. 1994. Genetic variation in male song and female song preference in the grasshopper *Chorthippus brunneus* (Orthoptera: Acrididae). Animal Behaviour 47: 399–411.

Charlton, R.E. and Cardé, R.T. 1982. Rate and diel periodicity of pheromone emission from female gypsy moths (*Lymantria dispar*) determined with a glass adsorption collection system. Journal of Insect Physiology 28: 423–430.

Cherry, E.C. 1953. Some experiments on the recognition of speech, with one and with two ears. Journal of the Acoustical Society of America 25: 293–324.

Christensen, T.A. and Carlson, A.D. 1981. Symmetrically organized dorsal unpaired median (dum) neurons and flash control in the male firefly, *Photuris versicolor*. Journal of Experimental Biology 93: 133–147.

Christensen, T.A. and Carlson, A.D. 1982. The neurophysiology of larval firefly luminescence: direct activation through 4 bifurcating (DUM) neurons. Journal of Comparative Physiology A 148: 503–514.

Christensen, T.A. and Hildebrand, J.G. 1987. Male-specific, sex pheromone-selective projection neurons in the antennal lobes of the moth *Manduca sexta*. Journal of Comparative Physiology A 160: 553–569.

Christensen, T.A. and Hildebrand, J.G. 1995. Neural regulation of sex-pheromone glands in Lepidoptera. Invertebrate Neuroscience 1: 97–103.

Christensen, T.A., Mustaparta, H. and Hildebrand, J.G. 1991. Chemical communication in heliothine moths. 2. Central processing of intraspecific and interspecific olfactory messages in the male corn earworm moth *Helicoverpa zea*. Journal of Comparative Physiology A 169: 259–274.

Christy, J.H. 1988a. Pillar function in the fiddler crab, *Uca beebei*. 1. Effects on male spacing and aggression. Ethology 78: 53–71.

Christy, J.H. 1988b. Pillar function in the fiddler crab, *Uca beebei*. 2. Competitive courtship signaling. Ethology 78: 113–128.

Christy, J.H. 1995. Mimicry, mate choice, and the sensory trap hypothesis. American Naturalist 146: 171–181.

Claridge, M. F. 1985. Acoustic signals in the Homoptera: behavior, taxonomy, and evolution. Annual Review of Entomology 30: 297–317.

Claridge, M.F., Morgan, J.C. and Moulds, M.S. 1999. Substrate-transmitted acoustic signals of the primitive cicada, *Tettigarcta crinita* Distant (Hemiptera: Cicadoidea: Tettigarctidae). Journal of Natural History 33: 1831–1834.

Clark, D.C., Beshear, D.D. and Moore, A.J. 1995. Role of familiarity in structuring male–male social interactions in the cockroach *Gromphadorhina portentosa* (Dictyoptera: Blaberidae). Annals of the Entomological Society of America 88: 554–561.

Clark, D.L. and Uetz, G.W. 1990. Video image recognition by the jumping spider, *Maevia inclemens* (Aranea, Salticidae). Animal Behaviour 40: 884–890.

Clutton-Brock, T.H. (ed.) 1988. Reproductive Success: Studies of Individual Variation in Contrasting Breeding Systems. University of Chicago Press, Chicago, IL.

Clutton-Brock, T.H. and Albon, S.D. 1979. The roaring of red deer and the evolution of honest advertisement. Behaviour 69: 145–170.

Clyne, P.J., Warr, C.G., Freeman, M.R., Lessing, D., Kim, J.H. and Carlson, J.R. 1999. A novel family of divergent seven-transmembrane proteins: candidate odorant receptors in Drosophila. Neuron 22: 327–338.

Clyne, P.J., Warr, C.G. and Carlson, J.R. 2000. Candidate taste receptors in *Drosophila*. Science 287: 1830–1834.

Cocroft, R.B. 1996. Insect vibrational defence signals. Nature 382: 679–680.

Cocroft, R.B. 1999a. Offspring–parent communication in a subsocial treehopper (Hemiptera: Membracidae: *Umbonia crassicornis*). Behaviour 136: 1–21.

Cocroft, R.B. 1999b. Parent–offspring communication in response to predators in a subsocial treehopper (Hemiptera: Membracidae: *Umbonia crassicornis*). Ethology 105: 553–568.

Cocroft, R.B., Tieu, T.D., Hoy, R.R. and Miles, R.N. 2000. Directionality in the mechanical response to substrate vibration in a treehopper (Hemiptera: Membracidae: *Umbonia crassicornis*). Journal of Comparative Physiology A 186: 695–705.

Cody, M.L. 1971. Finch flocks in the Mohave Desert. Theoretical Population Biology 2: 142–158.

Cohen, A.C. and Morin, J.G. 1990. Patterns of reproduction in ostracodes: a review. Journal of Crustacean Biology 10: 184–211.

Čokl, A. 1983. Functional properties of vibroreceptors in the legs of *Nezara viridula* (L.) (Heteroptera, Pentatomidae). Journal of Comparative Physiology A 150: 261–269.

Čokl, A., Virant-Doberlet, M. and Stritih, N. 2000. The structure and function of songs emitted by southern green stink bugs from Brazil, Florida, Italy and Slovenia. Physiological Entomology 25: 196–205.

Colepicolo-Neto, P., Costa, C. and Bechara, E.J.H. 1986. Brazilian species of luminescent Elateridae: luciferin identification and bioluminescence spectra. Insect Biochemistry 16: 803–810.

Collett, M., Despland, E., Simpson, S.J. and Krakauer, D.C. 1998. Spatial scales of desert locust gregarization. Proceedings of the National Academy of Sciences, U.S.A. 95: 13052–13055.

Collins, R.D. and Cardé, R.T. 1989. Selection for altered pheromone component ratios in the pink bollworm moth, *Pectinophora gossypiella* (Lepidoptera: Gelechiidae). Journal of Insect Behavior 2: 609–621.

Collins, R.D., Rosenblum, S.L. and Cardé, R.T. 1990. Selection for increased pheromone titer in the pink bollworm moth, *Pectinophora gossypiella* (Lepidoptera: Gelechiidae). Physiological Entomology 15: 141–147.

Collins, R.D., Jang, Y., Reinhold, K. and Greenfield, M.D. 1999. Quantitative genetics of ultrasonic advertisement signalling in the lesser waxmoth *Achroia grisella* (Lepidoptera: Pyralidae). Heredity 83: 644–651.

Conner, W.E. 1999. "Un chant d'appel amoureux": acoustic communication in moths. Journal of Experimental Biology 202: 1711–1723.

Conner, W.E. and Best, B.A. 1988. Biomechanics of the release of sex pheromone in moths: effects of body posture on local air flow. Physiological Entomology 13: 15–20.

Conner, W.E., Eisner, T., Vander Meer, R.K., Guerrero, A., Ghiringelli, D. and Meinwald, J. 1980. Sex attractant of an arctiid moth (*Utetheisa ornatrix*): a pulsed chemical signal. Behavioral Ecology and Sociobiology 7: 55–63.

Conner, W.E., Eisner, T., Vander Meer, R.K., Guerrero, A. and Meinwald, J. 1981. Pre-copulatory sexual interaction in an arctiid moth (*Utetheisa ornatrix*): role of a pheromone derived from dietary alkaloids. Behavioral Ecology and Sociobiology 9: 227–235.

Conner, W.E., Roach, B., Benedict, E., Meinwald, J. and Eisner, T. 1990. Courtship pheromone production and body size as correlates of larval diet in males of the arctiid moth, *Utetheisa ornatrix*. Journal of Chemical Ecology 16: 543–552.

Consi, T.R., Grasso, F., Mountain, D. and Atema, J. 1995. Explorations of turbulent odor plumes with an autonomous underwater robot. Biological Bulletin 189: 231–232.

Conway Morris, S. 1998. The Crucible of Creation: The Burgess Shale and the Rise of Animals. Oxford University Press, Oxford.

Costa, J.T. and Pierce, N.E. 1997. Social evolution in the Lepidoptera: ecological context and communication in larval societies. In The Evolution of Social Behavior in Insects and Arachnids (eds J.C. Choe and B.J. Crespi), pp. 407–442. Cambridge University Press, Cambridge, U.K.

Crane, J. 1975. Fiddler Crabs of the World, Ocypodidae: Genus *Uca*. Princeton University Press, Princeton, NJ.

Cremer, S. and Greenfield, M.D. 1998. Partitioning the components of sexual selection: attractiveness and agonistic behaviour in male wax moths, *Achroia grisella* (Lepidoptera: Pyralidae). Ethology 104: 1–9.

Cronin, T.W. and Marshall, N.J. 1989. A retina with at least 10 spectral types of photoreceptors in a mantis shrimp. Nature 339: 137–140.

Cronin, T.W., Jarvilehto, M., Weckstrom, M. and Lall, A.B. 2000. Tuning of photoreceptor spectral sensitivity in fireflies (Coleoptera: Lampyridae). Journal of Comparative Physiology A 186: 1–12.

Cronin, T.W., Caldwell, R.L. and Marshall, J. 2001. Tunable colour vision in a mantis shrimp. Nature 411: 547–548.

Culvenor, C.C.J. and Edgar, J.A. 1972. Dihydropyrrolizine secretions associated with coremata of *Utetheisa* moths (family Arctiidae). Experientia 28: 627–628.

Cusson, M. and McNeil, J.N. 1989. Involvement of juvenile hormone in the regulation of pheromone release activities in a moth. Science 243: 210–211.

Cuthill, I.C., Partridge, J.C., Bennett, A.T.D., Church, S.C., Hart, N.S. and Hunt, S. 2000. Ultraviolet vision in birds. Advances in the Study of Behavior 29: 159–214.

Dabelsteen, T. 1992. Interactive playback: a finely tuned response. In NATO ASI (Advanced Science Institutes) Series, Series A, Life Sciences, no. 228, pp. 97–109.

Darwin, C. 1871. The Descent of Man and Selection in Relation to Sex (2nd edn, 1874, John Murray, London).

David, C.T. and Kennedy, J.S. 1987. The steering of zigzagging flight by male gypsy moths. Naturwissenschaften 74: 194–196.

Davis, D.R. 1998. A world classification of the Harmaclonimae, a new subfamily of Tineidae (Lepidoptera: Tineoidea). Smithsonian Contributions to Zoology 597: 1–57.

D'Eath, R.B. 1998. Can video images imitate real stimuli in animal behaviour experiments? Biological Reviews of the Cambridge Philosophical Society 28: 267–292.

DeCock, R. and Matthysen, E. 1999. Aposematism and bioluminescence: experimental evidence from glow-worm larvae (Coleoptera: Lampyridae). Evolutionary Ecology 13: 619–639.

DeJong, M.C.M. and Sabelis, M.W. 1991. Limits to runaway sexual selection: the wallflower paradox. Journal of Evolutionary Biology 4: 637–656.

DeLuca, P.A. and Morris, G.K. 1998. Courtship communication in meadow katydids: female preferences for large male vibrations. Behaviour 135: 777–794.

Deneubourg, J.L. and Goss, S. 1989. Collective patterns and decision making. Ethology, Ecology and Evolution 1: 295–311.

Deneubourg, J.L., Aron, S., Goss, S. and Pasteels, J.M. 1990. The self organizing exploratory pattern of the Argentine ant. Journal of Insect Behavior 3: 159–168.

Denny, M.W. 1976. The physical properties of spider's silk and their role in the design of orb webs. Journal of Experimental Biology 65: 483–506.

Denny, M.W. 1988. Biology and the Mechanics of the Wave-Swept Environment. Princeton University Press, Princeton, NJ.

Denny, M.W. 1993. Air and Water. Princeton University Press, Princeton, NJ.

DeVries, P.J. 1990. Enhancement of symbioses between butterfly caterpillars and ants by vibrational communication. Science 248: 1104–1106.

DeVries, P.J. 1992. Singing caterpillars, ants and symbiosis. Scientific American 267(4): 76–82.

DeVries, P.J., Cocroft, R.B. and Thomas, J. 1993. Comparison of acoustical signals in *Maculinea* butterfly caterpillars and their obligate host *Myrmica* ants. Biological Journal of the Linnean Society 49: 229–238.

DeWinter, A.J. and Rollenhagen, T. 1990. The importance of male and female acoustic behavior for reproductive isolation in *Ribautodelphax* planthoppers

(Homoptera, Delphacidae). Biological Journal of the Linnean Society 40: 191–206.

Dickens, J.C. 1997. Neurobiology of pheromonal signal processing in insects. In Insect Pheromone Research: New Directions (eds R.T. Cardé and A.K. Minks), pp. 210–217. Chapman and Hall, New York.

Dierkes, S. and Barth, F.G. 1995. Mechanism of signal production in the vibratory communication of the wandering spider *Cupiennius getazi* (Arachnida, Araneae). Journal of Comparative Physiology A 176: 31–44.

Dill, M. and Heisenberg, M. 1995. Visual pattern memory without shape recognition. Philosophical Transactions of the Royal Society of London B 349: 143–152.

Dill, M., Wolf, R. and Heisenberg, M. 1993. Visual pattern recognition in *Drosophila* involves retinotopic matching. Nature 365: 751–753.

Dillon, R.J., Vennard, C.T. and Charnley, A.K. 2000. Pheromones: exploitation of gut bacteria in the locust. Nature 403: 851.

Dobler, S., Heller, K.-G. and von Helversen, O. 1994a. Song pattern recognition and an auditory time window in the female bush cricket *Ancistrura nigrovittata* (Orthoptera: Phaneropteridae). Journal of Comparative Physiology A 175: 67–74.

Dobler, S., Stumpner, A. and Heller, K.-G. 1994b. Sex-specific spectral tuning for the partners song in the duetting bush cricket *Ancistrura nigrovittata* (Orthoptera: Phaneropteridae). Journal of Comparative Physiology A 175: 303–310.

Doherty, J.A. 1982. Stereotypy and the effects of temperature on some spatio-temporal sub-components of the courtship wave in the fiddler crabs *Uca minax* (Leconte) and *Uca pugnax* (Smith) (Brachyura: Ocypodidae). Animal Behaviour 30: 352–363.

Doherty, J.A. 1985a. Temperature coupling and trade-off phenomena in the acoustic communication system of the cricket, *Gryllus bimaculatus* deGeer (Gryllidae). Journal of Experimental Biology 114: 17–35.

Doherty, J.A. 1985b. Trade-off phenomena in calling song recognition and phonotaxis in the cricket, *Gryllus bimaculatus* (Orthoptera: Gryllidae). Journal of Comparative Physiology A 156: 787–801.

Doherty, J.A. 1985c. Phonotaxis in the cricket, *Gryllus bimaculatus* deGeer: comparisons of choice and no-choice paradigms. Journal of Comparative Physiology A 157: 279–289.

Doherty, J.A. 1991. Song recognition and localization in the phonotaxis behavior of the field cricket, *Gryllus bimaculatus* (Orthoptera: Gryllidae). Journal of Comparative Physiology A 168: 213–222.

Doherty, J.A. and Gerhardt, H.C. 1983. Hybrid tree frogs: vocalizations of males and selective phonotaxis of females. Science 220: 1078–1080.

Doherty, J.A. and Gerhardt, H.C. 1984. Acoustic communication in hybrid treefrogs: sound production by males and selective phonotaxis by females. Journal of Comparative Physiology A 154: 319–330.

Douglass, J.K., Wilkens, L., Pantazelou, E. and Moss, F. 1993. Noise enhancement of information transfer in crayfish mechanoreceptors by stochastic resonance. Nature 365: 337–340.

Dreller, C. and Kirchner, W.H. 1993a. Hearing in honeybees: localization of the auditory sense organ. Journal of Comparative Physiology A 173: 275–279.

Dreller, C. and Kirchner, W.H. 1993b. How honeybees perceive the information of the dance language. Naturwissenschaften 80: 319–321.

Dreller, C. and Kirchner, W.H. 1994. Hearing in the Asian honeybees *Apis dorsata* and *Apis florea*. Insectes Sociaux 41: 291–299.

Dumont, J.P.C. and Robertson, R.M. 1986. Neuronal circuits: an evolutionary perspective. Science 233: 849–853.

Dumortier, B. 1963. Morphology of sound emission apparatus in Arthropoda. In Acoustic Behaviour of Animals (ed. R.-G. Busnel), pp. 277–345. Elsevier Publishing, Amsterdam.

Dunning, D.C. and Roeder, K.D. 1965. Moth sounds and insect catching behavior of bats. Science 147: 173–174.

Dunning, D.C., Acharya, L., Merriman, C.B. and Dalferro, L. 1992. Interactions between bats and arctiid moths. Canadian Journal of Zoology 70: 2218–2223.

Dusenbery, D.B. 1989. Calculated effect of pulsed pheromone release on range of attraction. Journal of Chemical Ecology 15: 971–977.

Dusenbery, D.B. 1992. Sensory Ecology: How Organisms Acquire and Respond to Information. W.H. Freeman and Company, New York.

Dussourd, D.E., Ubik, K., Harvis, C., Resch, J., Meinwald, J. and Eisner, T. 1988. Defense mechanisms of arthropods. 86. biparental defensive endowment of eggs with acquired plant alkaloid in the moth *Utetheisa ornatrix*. Proceedings of the National Academy of Sciences, U.S.A. 85: 5992–5996.

Dussourd, D.E., Harvis, C.A., Meinwald, J. and Eisner, T. 1989. Paternal allocation of sequestered plant pyrrolizidine alkaloid to eggs in the danaine butterfly, *Danaus gilippus*. Experientia 45: 896–898.

Dussourd, D.E., Harvis, C.A., Meinwald, J. and Eisner, T. 1991. Pheromonal advertisement of a nuptial gift by a male moth (*Utetheisa ornatrix*). Proceedings of the National Academy of Sciences, U.S.A. 88: 9224–9227.

Dyson, M.L. and Passmore, N.I. 1988. Two-choice phonotaxis in *Hyperolius marmoratus* (Anura: Hyperoliidae): the effect of temporal variation in presented stimuli. Animal Behaviour 36: 648–652.

Eberhard, W.G. 1977. Aggressive chemical mimicry by a bolas spider. Science 198: 1173–1175.

Eberhard, W.G. 1996. Female Control: Sexual Selection by Cryptic Female Choice. Princeton University Press, Princeton, NJ.

Edelstein-Keshet, L. 1994. Simple models for trail-following behavior: trunk trails versus individual foragers. Journal of Mathematical Biology 32: 303–328.

Edgar, J.A., Boppré, M. and Schneider, D. 1979. Pyrrolizidine alkaloid storage in African and Australian danaid butterflies. Experientia 35: 1447–1448.

Eggert, A.-K. and Müller, J.K. 1989a. Pheromone-mediated attraction in burying beetles. Ecological Entomology 14: 235–237.

Eggert, A.-K. and Müller, J.K. 1989b. Mating success of pheromone-emitting *Necrophorus* males: Do attracted females discriminate against resource owners? Behaviour 110: 248–257.

Eiríksson, T. 1992. Density dependent song duration in the grasshopper *Omocestus viridulus*. Behaviour 122: 121–132.

Eiríksson, T. 1993. Female preference for specific pulse duration of male songs in the grasshopper, *Omocestus viridulus*. Animal Behaviour 45: 471–477.

Eiríksson, T. 1994. Song duration and female response behaviour in the grasshopper *Omocestus viridulus*. Animal Behaviour 47: 707–712.

Eisner, T. and Meinwald, J. 1995. Defense mechanisms of arthropods. 129. the chemistry of sexual selection. Proceedings of the National Academy of Sciences, U.S.A. 92: 50–55.

Eisner, T., Goetz, M.A., Hill, D.E., Smedley, S.R. and Meinwald, J. 1997. Firefly "femmes fatales" acquire defensive steroids (lucibufagins) from their firefly prey. Proceedings of the National Academy of Sciences, U.S.A. 94: 9723–9728.

Eisner, T., Eisner, M., Rossini, C., Iyengar, V.K., Roach, B.L., Benedikt, E. and Meinwald, J. 2000. Chemical defense against predation in an insect egg. Proceedings of the National Academy of Sciences, U.S.A. 97: 1634–1639.

Elkinton, J.S. and Cardé, R.T. 1983. Appetitive flight behavior of male gypsy moths (*Lymantria dispar* L.) (Lepidoptera: Lymantriidae). Environmental Entomology 12: 1702–1707.

Elliot, C.J.H. 1983. Wing hair plates in crickets: physiological characteristics and connections with stridulatory motor neurones. Journal of Experimental Biology 107: 21–47.

Elliot, C.J.H. and Koch, U.T. 1985. The clockwork cricket. Naturwissenschaften 72: 150–152.

Ellner, S. and Hairston Jr., N.G. 1994. Role of overlapping generations in maintaining genetic variation in a fluctuating environment. American Naturalist 143: 403–417.

Elsner, N. 1974. Neuroethology of sound production in gomphocerine grasshoppers (Orthoptera: Acrididae). 1. Song patterns and stridulatory movements. Journal of Comparative Physiology 88: 67–102.

Elsner, N. 1983. A neuroethological approach to the phylogeny of leg stridulation in gomphocerine grasshoppers. In Neuroethology and Behavioral Physiology (eds F. Huber and H. Markl), pp. 54–68. Springer-Verlag, Berlin.

Endler, J.A. 1992. Signals, signal conditions, and the direction of evolution. American Naturalist 139: S125–S153.

Endler, J.A. 1993a. Some general comments on the evolution and design of animal communication systems. Philosophical Transactions of the Royal Society of London B 340: 215–225.

Endler, J.A. 1993b. The color of light in forests and its implications. Ecological Monographs 63: 1–27.

Endler, J.A. and Basolo, A.L. 1998. Sensory ecology, receiver biases and sexual selection. Trends in Ecology and Evolution 13: 415–420.

Engel, M.S. and Schultz, T.R. 1997. Phylogeny and behavior in honey bees (Hymenoptera: Apidae). Annals of the Entomological Society of America 90: 43–53.

Enquist, M. and Arak, A. 1993. Selection of exaggerated male traits by female aesthetic senses. Nature 361: 446–448.

Enquist, M. and Arak, A. 1994. Symmetry, beauty and evolution. Nature 372: 169–172.

Ermentrout, G.B. 1991. An adaptive model for synchrony in the firefly *Pteroptyx malaccae*. Journal of Mathematical Biology 29: 571–585.

Ernst, R. and Heisenberg, M. 1999. The memory template in *Drosophila* pattern vision at the flight simulator. Vision Research 39: 3920–3933.

Esch, H., Esch, I. and Kerr, W.E. 1965. Sound: an element common to communication of stingless bees and to dances of the honey bee. Science 149: 320–321.

Esch, H.E., Zhang, S., Srinivasan, M.V. and Tautz, J. 2001. Honeybee dances communicate distances measured by optic flow. Nature 411: 581–583.

Evans, C.S. and Marler, P. 1991. On the use of video images as social stimuli in birds: audience effects on alarm calling. Animal Behaviour 41: 17–26.

Evans, H.E. and O'Neill, K.M. 1988. The Natural History and Behavior of North American Beewolves. Comstock Publishing Associates, Ithaca, NY.

Ewing, A.W. 1989. Arthropod Bioacoustics: Neurobiology and Behavior. Cornell University Press, Ithaca, NY.

Fabre, J.H.C. 1879. Souvenirs Entomologiques, sér. VII. (6th edn, 1912, Librairie Ch. Delagrave, Paris).

Fadamiro, H.Y., Wyatt, T.D. and Hall, D.R. 1996. Behavioural response of *Prostephanus truncatus* (Horn) (Coleoptera: Bostrichidae) to the individual components of its pheromone in a flight tunnel: discrimination between two odour sources. Journal of Stored Products Research 32: 163–170.

Fadamiro, H.Y., Wyatt, T.D. and Birch, M.C. 1998. Flying beetles respond as moths predict: optomotor anemotaxis to pheromone plumes at different heights. Journal of Insect Behavior 11: 549–557.

Fadamiro, H.Y., Cosse, A.A. and Baker, T.C. 1999. Fine-scale resolution of closely spaced pheromone and antagonist filaments by flying male *Helicoverpa zea*. Journal of Comparative Physiology A 185: 131–141.

Farris, H.E., Forrest, T.G. and Hoy, R.R. 1997. The effects of calling song spacing and intensity on the attraction of flying crickets (Orthoptera: Gryllidae: Nemobiinae). Journal of Insect Behavior 10: 639–653.

Farris, H.E., Forrest, T.G. and Hoy, R.R. 1998. The effect of ultrasound on the attractiveness of acoustic mating signals. Physiological Entomology 23: 322–328.

Faure, P.A. and Hoy, R.R. 2000a. The sounds of silence: cessation of singing and song pausing are ultrasound-induced acoustic startle behaviors in the katydid *Neoconocephalus ensiger* (Orthoptera: Tettigoniidae). Journal of Comparative Physiology A 186: 129–142.

Faure, P.A. and Hoy, R.R. 2000b. Auditory symmetry analysis. Journal of Experimental Biology 203: 3209–3223.

Fenton, M.B. and Fullard, J.H. 1981. Moth hearing and the feeding strategies of bats. American Scientist 69: 266–275.

Ferkovich, S.M., Oliver, J.E. and Dillard, C. 1982. Pheromone hydrolysis by cuticular and interior esterases of the antennae, legs, and wings of the cabbage looper moth, *Trichoplusia ni* (Hubner). Journal of Chemical Ecology 8: 859–866.

Field, L.H. 1993. Structure and evolution of stridulatory mechanisms in New Zealand wetas (Orthoptera: Stenopelmatidae). International Journal of Insect Morphology and Embryology 22: 163–183.

Field, L.H. 2001. Stridulatory mechanisms and associated behaviour in New Zealand wetas. In The Biology of Wetas, King Crickets and Their Allies (ed. L.H. Field), pp. 271–295. CAB International, Wallingford, U.K.

Field, L.H. and Rind, F.C. 1992. Stridulatory behavior in a New Zealand weta, *Hemideina crassidens*. Journal of Zoology 228: 371–394.

Finelli, C.M., Pentcheff, N.D., Zimmer-Faust, R.K. and Wethey, D.S. 1999. Odor transport in turbulent flows: constraints on animal navigation. Limnology and Oceanography 44: 1056–1071.

Fisher, R.A. 1915. The evolution of sexual preference. Eugenics Review 7: 184–192.

Fisher, R.A. 1958. The Genetical Theory of Natural Selection, 2nd edn. Dover, New York.

Fitzgerald, T.D. 1995. The Tent Caterpillars. Cornell University Press, Ithaca, NY.

Fitzpatrick, S.M., McNeil, J.N. and Dumont, S. 1988. Does male pheromone effectively inhibit competition among courting true armyworm males (Lepidoptera: Noctuidae)? Animal Behaviour 36: 1831–1835.

Fleishman, L.J., McClintock, W.J., D'Eath, R.B., Brainard, D.H. and Endler, J.A. 1998. Colour perception and the use of video playback experiments in animal behaviour. Animal Behaviour 56: 1035–1040.

Fletcher, D.J.C. and Michener, C.D. (eds) 1987. Kin Recognition in Animals. John Wiley & Sons, Chichester, U.K.

Fletcher, N.J.C. 1992. Acoustic Systems in Biology. Oxford University Press, Oxford.

Flook, P.K., Klee, S. and Rowell, C.H.F. 1999. Combined molecular phylogenetic analysis of the Orthoptera (Arthropoda: Insecta) and implications for their higher systematics. Systematic Biology 48: 233–253.

Fonseca, P.J. 1993. Directional hearing of a cicada: biophysical aspects. Journal of Comparative Physiology A 172: 767–774.

Fonseca, P.J. and Popov, A.V. 1997. Directionality of the tympanal vibrations in a cicada: a biophysical analysis. Journal of Comparative Physiology A 180: 417–427.

Fonseca, P.J., Münch, D. and Hennig, R.M. 2000. Auditory perception: how cicadas interpret acoustic signals. Nature 405: 297–298.

Ford, N.B. and Low, J.R. 1984. Sex pheromone source location by garter snakes: a mechanism for detection of direction in nonvolatile trails. Journal of Chemical Ecology 10: 1193–1199.

Fordyce, D. and Cronin, T.W. 1993. Trilobite vision: a comparison of schizochroal and holochroal eyes with the compound eyes of modern arthropods. Paleobiology 19: 288–303.

Forel, A. 1908. The Senses of Insects (English translation by Macleod Yearsley). Methuen, London.

Forrest, T.G. 1982. Acoustic communication and baffling behaviors of crickets. Florida Entomologist 65: 33–44.

Forrest, T.G. 1991. Power output and efficiency of sound production by crickets. Behavioral Ecology 2: 327–338.

Forrest, T.G. 1994. From sender to receiver: propagation and environmental effects on acoustic signals. American Zoologist 34: 644–654.

Forrest, T.G. and Raspet, R. 1994. Models of female choice in acoustic communication. Behavioral Ecology 5: 293–303.

Forrest, T.G., Read, M.P., Farris, H.E. and Hoy, R.R. 1997. A tympanal hearing organ in scarab beetles. Journal of Experimental Biology 200: 601–606.

Forster, L. 1982. Visual communication in jumping spiders (Salticidae). In Spider Communication: Mechanisms and Ecological Significance (eds P.N. Witt and J.S. Rovner), pp. 161–212. Princeton University Press, Princeton, NJ.

Franks, N.R., Gomez, N., Goss, S. and Deneubourg, J.L. 1991. The blind leading the blind in army ant raid patterns: testing a model of self organization (Hymenoptera: Formicidae). Journal of Insect Behavior 4: 583–607.

Freeland, W.J. 1980. Insect flight times and atmospheric hydrocarbons. American Naturalist 116: 736–742.

Fretwell, S.D. 1972. Populations in a Seasonal Environment. Princeton University Press, Princeton, NJ.

Friedrich, M. and Tautz, D. 1995. Ribosomal DNA phylogeny of the major extant arthropod classes and the evolution of myriapods. Nature 376: 165–167.

Fullard, J.H. 1994. Auditory changes in noctuid moths endemic to a bat-free habitat. Journal of Evolutionary Biology 7: 435–445.

Fullard, J.H. and Yack, J.E. 1993. The evolutionary biology of insect hearing. Trends in Ecology and Evolution 8: 248–252.

Fullard, J.H., Dawson, J.W., Otero, L.D. and Surlykke, A. 1997. Bat-deafness in day-flying moths (Lepidoptera: Notodontidae: Dioptinae). Journal of Comparative Physiology A 181: 477–483.

Futrelle, R.P. 1984. How molecules get to their detectors: the physics of diffusion of insect pheromones. Trends in Neurosciences 7: 116–120.

Gadagkar, R. 1996. What's the essence of royalty—one keto group? Current Science 71: 975–980.

Gadagkar, R. 1997a. Social evolution—has nature ever rewound the tape? Current Science 72: 950–956.

Gadagkar, R. 1997b. The evolution of communication and the communication of evolution: the case of the honey bee queen pheromone. In Orientation and Communication in Arthropods (ed. M. Lehrer), pp. 375–395. Birkhäuser Verlag, Basel.

Galliart, P.L. and Shaw, K.C. 1991a. Effect of intermale distance and female presence on the nature of chorusing by paired *Amblycorypha parvipennis* (Orthoptera: Tettigoniidae) males. Florida Entomologist 74: 559–569.

Galliart, P.L. and Shaw, K.C. 1991b. Role of weight and acoustic parameters, including nature of chorusing, in the mating success of males of the katydid, *Amblycorypha parvipennis* (Orthoptera: Tettigoniidae). Florida Entomologist 74: 453–464.

Galliart, P.L. and Shaw, K.C. 1996. The effect of variation in parameters of the male calling song of the katydid, *Amblycorypha parvipennis* (Orthoptera:

Tettigonndae), on female phonotaxis and phonoresponse. Journal of Insect Behavior 9: 841–855.

Gamboa, G.J., Reeve, H.K. and Pfennig, D.W. 1986. The evolution and ontogeny of nestmate recognition in social wasps. Annual Review of Entomology 31: 431–454.

Gemeno, C. and Haynes, K.F. 2000. Periodical and age-related variation in chemical communication system of black cutworm moth, *Agrotis ipsilon*. Journal of Chemical Ecology 26: 329–342.

Gemeno, C., Yeargan, K.V. and Haynes, K.F. 2000. Aggressive chemical mimicry by the bolas spider *Mastophora hutchinsoni*: identification and quantification of a major prey's sex pheromone components in the spider's volatile emissions. Journal of Chemical Ecology 26: 1235–1243.

Gerhardt, H.C. 1978. Temperature coupling in the vocal communication system of the gray treefrog, *Hyla versicolor*. Science 199: 992–994.

Gerhardt, H.C. 1991. Female mate choice in treefrogs: static and dynamic acoustic criteria. Animal Behaviour 42: 615–635.

Gerhardt, H.C. 1994. Selective responsiveness to long-range acoustic signals in insects and anurans. American Zoologist 34: 706–714.

Gerhardt, H.C., Roberts, J.D., Bee, M.A. and Schwartz, J.J. 2000. Call matching in the quacking frog (*Crinia georgiana*). Behavioral Ecology and Sociobiology 48: 243–251.

Getz, W.M. 1981. Genetically based kin recognition systems. Journal of Theoretical Biology 92: 209–226.

Getz, W.M. and Chapman, R.F. 1987. An odor discrimination model with application to kin recognition in social insects. International Journal of Neuroscience 32: 963–978.

Getz, W.M. and Page, R.E. 1991. Chemosensory kin communication systems and kin recognition in honeybees. Ethology 87: 298–315.

Getz, W.M. and Smith, K.B. 1983. Genetic kin recognition: honey bees discriminate between full and half sisters. Nature 302: 147–148.

Getz, W.M, Brückner, D. and Smith, K.B. 1986. Conditioning honeybees to discriminate between heritable odors from full and half sisters. Journal of Comparative Physiology A 159: 251–256.

Ghiradella, H. 1977a. Fine structure of tracheoles of lantern of a photurid firefly. Journal of Morphology 153: 187–203.

Ghiradella, H. 1977b. Implication of respiratory system in flash control in a photinid firefly. Journal of Cell Biology. 75: A106.

Ghiradella, H. 1984. Structure of iridescent lepidopteran scales: variations on several themes. Annals of the Entomological Society of America 77: 637–645.

Ghiradella, H. 1998. The anatomy of light production: the fine structure of the firefly lantern. In Microscopic Anatomy of Invertebrates, vol. 11A, Insecta (eds F.W. Harrison and M. Locke), pp. 363–381. Wiley-Liss, New York.

Ghiradella, H., Eisner, T., Silberglied, R.E., Aneshansley, D. and Hinton, H.E. 1972. Ultraviolet reflection of a male butterfly: interference color caused by thin-layer elaboration of wing scales. Science 178: 1214–1217.

Gibson, R.M. and Höglund, J. 1992. Copying and sexual selection. Trends in Ecology and Evolution 7: 229–232.

Giger, A.D. and Srinivasan, M.V. 1996. Pattern recognition in honeybees: chromatic properties of orientation analysis. Journal of Comparative Physiology A 178: 763–769.

Gil, D., Graves, J., Hazon, N. and Wells, A. 1999. Male at attractiveness and differential testosterone investment in zebra finch eggs. Science 286: 126–128.

Gilbert, C. 1994. Form and function of stemmata in larvae of holometabolous insects. Annual Review of Entomology 39: 323–349.

Gilburn, A.S. and Day, T.H. 1994a. Evolution of female choice in seaweed flies: Fisherian and good genes mechanisms operate in different populations. Proceedings of the Royal Society of London B 255: 159–165.

Gilburn, A.S. and Day, T.H. 1994b. The inheritance of female mating behavior in the seaweed fly, *Coelopa frigida*. Genetical Research 64: 19–25.

Gilburn, A.S., Foster, S.P. and Day, T.H. 1993. Genetic correlation between a female mating preference and the preferred male character in seaweed flies (*Coelopa frigida*). Evolution 47: 1788–1795.

Giurfa, M., Eichmann, B. and Menzel, R. 1996. Symmetry perception in an insect. Nature 382: 458–461.

Gleason, J.M. and Ritchie, M.G. 1998. Evolution of courtship song and reproductive isolation in the *Drosophila willistoni* species complex: Do sexual signals diverge the most quickly? Evolution 52: 1493–1500.

Gnatzy, W. and Heusslein, R. 1986. Digger wasp against crickets. 1. receptors involved in the antipredator strategies of the prey. Naturwissenschaften 73: 212–215.

Gnatzy, W. and Kämper, G. 1990. Digger wasp against crickets. 2. an airborne signal produced by a running predator. Journal of Comparative Physiology A 167: 551–556.

Godfray, H.C.J. 1994. Parasitoids: Behavioral and Evolutionary Ecology. Princeton University Press, Princeton, NJ.

Gogala, M. 1970. Artspezifität der Lautäußerungen bei Erdwanzen (Heteroptera, Cydnidae). Zeitschrift für Vergleichende Physiologie 70: 20–28.

Gogala, M. and Čokl, A. 1983. The acoustic behavior of the bug *Phymata crassipes* (F.) (Heteroptera). Revue Canadienne de Biologie Experimentale 42: 249–256.

Gogala, M., Čokl, A., Draslar, K. and Blazevic, A. 1974. Substrate-borne sound communication in Cydnidae (Heteroptera). Journal of Comparative Physiology 94: 25–31.

Goldberg, R.L. and Henson, O.W. 1998. Changes in cochlear mechanics during vocalization: evidence for a phasic medial efferent effect. Hearing Research 122: 71–81.

Goldsmith, T.H. 1991. Photoreception and vision. In Neural and Integrative Animal Physiology: Comparative Animal Physiology, 4th edn (ed. C.L. Prosser), pp. 171–245. Wiley-Liss, New York.

Gomez, G. and Atema, J. 1996. Temporal resolution in olfaction: stimulus integration time of lobster chemoreceptor cells. Journal of Experimental Biology 199: 1771–1779.

Gomez, G., Voigt, R. and Atema, J. 1999. Temporal resolution in olfaction. III. flicker fusion and concentration-dependent synchronization with stimulus pulse trains of antennular chemoreceptor cells in the American lobster. Journal of Comparative Physiology A 185: 427–436.

Gonzalez, A., Rossini, C., Eisner, M. and Eisner, T. 1999. Sexually transmitted chemical defense in a moth (*Utetheisa ornatrix*). Proceedings of the National Academy of Sciences, U.S.A. 96: 5570–5574.

Göpfert, M.C. and Robert, D. 2000. Nanometre-range acoustic sensitivity in male and female mosquitoes. Proceedings of the Royal Society of London B 267: 453–457.

Göpfert, M.C. and Robert, D. 2001. Turning the key on *Drosophila* audition. Nature 411: 908.

Göpfert, M.C. and Wasserthal, L.T. 1999. Hearing with the mouthparts: behavioural responses and the structural basis of ultrasound perception in acherontiine hawkmoths. Journal of Experimental Biology 202: 909–918.

Göpfert, M.C., Briegel, H. and Robert, D. 1999. Mosquito hearing: Sound-induced antennal vibrations in male and female *Aedes aegypti*. Journal of Experimental Biology 202: 2727–2738.

Gotwald, W.H. Jr. 1995. Army Ants: The Biology of Social Predation. Cornell University Press, Ithaca, NY.

Gould, J.L. and Towne, W.F. 1987. Evolution of the dance language. American Naturalist 130: 317–338.

Gould, J.L., Dyer, F.C. and Towne, W.F. 1985. Recent progress in the study of the dance language. Fortschritte der Zoologie 31: 141–161.

Grafe, T.-U. 1999. A function of synchronous chorusing and a novel female preference shift in an anuran. Proceedings of the Royal Society of London B 266: 2331–2336.

Grafen, A. 1990. Sexual selection unhandicapped by the Fisher process. Journal of Theoretical Biology 144: 517–546.

Gray, D.A. 1997. Female house crickets, *Acheta domesticus*, prefer the chirps of large males. Animal Behaviour 54: 1553–1562.

Gray, D.A. and Cade, W.H. 1999a. Quantitative genetics of sexual selection in the field cricket, *Gryllus integer*. Evolution 53: 848–854.

Gray, D.A. and Cade, W.H. 1999b. Sex, death and genetic variation: natural and sexual selection on cricket song. Proceedings of the Royal Society of London B 266: 707–709.

Green, S. and Marler, P.M. 1979. The analysis of animal communication. In Handbook of Behavioral Neurobiology, Vol. 3. Social Behavior and Communication (eds P. Marler and J.G. Vandebergh), pp. 73–158. Plenum Press, New York.

Greenberg, L. 1979. Genetic component of bee odor in kin recognition. Science 206: 1095–1097.

Greenfield, M.D. 1981. Moth sex pheromones: an evolutionary perspective. Florida Entomologist 64: 4–17.

Greenfield, M.D. 1982. The question of paternal investment in Lepidoptera: male contributed proteins in *Plodia interpunctella*. International Journal of Invertebrate Reproduction 5: 323–330.

Greenfield, M.D. 1983. Reproductive isolation in clearwing moths (Lepidoptera: Sesiidae): a tropical–temperate comparison. Ecology 64: 362–375.

Greenfield, M.D. 1988. Interspecific acoustic interactions among katydids, Neoconocephalus: inhibition-induced shifts in diel periodicity. Animal Behaviour 36: 684–695.

Greenfield, M.D. 1990. Evolution of acoustic communication in the genus Neoconocephalus: discontinuous songs, synchrony, and interspecific interactions. In The Tettigoniidae: Biology, Systematics and Evolution (eds W.J. Bailey and D.C.F. Rentz), pp. 71–97. Springer-Verlag, Berlin.

Greenfield, M.D. 1992. The evening chorus of the desert clicker, *Ligurotettix coquilletti* (Orthoptera: Acrididae): mating investment with delayed returns. Ethology 91: 265–278.

Greenfield, M.D. 1993. Inhibition of male calling by heterospecific signals: Artifact of chorusing or abstinence during suppression of female phonotaxis? Naturwissenschaften 80: 570–573.

Greenfield, M.D. 1994a. Cooperation and conflict in the evolution of signal interactions. Annual Review of Ecology and Systematics 25: 97–126.

Greenfield, M.D. 1994b. Synchronous and alternating choruses in insects and anurans: common mechanisms and diverse functions. American Zoologist 34: 605–615.

Greenfield, M.D. 1997a. Acoustic communication in Orthoptera. In The Bionomics of Grasshoppers, Katydids, and their Kin (eds S.K. Gangwere, M.C. Muralirangan, and Meera Muralirangan), pp. 197–230. CAB International, Wallingford, U.K.

Greenfield, M.D. 1997b. Sexual selection and the evolution of advertisement signals. In Perspectives in Ethology, Vol. 5. Communication (eds D.H. Owings, M.D. Beecher, and N.S. Thompson), pp. 145–177. Plenum Press, New York.

Greenfield, M.D. 1997c. Sexual selection in resource defense polygyny: lessons from territorial grasshoppers. In The Evolution of Mating Systems in Insects and Arachnids (eds J.C. Choe and B.J. Crespi), pp. 75–88. Cambridge University Press, Cambridge, U.K.

Greenfield, M.D. and Coffelt, J.A. 1983. Reproductive behavior of the lesser waxmoth, *Achroia grisella* (Pyralidae: Galleriinae): signaling, pair formation, male interactions, and mate guarding. Behaviour 84: 287–315.

Greenfield, M.D. and Karandinos, M.G. 1979. Resource partitioning of the sex communication channel in clearwing moths (Lepidoptera: Sesiidae) of Wisconsin. Ecological Monographs 49: 403–426.

Greenfield, M.D. and Minckley, R.L. 1993. Acoustic dueling in tarbush grasshoppers: settlement of territorial contests via alternation of reliable signals. Ethology 95: 309–326.

Greenfield, M.D. and Rand, A.S. 2000. Frogs have rules: selective attention algorithms regulate chorusing in *Physalaemus pustulosus* (Leptodactylidae). Ethology 106: 331–347.

Greenfield, M.D. and Roizen, I. 1993. Katydid synchronous chorusing is an evolutionarily stable outcome of female choice. Nature 364: 618–620.

Greenfield, M.D. and Shelly, T.E. 1985. Alternative mating strategies in a desert grasshopper: evidence of density-dependence. Animal Behaviour 33: 1192–1210.

Greenfield, M.D. and Weber, T. 2000. Evolution of ultrasonic signalling in wax moths: discrimination of ultrasonic mating calls from bat echolocation signals and the exploitation of an anti-predator receiver bias by sexual advertisement. Ethology, Ecology and Evolution 12: 259–279.

Greenfield, M.D., Shelly, T.E. and Gonzalez-Coloma, A. 1989. Territory selection in a desert grasshopper: the maximization of conversion efficiency on a chemically defended shrub. Journal of Animal Ecology 58: 761–771.

Greenfield, M.D., Tourtellot, M.K. and Snedden, W.A. 1997. Precedence effects and the evolution of chorusing. Proceedings of the Royal Society of London B 264: 1355–1361.

Greenwood, P.E., Ward, L.M., Russell, D.F., Neiman, A. and Moss, F. 2000. Stochastic resonance enhances the electrosensory information available to paddlefish for prey capture. Physical Review Letters 84: 4773–4776.

Griffin, D.R. 1971. Importance of atmospheric attenuation for echolocation of bats (Chiroptera). Animal Behaviour 19: 55–61.

Gullan, P.J. and Cranston, P.S. 2000. The Insects: An Outline of Entomology, 2nd edn. Blackwell Science, Oxford.

Gwynne, D.T. 1984. Courtship feeding increases female reproductive success in bush crickets. Nature 307: 361–363.

Gwynne, D.T. 1987. Sex-biased predation and the risky mate-locating behavior of male tick-tock cicadas (Homoptera: Cicadidae). Animal Behaviour 35: 571–576.

Gwynne, D.T. 1995. Phylogeny of the Ensifera (Orthoptera): a hypothesis supporting multiple origins of acoustical signalling, complex spermatophores, and maternal care in crickets, katydids, and weta. Journal of Orthoptera Research 4: 203–218.

Gwynne, D.T. and Bailey, W.J. 1988. Mating system, mate choice and ultrasonic calling in a zaprochiline katydid (Orthoptera: Tettigoniidae). Behaviour 105: 202–223.

Gwynne, D.T. and Bailey, W.J. 1999. Female–female competition in katydids: sexual selection for increased sensitivity to a male signal? Evolution 53: 546–551.

Gwynne, D.T. and Edwards, E.D. 1986. Ultrasound production by genital stridulation in *Syntonarcha iriastis* (Lepidoptera: Pyralidae): long-distance signaling by male moths. Zoological Journal of the Linnean Society 88: 363–376.

Gwynne, D.T. and Morris, G.K. 1986. Heterospecific recognition and behavioral isolation in acoustic Orthoptera (Insecta). Evolutionary Theory 8: 33–38.

Hack, M.A. 1997a. Assessment strategies in the contests of male crickets, *Acheta domesticus* (L.). Animal Behaviour 53: 733–747.

Hack, M.A. 1997b. The energetic costs of fighting in the house cricket, *Acheta domesticus* L. Behavioral Ecology 8: 28–36.

Hack, M.A. 1998. The energetics of male mating strategies in field crickets (Orthoptera: Gryllinae: Gryllidae). Journal of Insect Behavior 11: 853–867.

Hailman, J.P. 1977. Optical Signals: Animal Communication and Light. Indiana University Press, Bloomington.

Hamilton, W.D. 1971. Geometry for the selfish herd. Journal of Theoretical Biology 31: 295–311.

Hamner, P. and Hamner, W.M. 1977. Chemosensory tracking of scent trails by planktonic shrimp *Acetes sibogae australis*. Science 195: 886–888.

Hansen, K., Schneider, D. and Boppré, M. 1983. Chiral pheromone and reproductive isolation between the gypsy- and nun moth. Naturwissenschaften 70: 466–467.

Hanson, F.E. 1978. Comparative studies of firefly pacemakers. Federation Proceedings 37: 2158–2164.

Hanson, F.E., Case, J.F., Buck, E. and Buck, J. 1971. Synchrony and flash entrainment in a New Guinea firefly. Science. 174: 161–164.

Hansson, B.S. 1995. Olfaction in Lepidoptera. Experientia 51: 1003–1027.

Hansson, B.S. 1999. Insect Olfaction. Springer-Verlag, Berlin.

Hansson, B.S. and Christensen, T.A. 1999. Functional characteristics of the antennal lobe. In Insect Olfaction (ed. B.S. Hansson), pp. 125–161. Springer-Verlag, Berlin.

Harris, M.O. and Miller, J.R. 1983. Color stimuli and oviposition behavior of the onion fly, *Delia antiqua* (Meigen) (Diptera: Anthomyiidae). Annals of the Entomological Society of America 76: 766–771.

Harris, V.E., Todd, J.W., Webb, J.C. and Benner, J.C. 1982. Acoustical and behavioral analysis of the songs of the southern green stink bug, *Nezara viridula* (Hemiptera, Pentatomidae). Annals of the Entomological Society of America 75: 234–249.

Hartley, J.C. 1993. Acoustic behaviour and phonotaxis in the duetting ephippigerines, *Steropleurus nobrei* and *Steropleurus stali* (Tettigoniidae). Zoological Journal of the Linnean Society 107: 155–167.

Hartley, J.C. and Stephen, R.O. 1992. A paradoxical problem in insect communication: Can bush crickets discriminate frequency? Journal of Experimental Biology 163: 359–365.

Hasson, O. 1990. The role of amplifiers in sexual selection: an integration of the amplifying and the Fisherian mechanisms. Evolutionary Ecology 4: 277–289.

Hasson, O. 1991. Sexual displays as amplifiers: practical examples with an emphasis on feather decorations. Behavioral Ecology 2: 189–197.

Hasson, O. 1994. Cheating signals. Journal of Theoretical Biology 167: 223–238.

Hasson, O. 1997. Towards a general theory of biological signaling. Journal of Theoretical Biology 185: 139–156.

Hastings, J.W. and Morin, J.G. 1991. Bioluminescence. In Neural and Integrative Animal Physiology: Comparative Animal Physiology, 4th edn (ed. C.L. Prosser), pp. 131–170. Wiley-Liss, New York.

Hauser, M.D. 1996. The Evolution of Communication. The MIT Press, Cambridge, MA.

Hawkins, A.D. and Myrberg, A.A.Jr. 1983. Hearing and sound communication under water. In Bioacoustics: A Comparative Approach (ed. B. Lewis), pp. 347–405. Academic Press, London.

Haynes, K.F. and Baker, T.C. 1989. An analysis of anemotactic flight in female moths stimulated by host odor and comparison with the males response to sex pheromone. Physiological Entomology 14: 279–289.

Haynes, K.F. and Birch, M.C. 1986. Temporal reproductive isolation between 2 species of plume moths (Lepidoptera: Pterophoridae). Annals of the Entomological Society of America 79: 210–215.

Haynes, K.F. and Hunt, R.E. 1990. A mutation in pheromonal communication system of the cabbage looper moth, *Trichoplusia ni*. Journal of Chemical Ecology 16: 1249–1257.

Hazlett, B.A. 1979. The meral spot of *Gonodactylus oerstedii* (Hansen) as a visual stimulus (Stomatopoda: Gonodactylidae). Crustaceana 36: 196–198.

Heath, R.R. and Tumlinson, J.H. 1984. Techniques for purifying, analyzing, and identifying pheromones. In Techniques in Pheromone Research (eds H.E. Hummel and T.A. Miller), pp. 287–322. Springer-Verlag, New York.

Hebets, E.A. and Uetz, G.W. 2000. Leg ornamentation and the efficacy of courtship display in four species of wolf spider (Araneae: Lycosidae). Behavioral Ecology and Sociobiology. 47: 280–286.

Hedrick, A.V. 1986. Female preferences for male calling bout duration in a field cricket. Behavioral Ecology and Sociobiology 19: 73–77.

Hedrick, A.V. 1988. Female choice and the heritability of attractive male traits: an empirical study. American Naturalist 132: 267–276.

Hedrick, A.V. 2000. Crickets with extravagant mating songs compensate for predation risk with extra caution. Proceedings of the Royal Society of London B 267: 671–675.

Hedrick, A.V. and Dill, L.M. 1993. Mate choice by female crickets is influenced by predation risk. Animal Behaviour 46: 193–196.

Hedrick, A. and Weber, T. 1998. Variance in female responses to the fine structure of male song in the field cricket, *Gryllus integer*. Behavioral Ecology 9: 582–591.

Hedwig, B. 1990. Modulation of auditory responsiveness in stridulating grasshoppers. Journal of Comparative Physiology A 167: 847–856.

Hedwig, B. 1992. On the control of stridulation in the acridid grasshopper *Omocestus viridulus* L. 1. interneurons involved in rhythm generation and bilateral coordination. Journal of Comparative Physiology A 171: 117–128.

Hedwig, B. 1996. A descending brain neuron elicits stridulation in the cricket *Gryllus bimaculatus* (de Geer). Naturwissenschaften 83: 428–429.

Hedwig, B. 2000. Control of cricket stridulation by a command neuron: efficacy depends on the behavioral state. Journal of Neurophysiology 83: 712–722.

Hedwig, B. and Heinrich, R. 1997. Identified descending brain neurons control different stridulatory motor patterns in an acridid grasshopper. Journal of Comparative Physiology A 180: 285–294.

Hedwig, B., Lang, F. and Elsner, N. 1990. Modulation of auditory information processing by motor activity and mechanical stimulation in grasshoppers and locusts. In Sensory Systems and Communication in Arthropods (eds F.G. Gribakin, K. Wiese, and A.V. Popov), pp. 193–198. Birkhäuser Verlag, Basel.

Heidelbach, J., Dambach, M. and Böhm, H. 1991. Processing wing flick-generated air-vortex signals in the african cave cricket *Phaeophilacris spectrum*. Naturwissenschaften 78: 277–278.

Heifetz, Y. and Applebaum, S.W. 1995. Density-dependent physiological phase in a non-migratory grasshopper *Aiolopus thalassinus*. Entomologia Experimentalis et Applicata 77: 251–262.

Heinrich, R. and Elsner, N. 1997. Central nervous control of hindleg coordination in stridulating grasshoppers. Journal of Comparative Physiology A 180: 257–269.

Heinrich, R., Jatho, M. and Kalmring, K. 1993. Acoustic transmission characteristics of the tympanal tracheae of bushcrickets (Tettigoniidae). II. comparative studies of the tracheae of seven species. Journal of the Acoustical Society of America 93: 3481–3489.

Heller, K.-G. 1986. Warm-up and stridulation in the bush cricket, *Hexacentrus unicolor* Serville (Orthoptera: Conocephalidae: Listroscelidinae). Journal of Experimental Biology 126: 97–109.

Heller, K.-G. 1988. Bioakustik der Europischen Laubheuschrecken. Verlag Josef Margraf, Weikersheim, Germany.

Heller, K.-G. 1992. Risk shift between males and females in the pair-forming behavior of bush crickets. Naturwissenschaften 79: 89–91.

Heller, K.-G. 1995. Acoustic signaling in palaeotropical bush crickets (Orthoptera: Tettigonioidea: Pseudophyllidae): Does predation pressure by eavesdropping enemies differ in the Palaeotropics and Neotropics? Journal of Zoology 237: 469–485.

Heller, K.-G. and Krahe, R. 1994. Sound production and hearing in the pyralid moth *Symmoracma minoralis*. Journal of Experimental Biology 187: 101–111.

Heller, K.-G. and von Helversen, D. 1986. Acoustic communication in phaneropterid bush crickets: species-specific delay of female stridulatory response and matching male sensory time window. Behavioral Ecology and Sociobiology 18: 189–198.

Heller, K.-G. and von Helversen, D. 1991. Operational sex ratio and individual mating frequencies in two bushcricket species (Orthoptera, Tettigonioidea, *Poecilimon*). Ethology 89: 211–228.

Heller, K.-G., Schul, J. and Ingrisch, S. 1997a. Sex specific differences in song frequency and tuning of the ears in some duetting bushcrickets (Orthoptera: Tettigonioidea: Phaneropteridae). Zoology: Analysis of Complex Systems 100: 110–118.

Heller, K.-G., von Helversen, O. and Sergejeva, M. 1997b. Indiscriminate response behaviour in a female bushcricket: sex role reversal in selectivity of acoustic mate recognition? Naturwissenschaften 84: 252–255.

Hendry, L.B. 1976. Insect pheromones: diet related. Science 192: 143–145.

Hendry, L.B., Wichmann, J.K., Hindenlang, D.M., Mumma, R.O. and Anderson, M.E. 1975. Evidence for origin of insect sex pheromones: presence in food plants. Science 188: 59–63.

Hennig, R.M. 1990. Neuronal control of the forewings in two different behaviors: stridulation and flight in the cricket, *Teleogryllus commodus*. Journal of Comparative Physiology A 167: 617–628.

Hennig, R.M., Weber, T., Huber, F., Kleindienst, H.U., Moore, T.E. and Popov, A.V. 1994. Auditory threshold change in singing cicadas. Journal of Experimental Biology 187: 45–55.

Henry, C.S. 1980. The importance of low frequency, substrate borne sounds in lacewing communication (Neuroptera, Chrysopidae). Annals of the Entomological Society of America 73: 617–621.

Henry, C.S. 1986. Good vibrations. Natural History 95(8): 46–53.

Henry, C.S. 1994. Singing and cryptic speciation in insects. Trends in Ecology and Evolution 9: 388–392.

Henry, C.S., Wells, M.L.M. and Simon, C.M. 1999. Convergent evolution of courtship songs among cryptic species of the *carnea* group of green lacewings (Neuroptera: Chrysopidae: *Chrysoperla*). Evolution 53: 1165–1179.

Henson, O.W. Jr. 1965. Activity and function of middle-ear muscles in echolocating bats. Journal of Physiology 180: 871–887.

Hill, G.E. 1990. Female house finches prefer colorful males: sexual selection for a condition-dependent trait. Animal Behaviour 40: 563–572.

Hill, G.E. 1991. Plumage coloration is a sexually selected indicator of male quality. Nature 350: 337–339.

Hill, K.G. and Boyan, G.S. 1977. Sensitivity to frequency and direction of sound in the auditory system of crickets (Gryllidae). Journal of Comparative Physiology A 121: 79–97.

Hill, K.G. and Oldfield, B.P. 1981. Auditory function in Tettigoniidae (Orthoptera: Ensifera). Journal of Comparative Physiology A 142: 169–180.

Hirai, K., Shorey, H.H. and Gaston, L.K. 1978. Competition among courting male moths: male-to-male inhibitory pheromone. Science 202: 644–645.

Hoback, W.W. and Wagner, W.E. 1997. The energetic cost of calling in the variable field cricket, *Gryllus lineaticeps*. Physiological Entomology 22: 286–290.

Höglund, J. and Alatalo, R.V. 1995. Leks. Princeton University Press, Princeton, NJ.

Hoikkala, A. and Suvanto, L. 1999. Male courtship song frequency as an indicator of male mating success in *Drosophila montana*. Journal of Insect Behavior 12: 599–609.

Holland, B. and Rice, W.R. 1998. Perspective: Chase-away sexual selection: antagonistic seduction versus resistance. Evolution 52: 1–7.

Hölldobler, B. 1971. Recruitment behavior in *Camponotus socius* (Hym. Formicidae). Zeitschrift für Vergleichende Physiologie 75: 123–142.

Hölldobler, B. 1999. Multimodal signals in ant communication. Journal of Comparative Physiology A 184: 129–141.

Hölldobler, B., Moglich, M. and Maschwitz, U. 1974. Communication by tandem running in the ant *Camponotus sericeus*. Journal of Comparative Physiology 90: 105–127.

Hölldobler, B., Obermayer, M. and Wilson, E.O. 1992. Communication in the primitive cryptobiotic ant *Prionopelta amabilis* (Hymenoptera: Formicidae). Journal of Comparative Physiology A 171: 9–16.

Hollocher, H., Ting, C.T., Pollack, F. and Wu, C.I. 1997a. Incipient speciation by sexual isolation in *Drosophila melanogaster*: variation in mating preference and correlation between sexes. Evolution 51: 1175–1181.

Hollocher, H., Ting, C.T., Wu, M.L. and Wu, C.I. 1997b. Incipient speciation by sexual isolation in *Drosophila melanogaster*: extensive genetic divergence without reinforcement. Genetics 147: 1191–1201.

Horch, K.W. 1971. An organ for hearing and vibration sense in the ghost crab *Ocypode*. Zeitschrift für Vergleichende Physiologie 73: 1–21.

Horch, K.W. 1974. Barth's myochordotonal organ as an acoustic sensor in the ghost crab, *Ocypode*. Experientia 30: 630–631.

Horch, K.W. 1975. Acoustic behavior of the ghost crab *Ocypode cordimana* Latrielle, 1818 (Decapoda, Brachyura). Crustaceana 29: 193–205.

Horridge, G.A., Zhang, S.W. and Lehrer, M. 1992. Bees can combine range and visual angle to estimate absolute size. Philosophical Transactions of the Royal Society of London B 337: 49–57.

Howard, R.W. and Akre, R.D. 1995. Propaganda, crypsis, and slave-making. In Chemical Ecology of Insects, Vol. 2 (eds R.T. Cardé and W.J. Bell), pp. 364–410. Chapman and Hall, New York.

Hoy, R.R. 1992. The evolution of hearing in insects as an adaptation to predation from bats. In The Evolutionary Biology of Hearing (eds D.B. Webster, R.R. Fay and A.N. Popper), pp. 115–129. Springer-Verlag, New York.

Hoy, R.R. 1994. Ultrasound acoustic startle in flying insects: some neuroethological and comparative aspects. Fortschritte der Zoologie 39: 227–241.

Hoy, R.R. and Robert, D. 1996. Tympanal hearing in insects. Annual Review of Entomology 41: 433–450.

Hoy, R.R., Hahn, J. and Paul, R.C. 1977. Hybrid cricket auditory behavior: evidence for genetic coupling in animal communication. Science 195: 82–84.

Hoy, R.R., Hoikkala, A. and Kaneshiro, K. 1988. Hawaiian courtship songs: evolutionary innovation in communication signals of *Drosophila*. Science 240: 217–219.

Hrncir, M., Jarau, S., Zucchi, R. and Barth, F.G. 2000. Recruitment behavior in stingless bees, *Melipona scutellaris* and *M. quadrifasciata*. II. possible mechanisms of communication. Apidologie 31: 93–113.

Huber, F. 1990. Nerve cells and insect behavior: studies on crickets. American Zoologist 30: 609–627.

Huber, F. 1992. Behavior and neurobiology of acoustically oriented insects. Naturwissenschaften 79: 393–406.

Huber, F. and Thorson, J. 1985. Cricket auditory communication. Scientific American 253(6): 60–68.

Imaizumi, K. and Pollack, G.S. 1999. Neural coding of sound frequency by cricket auditory receptors. Journal of Neuroscience 19: 1508–1516.

Ishida, H., Hayashi, K., Takakusaki, M., Nakamoto, T., Moriizumi, T. and Kanzaki, R. 1995. Odour source localization system mimicking behaviour of silkworm moth. Sensors and Actuators A-Physical 51: 225–230.

Isingrini, M., Lenoir, A. and Jaisson, P. 1985. Preimaginal learning as a basis of colony-brood recognition in the ant *Cataglyphis cursor*. Proceedings of the National Academy of Sciences, U.S.A. 82: 8545–8547.

Itagaki, H. and Conner, W.E. 1987. Neural control of rhythmic pheromone gland exposure in *Utetheisa ornatrix* (Lepidoptera: Arctiidae). Journal of Insect Physiology 33: 177–181.

Itagaki, H. and Conner, W.E. 1988. Calling behavior of *Manduca sexta* (L.) (Lepidoptera: *Sphingidae*) with notes on the morphology of the female sex pheromone gland. Annals of the Entomological Society of America 81: 798–807.

Iwasaki, M., Itoh, T. and Tominaga, Y. 1999. Mechano- and phonoreceptors. In Atlas of Arthropod Sensory Receptors (eds E. Eguchi and Y. Tominaga), pp. 177–190. Springer-Verlag, Tokyo.

Iyengar, V.K. and Eisner, T. 1999a. Heritability of body mass, a sexually selected trait, in an arctiid moth (*Utetheisa ornatrix*). Proceedings of the National Academy of Sciences, U.S.A. 96: 9169–9171.

Iyengar, V.K. and Eisner, T. 1999b. Female choice increases offspring fitness in an arctiid moth (*Utetheisa ornatrix*). Proceedings of the National Academy of Sciences, U.S.A. 96: 15013–15016.

Jackson, R.R. and Blest, A.D. 1982. The distances at which a primitive jumping spider, *Portia fimbriata*, makes visual discriminations. Journal of Experimental Biology 97: 441–445.

Jacquin, E., Nagnan, P. and Frerot, B. 1991. Identification of hairpencil secretion from male *Mamestra brassicae* (L.) (Lepidoptera: Noctuidae) and electroantennogram studies. Journal of Chemical Ecology 17: 239–246.

Jang, Y. and Greenfield, M.D. 1996. Ultrasonic communication and sexual selection in wax moths: female choice based on energy and asynchrony of male signals. Animal Behaviour 51: 1095–1106.

Jang, Y. and Greenfield, M.D. 1998. Absolute versus relative measurements of sexual selection: assessing the contributions of ultrasonic signal characters to mate attraction in lesser wax moths, *Achroia grisella* (Lepidoptera: Pyralidae). Evolution 52: 1383–1393.

Jang, Y. and Greenfield, M.D. 2000. Quantitative genetics of female choice in an ultrasonic pyralid moth, *Achroia grisella*: variation and evolvability of preference along multiple dimensions of the male advertisement signal. Heredity 84: 73–80.

Jang, Y., Collins, R.D. and Greenfield, M.D. 1997. Variation and repeatability of ultrasonic sexual advertisement signals in *Achroia grisella* (Lepidoptera: Pyralidae). Journal of Insect Behavior 10: 87–98.

Jansson, A. 1973. Stridulation and its significance in genus *Cenocorixa* (Hemiptera, Corixidae). Behaviour 46: 1–36.

Jarau, S., Hrncir, M., Zucchi, R. and Barth, F.G. 2000. Recruitment behavior in stingless bees, *Melipona scutellaris* and *M. quadrifasciata*. I. foraging at food sources differing in direction and distance. Apidologie 31: 81–91.

Jennions, M.D. and Backwell, P.R.Y. 1998. Variation in courtship rate in the fiddler crab *Uca annulipes*: Is it related to male attractiveness? Behavioral Ecology 9: 605–611.

Jennions, M.D. and Petrie, M. 1997. Variation in mate choice and mating preferences: a review of causes and consequences. Biological Reviews of the Cambridge Philosophical Society 72: 283–327.

Jennions, M.D. and Petrie, M. 2000. Why do females mate multiply? A review of the genetic benefits. Biological Reviews of the Cambridge Philosophical Society 75: 21–64.

Jennions, M.D., Møller, A.P. and Petrie, M. 2001. Sexually selected traits and adult survival: a meta-analysis. Quarterly Review of Biology. 76: 3–36.

Jia, F.-Y. and Greenfield, M.D. 1997. When are good genes good? Variable outcomes of female choice in wax moths. Proceedings of the Royal Society of London B 264: 1057–1063.

Jia, F.-Y., Greenfield, M.D. and Collins, R.D. 2000. Genetic variance of sexually selected traits in waxmoths: maintenance by genotype × environment interaction. Evolution 54: 953–967.

Jia, F.-Y., Greenfield, M.D. and Collins, R.D. 2001. Ultrasonic signal competition among wax moths. Journal of Insect Behavior 14: 19–33.

Jiggins, C.D., Naisbit, R.E., Coe, R.L. and Mallet, J. 2001. Reproductive isolation caused by colour pattern mimicry. Nature 411: 302–305.

Johnston, C. 1855. Auditory apparatus of the *Culex* mosquito. Quarterly Journal of Microscopical Science 3: 97–102.

Johnstone, R.A. 1994. Honest signaling, perceptual error and the evolution of all-or-nothing displays. Proceedings of the Royal Society of London B 256: 169–175.

Johnstone, R.A. 1995a. Honest advertisement of multiple qualities using multiple signals. Journal of Theoretical Biology 177: 87–94.

Johnstone, R.A. 1995b. Sexual selection, honest advertisement and the handicap principle: reviewing the evidence. Biological Reviews of the Cambridge Philosophical Society 70: 1–65.

Johnstone, R.A. 1996. Multiple displays in animal communication: 'backup signals' and 'multiple messages'. Philosophical Transactions of the Royal Society of London B 351: 329–338.

Johnstone, R.A. 1997. Recognition and the evolution of distinctive signatures: when does it pay to reveal identity? Proceedings of the Royal Society of London B 264: 1547–1553.

Johnstone, R.A. 1998a. Efficacy and honesty in communication between relatives. American Naturalist 152: 45–58.

Johnstone, R.A. 1998b. Conspiratorial whispers and conspicuous displays: games of signal detection. Evolution 52: 1554–1563.

Jones, M.D.R. 1966. The acoustical behaviour of the bushcricket *Pholidoptera griseoaptera*. 1. alternation, synchronism, and rivalry between males. Journal of Experimental Biology 45: 15–30.

Jones, T.M., Quinnell, R.J. and Balmford, A. 1998. Fisherian flies: benefits of female choice in a lekking sandfly. Proceedings of the Royal Society of London B 265: 1651–1657.

Josephson, R.K. 1985. The mechanical power output of a tettigoniid wing muscle during singing and flight. Journal of Experimental Biology 117: 357–368.

Josephson, R.K. and Young, D. 1979. Body temperature and singing in the bladder cicada, *Cystosoma saundersii*. Journal of Experimental Biology 80: 69–81.

Jurenka, R.A., Haynes, K.F., Adlof, R.O., Bengtsson, M. and Roelofs, W.L. 1994. Sex pheromone component ratio in the cabbage looper moth altered by a mutation affecting the fatty-acid chain-shortening reactions in the pheromone biosynthetic pathway. Insect Biochemistry and Molecular Biology 24: 373–381.

Kaissling, K.-E. and Priesner, E. 1970. Die Riechschwelle des Seidenspinngers. Naturwissenschaften 57: 23–28.

Kaib, M. and Dittebrand, H. 1990. Foraging and food recruitment communication in the ant Myrmicaria eumenoides (Hymenoptera: Myrmicinae). In Social Insects and the Environment (eds G.K. Veeresh, B. Mallik and C.A. Viraktamath), pp. 560–561. Proceedings of the 11th International Congress of IUSSI. E.J. Brill, Leiden, The Netherlands.

Kalmring, K., Keuper, A. and Kaiser, W. 1990. Aspects of acoustic and vibratory communication in seven European bushcrickets. In The Tettigoniidae: Biology, Systematics, and Evolution (eds W.J. Bailey and D.C.F. Rentz), pp. 191–216. Springer-Verlag, Berlin.

Kalmring, K., Jatho, M., Rossler, W. and Sickmann, T. 1997. Acousto-vibratory communication in bushcrickets (Orthoptera: Tettigoniidae). Entomologia Generalis 21: 265–291.

Kamimura, M. and Tatsuki, S. 1993. Diel rhythms of calling behavior and pheromone production of oriental tobacco budworm moth, *Helicoverpa assulta* (Lepidoptera: Noctuidae). Journal of Chemical Ecology 19: 2953–2963.

Kämper, G. 1985. Processing of species-specific low-frequency song components by interneurons in crickets. In Acoustic and Vibrational Communication in Insects (eds K. Kalmring and N. Elsner), pp. 169–176. Verlag Paul Parey, Berlin.

Kämper, G. and Kleindienst, H.U. 1990. Oscillation of cricket sensory hairs in a low-frequency sound field. Journal of Comparative Physiology A 167: 193–199.

Kanzaki, R. 1996. Behavioral and neural basis of instinctive behavior in insects: odor-source searching strategies without memory and learning. Robotics and Autonomous Systems 18: 33–43.

Kanzaki, R. 1998. Coordination of wing motion and walking suggests common control of zigzag motor program in a male silkworm moth. Journal of Comparative Physiology A 182: 267–276.

Kanzaki, R. and Mishima, T. 1996. Pheromone-triggered 'flipflopping' neural signals correlate with activities of neck motor neurons of a male moth, *Bombyx mori*. Zoological Science 13: 79–87.

Kanzaki, R., Sugi, N. and Shibuya, T. 1992. Self-generated zigzag turning of *Bombyx mori* males during pheromone-mediated upwind walking. Zoological Science 9: 515–527.

Karavanich, C. and Atema, J. 1998. Individual recognition and memory in lobster dominance. Animal Behaviour 56: 1553–1560.

Karlson, P. and Lüscher, M. 1959. "Pheromones": a new term for a class of biologically active substances. Nature 183: 55–56.

Karlson, P. and Schneider, D. 1973. Sexual pheromones of Lepidoptera as model systems for chemical communication. Naturwissenschaften 60: 113–121.

Kasang, G., Kaissling, K.E., Vostrowsky, O. and Bestmann, H.J. 1978. Bombykal, a 2nd pheromone component of the silkworm moth *Bombyx mori* L.. Angewandte Chemie–International Edition in English 17: 60.

Katzav-Gozansky, T., Soroker, V., Hefetz, A., Cojocaru, M., Erdmann, D.H. and Francke, W. 1997. Plasticity of caste-specific Dufour's gland secretion in the honey bee (*Apis mellifera* L.). Naturwissenschaften 84: 238–241.

Kavanagh, M.W. 1987. The efficiency of sound production in two cricket species, *Gryllotalpa australis* and *Teleogryllus commodus* (Orthoptera: Grylloidea). Journal of Experimental Biology 130: 107–119.

Kavanagh, M.W. and Young, D. 1989. Bilateral symmetry of sound production in the mole cricket, *Gryllotalpa australis*. Journal of Comparative Physiology A 166: 43–49.

Kelber, A. 1999. Why "false" colors are seen by butterflies. Nature 402: 251.

Kelber, A. and Pfaff, M. 1999. True colour vision in the orchard butterfly, *Papilio aegeus*. Naturwissenschaften 86: 221–224.

Keller, L. 1997. Indiscriminate altruism: unduly nice parents and siblings. Trends in Ecology and Evolution 12: 99–103.

Keller, L. and Nonacs, P. 1993. The role of queen pheromones in social insects: queen control or queen signal. Animal Behaviour 45: 787–794.

Kelly, J.K. and Noor, M.A.F. 1996. Speciation by reinforcement: a model derived from studies of *Drosophila*. Genetics 143: 1485–1497.

Kennedy, J.S. 1939. The visual responses of flying mosquitoes. Proceedings of the Zoological Society of London A 109: 221–242.

Kessel, E.L. 1955. Mating activities of balloon flies. Systematic Zoology 4: 97–104.

Keuper, A. and Kuhne, R. 1983. The acoustic behavior of the bushcricket *Tettigonia cantans*. 2. transmission of airborne sound and vibration signals in the biotope. Behavioural Processes 8: 125–145.

Kiflawi, M. and Gray, D.A. 2000. Size-dependent response to conspecific mating calls by male crickets. Proceedings of the Royal Society of London B 267: 2157–2161.

Kingan, T.G., Bodnar, W.M., Raina, A.K., Shabanowitz, J. and Hunt, D.F. 1995. The loss of female sex pheromone after mating in the corn earworm moth *Helicoverpa zea*: identification of a male pheromonostatic peptide. Proceedings of the National Academy of Sciences, U.S.A. 92: 5082–5086.

Kingett, P.D., Lambert, D.M. and Telford, S.R. 1981. Does mate choice occur in *Drosophila melanogaster*. Nature 293: 492.

Kinsler, L.E. and Frey, A.R. 1962. Fundamentals of Acoustics, 2nd edn, Wiley, New York.

Kirchner, W.H. 1993. Acoustical communication in honeybees. Apidologie 24: 297–307.

Kirchner, W.H. 1994. Hearing in honeybees: the mechanical response of the bee's antenna to near-field sound. Journal of Comparative Physiology A 175: 261–265.

Kirchner, W.H. 1997. Acoustical communication in social insects. In Orientation and Communication in Arthropods (ed. M. Lehrer), pp. 273–300. Birkhäuser Verlag, Basel.

Kirchner, W.H. and Dreller, C. 1993. Acoustical signals in the dance language of the giant honeybee, *Apis dorsata*. Behavioral Ecology and Sociobiology 33: 67–72.

Kirchner, W.H. and Towne, W.F. 1994. The sensory basis of the honeybees dance language. Scientific American 270: 74–80.

Kirchner, W.H., Broecker, I. and Tautz, J. 1994. Vibrational alarm communication in the dampwood termite *Zootermopsis nevadensis*. Physiological Entomology 19: 187–190.

Kirchner, W.H., Dreller, C., Grasser, A. and Baidya, D. 1996. The silent dances of the Himalayan honeybee, *Apis laboriosa*. Apidologie 27: 331–339.

Kirkpatrick, M. 1982. Sexual selection and the evolution of female choice. Evolution 36: 1–12.

Kirschfeld, K. 1974a. The absolute sensitivity of lens and compound eyes. Zeitschrift für Naturforschung 29C: 592–596.

Kirschfeld, K. 1974b. Lateral inhibition in the compound eye of the fly, *Musca*. Zeitschrift für Naturforschung 29C: 95–97.

Kirschfeld, K. 1976. The resolution of lens and compound eyes. In Neural Principles in Vision (eds F. Zettler and R. Weiler), pp. 354–370. Springer-Verlag, Berlin.

Kitamoto, J., Sakamoto, K., Ozaki, K., Mishina, Y. and Arikawa, K. 1998. Two visual pigments in a single photoreceptor cell: identification and histological localization of three mRNAs encoding visual pigment opsins in the retina of the butterfly *Papilio xuthus*. Journal of Experimental Biology 201: 1255–1261.

Kittredge, J.S. and Takahashi, F.T. 1972. Evolution of sex pheromone communication in Arthropoda. Journal of Theoretical Biology 35: 467–471.

Kleindienst, H.-U., Wohlers, D.W. and Larsen, O.N. 1983. Tympanal membrane motion is necessary for hearing in crickets. Journal of Comparative Physiology A 151: 397–400.

Klun, J.A. and Maini, S. 1979. Genetic basis of an insect chemical communication system: the European corn borer (*Ostrinia nubilalis*). Environmental Entomology 8: 423–426.

Knoll, A.H. and Carroll, S.B. 1999. Early animal evolution: emerging views from comparative biology and geology. Science 284: 2129–2137.

Koch, U. 1980. Analysis of cricket stridulation using miniature angle detectors. Journal of Comparative Physiology A 136: 247–256.

Koehl, M.A.R. 1996. Small-scale fluid dynamics of olfactory antennae. Marine and Freshwater Behaviour and Physiology 27: 127–141.

Koga, T., Backwell, P.R.Y., Jennions, M.D. and Christy, J.H. 1998. Elevated predation risk changes mating behaviour and courtship in a fiddler crab. Proceedings of the Royal Society of London B 265: 1385–1390.

Kolluru, G.R. 1999. Variation and repeatability of calling behavior in crickets subject to a phonotactic parasitoid. Journal of Insect Behavior 12: 611–626.

Kolmes, S.A. 1985. Surface vibrational cues in the precopulatory behavior of whirligig beetles. Journal of the New York Entomological Society 93: 1137–1140.

Kotiaho, J., Alatalo, R.V., Mappes, J. and Parri, S. 1996. Sexual selection in a wolf spider: male drumming activity, body size, and viability. Evolution 50: 1977–1981.

Kotiaho, J.S., Alatalo, R.V., Mappes, J., Nielsen, M.G., Parri, S. and Rivero, A. 1998a. Energetic costs of size and sexual signalling in a wolf spider. Proceedings of the Royal Society of London B 265: 2203–2209.

Kotiaho, J., Alatalo, R.V., Mappes, J., Parri, S. and Rivero, A. 1998b. Male mating success and risk of predation in a wolf spider: a balance between sexual and natural selection. Journal of Animal Ecology 67: 287–291.

Kotiaho, J.S., Alatalo, R.V., Mappes, J. and Parri, S. 1999. Sexual signalling and viability in a wolf spider (*Hygrolycosa rubrofasciata*): measurements under laboratory and field conditions. Behavioral Ecology and Sociobiology 46: 123–128.

Krahe, R. and Ronacher, B. 1993. Long rise times of sound pulses in grasshopper songs improve the directionality cues received by the CNS from the auditory receptors. Journal of Comparative Physiology A 173: 425–434.

Krakauer, D.C. and Johnstone, R.A. 1995. The evolution of exploitation and honesty in animal communication: a model using artificial neural networks. Philosophical Transactions of the Royal Society of London B 348: 355–361.

Kramer, E. 1975. Orientation of male silkmoth to the sex attractant bombykol. In Olfaction and Taste (eds D.A. Denton and J.P. Coghlan), pp. 329–355. Academic Press, New York.

Krasnoff, S.B. and Dussourd, D.E. 1989. Dihydropyrrolizine attractants for arctiid moths that visit plants containing pyrrolizidine alkaloids. Journal of Chemical Ecology 15: 47–60.

Krasnoff, S.B. and Roelofs, W.L. 1988. Sex pheromone released as an aerosol by the moth *Pyrrharctia isabella*. Nature 333: 263–265.

Krasnoff, S.B. and Roelofs, W.L. 1989. Quantitative and qualitative effects of larval diet on male scent secretions of *Estigmene acrea*, *Phragmatobia fuliginosa*, and *Pyrrharctia isabella* (Lepidoptera: Arctiidae). Journal of Chemical Ecology 15: 1077–1093.

Krasnoff, S.B. and Roelofs, W.L. 1990. Evolutionary trends in the male pheromone systems of arctiid moths: evidence from studies of courtship in *Phragmatobia fuliginosa* and *Pyrrharctia isabella* (Lepidoptera: Arctiidae). Zoological Journal of the Linnean Society 99: 319–338.

Krasnoff, S.B., Bjostad, L.B. and Roelofs, W.L. 1987. Quantitative and qualitative variation in male pheromones of *Phragmatobia fuliginosa* and *Pyrrharctia isabella* (Lepidoptera: Arctiidae). Journal of Chemical Ecology 13: 807–822.

Kraus, W.F. 1989. Surface wave communication during courtship in the giant water bug, *Abedus identatus* (Heteroptera: Belostomatidae). Journal of the Kansas Entomological Society 62: 316–328.

Kriegbaum, H. 1989. Female choice in the grasshopper *Chorthippus biguttulus*: mating success is related to song characteristics of the male. Naturwissenschaften 76: 81–82.

Krieger, J. and Breer, H. 1999. Olfactory reception in invertebrates. Science 286: 720–723.

Kristensen, N.P. 1999. Phylogeny of endopterygote insects, the most successful lineage of living organisms. European Journal of Entomology 96: 237–253.

Kroodsma, D.E. 1989. Suggested experimental designs for song playbacks. Animal Behaviour 37: 600–609.

Kuenen, L.P.S. and Baker, T.C. 1983. A non-anemotactic mechanism used in pheromone source location by flying moths. Physiological Entomology 8: 277–289.

Kunzler, R. and Bakker, T.C.M. 1998. Computer animations as a tool in the study of mating preferences. Behaviour 135: 1137–1159.

Lahav, S., Soroker, V., Hefetz, A. and Vander Meer, R.K. 1999. Direct behavioral evidence for hydrocarbons as ant recognition discriminators. Naturwissenschaften 86: 246–249.

Lakes-Harlan, R. and Heller, K.-G. 1992. Ultrasound sensitive ears in a parasitoid fly. Naturwissenschaften 79: 224–226.

Lakes-Harlan, R., Bailey, W.J. and Schikorski, T. 1991. The auditory system of an atympanate bushcricket, *Phasmodes ranatriformes* (Westwood) (Tettigoniidae: Orthoptera). Journal of Experimental Biology 158: 307–324.

Lakes-Harlan, R., Stolting, H. and Stumpner, A. 1999. Convergent evolution of insect hearing organs from a preadaptive structure. Proceedings of the Royal Society of London B 266: 1161–1167.

Lall, A.B., Seliger, H.H., Biggley, W.H. and Lloyd, J.E. 1980. Ecology of colors of firefly bioluminescence. Science 210: 560–562.

LaMunyon, C.W. 1997. Increased fecundity, as a function of multiple mating, in an arctiid moth, *Utetheisa ornatrix*. Ecological Entomology 22: 69–73.

LaMunyon, C.W. and Eisner, T. 1993. Defense mechanisms of arthropods. 115. postcopulatory sexual selection in an arctiid moth (*Utetheisa ornatrix*). Proceedings of the National Academy of Sciences, U.S.A. 90: 4689–4692.

LaMunyon, C.W. and Eisner, T. 1994. Spermatophore size as determinant of paternity in an arctiid moth (*Utetheisa ornatrix*). Proceedings of the National Academy of Sciences, U.S.A. 91: 7081–7084.

Land, M.F. 1981. Optics and vision in invertebrates. In Handbook of Sensory Physiology, Vol. VII/6B (ed. H. Autrum), pp. 472–592. Springer-Verlag, Berlin.

Land, M.F. 1989. Variations in the structure and design of compound eyes. In Facets of Vision (eds D.G. Stavenga and R.C. Hardie), pp. 90–111. Springer-Verlag, Berlin.

Land, M.F. 1997. The resolution of insect compound eyes. Israel Journal of Plant Sciences 45: 79–91.

Lande, R. 1981. Models of speciation by sexual selection on polygenic traits. Proceedings of the National Academy of Sciences, U.S.A. 78: 3721–3725.

Landolt, P.J. and Heath, R.R. 1990. Sexual role reversal in mate-finding strategies of the cabbage looper moth. Science 249: 1026–1028.

Landolt, P.J. and Phillips, T.W. 1997. Host plant influences on sex pheromone behavior of phytophagous insects. Annual Review of Entomology 42: 371–391.

Landolt, P.J., Heath, R.R. and Chambers, D.L. 1992. Oriented flight responses of female Mediterranean fruit flies to calling males, odor of calling males, and a synthetic pheromone blend. Entomologia Experimentalis et Applicata 65: 259–266.

Landolt, P.J., Jeanne, R.L. and Reed, H.C. 1998. Chemical communication in social wasps. In Pheromone Communication in Social Insects: Ants, Wasps, Bees and Termites (eds R.K. Vander Meer, M.D. Breed, K.E. Espelie and M.L. Winston), pp. 216–235. Westview Press, Boulder, CO.

Larsen, O.N. 1987. The cricket's anterior tympanum revisited. Naturwissenschaften 74: 92–94.

Larsen, O.N. and Michelsen, A. 1978. Biophysics of the ensiferan ear. III. the cricket ear as a four-input system. Journal of Comparative Physiology A 123: 217–227.

Larsen, O.N., Surlykke, A. and Michelsen, A. 1984. Directionality of the cricket ear: a property of the tympanal membrane. Naturwissenschaften 71: 538–540.

Latimer, W. and Broughton, W.B. 1984. Acoustic interference in bush crickets: a factor in the evolution of singing insects. Journal of Natural History 18: 599–616.

Latimer, W. and Sippel, M. 1987. Acoustic cues for female choice and male competition in *Tettigonia cantans*. Animal Behaviour 35: 887–900.

Lawrence, B.D. and Simmons, J.A. 1982. Measurements of atmospheric attenuation of ultrasonic frequencies and the significance for echolocation by bats. Journal of the Acoustical Society of America 71: 585–590.

Lazard, D., Zupko, K., Poria, Y., Nef, P., Lazarovits, J., Horn, S., Khen, M. and Lancet, D. 1991. Odorant signal termination by olfactory UDP glucuronosyl transferase. Nature 349: 790–793.

Leal, W.S. 1996. Chemical communication in scarab beetles: reciprocal behavioral agonist-antagonist activities of chiral pheromones. Proceedings of the National Academy of Sciences, U.S.A. 93: 12112–12115.

Leal, W.S. 1998. Chemical ecology of phytophagous scarab beetles. Annual Review of Entomology 43: 39–61.

Leal, W.S., Kawamura, F. and Ono, M. 1994. The scarab beetle *Anomala albopilosa* Sakishimana utilizes the same sex pheromone blend as a closely related and geographically isolated species, *Anomala cuprea*. Journal of Chemical Ecology 20: 1667–1676.

Leal, W.S., Kuwahara, S., Shi, X.W., Higuchi, H., Marino, C.E.B., Ono, M. and Meinwald, J. 1998. Male-released sex pheromone of the stink bug *Piezodorus hybneri*. Journal of Chemical Ecology 24: 1817–1829.

Lecomte, C., Thibout, E., Pierre, D. and Auger, J. 1998. Transfer, perception, and activity of male pheromone of *Acrolepiopsis assectella* with special reference to conspecific male sexual inhibition. Journal of Chemical Ecology 24: 655–671.

Lehrer, M. 1999. Shape perception in the honeybee: symmetry as a global framework. International Journal of Plant Sciences 160: S51-S65.

Lesna, I. and Sabelis, M.W. 1999. Diet-dependent female choice for males with "good genes" in a soil predatory mite. Nature 401: 581–584.

Leyrer, R.L. and Monroe, R.E. 1973. Isolation and identification of scent of the moth, *Galleria mellonella*, and a reevaluation of its sex pheromone. Journal of Insect Physiology 19: 2267–2271.

Libersat, F. and Hoy, R.R. 1991. Ultrasonic startle behavior in bush crickets (Orthoptera, Tettigoniidae). Journal of Comparative Physiology A 169: 507–514.

Lighthill, J. 1978. Waves in Fluids. Cambridge University Press, Cambridge, UK.

Lighton, J.R.B. 1987. Cost of tokking: the energetics of substrate communication in the tok-tok beetle, *Psammodes striatus*. Journal of Comparative Physiology B 157: 11–20.

Lin, Y., Kalmring, K., Jatho, M., Sickmann, T. and Rössler, W. 1993. Auditory receptor organs in the forelegs of *Gampsocleis gratiosa* (Tettigoniidae): morphology and function of the organs in comparison to the frequency parameters of the conspecific song. Journal of Experimental Zoology 267: 377–388.

Lindauer, M. 1961. Communication among Social Bees. Harvard University Press, Cambridge, MA.

Linn, C.E. and Roelofs, W.L. 1983. Effect of varying proportions of the alcohol component on sex pheromone blend discrimination in male oriental fruit moths. Physiological Entomology 8: 291–306.

Linn, C.E., Bjostad, L.B., Du, J.W. and Roelofs, W.L. 1984. Redundancy in a chemical signal—behavioral-responses of male *Trichoplusia ni* to a 6-component sex pheromone blend. Journal of Chemical Ecology 10: 1635–1658.

Linsenmair, K.-E. 1984. Comparative studies on the social behaviour of the desert isopod *Hemilepistus reaumuri* and a *Porcellio* species. Symposia of the Zoological Society of London 53: 423–453.

Linsenmair, K.-E. 1985. Individual and family recognition in subsocial arthropods, in particular in the desert isopod *Hemilepistus reaumuri*. Fortschritte der Zoologie 31: 411–436.

Linsenmair, K.-E. 1987. Kin recognition in subsocial arthropods, in particular in the desert isopod *Hemilepistus reaumuri*. In Kin Recognition in Animals (eds D.J.C. Fletcher and C.D. Michener), pp. 121–208. John Wiley and Sons, Chichester, U.K.

Liou, L.W. and Price, T.D. 1994. Speciation by reinforcement of premating isolation. Evolution 48: 1451–1459.

Liu, Y.B. and Haynes, K.F. 1994. Temporal and temperature-induced changes in emission rates and blend ratios of sex-pheromone components in *Trichoplusia ni*. Journal of Insect Physiology 40: 341–346.

Ljungberg, H., Anderson, P. and Hansson, B.S. 1993. Physiology and morphology of pheromone-specific sensilla on the antennae of male and female *Spodoptera littoralis* (Lepidoptera: Noctuidae). Journal of Insect Physiology 39: 253–260.

Lloyd, J.E. 1971. Bioluminescent communication in insects. Annual Review of Entomology. 16: 97–122.

Lloyd, J.E. 1975. Aggressive mimicry in *Photuris* fireflies: signal repertoires by femmes fatales. Science 187: 452–453.

Lloyd, J.E. 1989. Bat (Chiroptera) connections with firefly (Coleoptera, Lampyridae) luminescence. 1. Potential significance, historical evidence, and opportunity. Coleopterists Bulletin 43: 83–91.

Lloyd, J.E. 1997. Firefly mating ecology, selection and evolution. In The Evolution of Mating Systems in Insects and Arachnids (eds J.C. Choe and B.J. Crespi), pp. 184–192. Cambridge University Press, Cambridge, U.K.

Lloyd, J.E. and Wing, S.R. 1983. Nocturnal aerial predation of fireflies by light-seeking fireflies. Science 222: 634–635.

Lloyd, J.E., Wing, S.R. and Hongtrakul, T. 1989. Ecology, flashes, and behavior of congregating Thai fireflies. Biotropica 21: 373–376.

Lockwood, J.A. and Rentz, D.C.F. 1996. Nest construction and recognition in a gryllacridid: the discovery of pheromonally mediated autorecognition in an insect. Australian Journal of Zoology 44: 129–141.

Löfstedt, C. 1993. Moth pheromone genetics and evolution. Philosophical Transactions of the Royal Society of London B 340: 167–177.

Löfstedt, C. and Kozlov, M. 1997. A phylogenetic analysis of pheromone communication in primitive moths. In Insect Pheromone Research: New Directions (eds R.T. Cardé and A.K. Minks), pp. 473–489. Chapman and Hall, New York.

Löfstedt, C., Lofqvist, J., Lanne, B.S., Vander Pers, J.N.C. and Hansson, B.S. 1986. Pheromone dialects in European turnip moths *Agrotis segetum*. Oikos 46: 250–257.

Löfstedt, C., Hansson, B.S., Roelofs, W. and Bengtsson, B.O. 1989a. No linkage between genes controlling female pheromone production and male pheromone response in the European corn borer, *Ostrinia nubilalis* Hubner (Lepidoptera: Pyralidae). Genetics 123: 553–556.

Löfstedt, C., Vickers, N.J., Roelofs, W.L. and Baker, T.C. 1989b. Diet related courtship success in the oriental fruit moth, *Grapholita molesta* (Tortricidae). Oikos 55: 402–408.

Loudon, C. 1995. Insect morphology above the molecular level: biomechanics. Annals of the Entomological Society of America 88: 1–4.

Loudon, C. and Koehl. M.A.R. 2000. Sniffing by a silkworm moth: wing fanning enhances air penetration through and pheromone interception by antennae. Journal of Experimental Biology 203: 2977–2990.

Lythgoe, J.N. 1979. The Ecology of Vision. Clarendon Press of Oxford University Press, Oxford.

Lyytinen, A., Alatalo, R.V., Lindström, L. and Mappes, J. 2001. Can ultraviolet cues function as aposematic signals? Behavioral Ecology 12: 65–70.

MacNally, R.C. and Young, D. 1981. Song energetics of the bladder cicada, *Cystosoma saundersii*. Journal of Experimental Biology 90: 185–196.

Mafra-Neto, A. and Cardé, R.T. 1994. Fine-scale structure of pheromone plumes modulates upwind orientation of flying moths. Nature 369: 142–144.

Mafra-Neto, A. and Cardé, R.T. 1995. Effect of the fine-scale structure of pheromone plumes: pulse frequency modulates activation and upwind flight of almond moth males. Physiological Entomology 20: 229–242.

Mafra-Neto, A. and Cardé, R.T. 1996. Dissection of the pheromone-modulated flight of moths using single-pulse response as a template. Experientia 52: 373–379.

Magnus, D. 1958. Experimentalle Untersuchunger zur Bionomie und Ethologie des Kaisermantels *Argynnis paphia* L. (Lepidoptera: Nymphalidae). Zeitschrift für Tierpsychologie 15: 397–426.

Malakoff, D. 1999. Pentagon agency thrives on in-your-face science. Science 285: 1476–1479.

Mallock, A. 1894. Insect sight and the defining power of composite eyes. Proceedings of the Royal Society of London B 55: 85–90.

Mankin, R.W. and Hagstrum, D.W. 1995. Three-dimensional orientation of male *Cadra cautella* (Lepidoptera: Pyralidae) flying to calling females in a windless environment. Environmental Entomology 24: 1616–1626.

Mappes, J., Alatalo, R.V., Kotiaho, J. and Parri, S. 1996. Viability costs of condition-dependent sexual male display in a drumming wolf spider. Proceedings of the Royal Society of London B 263: 785–789.

Marshall, D.C. and Cooley, J.R. 2000. Reproductive character displacement and speciation in periodical cicadas, with description of a new species, 13-year *Magicicada neotredecim*. Evolution 54: 1313–1325.

Marshall, J., Cronin, T.W., Shashar, N. and Land, M. 1999a. Behavioural evidence for polarisation vision in stomatopods reveals a potential channel for communication. Current Biology 9: 755–758.

Marshall, J., Kent, J. and Cronin, T. 1999b. Visual adaptations in crustaceans: spectral sensitivity in diverse habitats. In Adaptive Mechanisms in the Ecology of Vision (eds S.N. Archer, M.B.A. Djamgoz, E.R. Loew, J.C. Partridge and S. Vallerga), pp. 285–327. Kluwer Academic Publishers, Dordrecht, The Netherlands.

Marshall, N.J., Land, M.F., King, C.A. and Cronin, T.W. 1991. The compound eyes of mantis shrimps (Crustacea, Hoplocarida, Stomatopoda). 2. Color pigments in the eyes of stomatopod crustaceans: polychromatic vision by serial and lateral filtering. Philosophical Transactions of the Royal Society of London B 334: 57–84.

Mason, A.C. 1996. Territoriality and the function of song in the primitive acoustic insect *Cyphoderris monstrosa* (Orthoptera: Haglidae). Animal Behaviour 51: 211–224.

Mason, A.C. and Bailey, W.J. 1998. Ultrasound hearing and male–male communication in Australian katydids (Tettigoniidae: Zaprochilinae) with sexually dimorphic ears. Physiological Entomology 23: 139–149.

Mason, A.C., Morris, G.K. and Wall, P. 1991. High ultrasonic hearing and tympanal slit function in rainforest katydids. Naturwissenschaften 78: 365–367.

Mason, A.C., Morris, G.K. and Hoy, R.R. 1999. Peripheral frequency mis-match in the primitive ensiferan *Cyphoderris monstrosa* (Orthoptera: Haglidae). Journal of Comparative Physiology A 184: 543–551.

Mason, A.C., Oshinsky, M.L. and Hoy, R.R. 2001. Hyperacute directional hearing in a microscale auditory system. Nature 410: 686–690.

Masters, W.M. 1979. Insect disturbance stridulation: its defensive role. Behavioral Ecology and Sociobiology 5: 187–200.

Masters, W.M. 1980. Insect disturbance stridulation: characterization of airborne and vibrational components of the sound. Journal of Comparative Physiology A 135: 259–268.

Masters, W.M. 1984a. Vibrations in the orbwebs of *Nuctenea sclopetaria* (Araneidae). I. transmission through the web. Behavioral Ecology and Sociobiology 15: 207–215.

Masters, W.M. 1984b. Vibrations in the orbwebs of *Nuctenea sclopetaria* (Araneidae). II. prey and wind signals and the spider's response threshold. Behavioral Ecology and Sociobiology 15: 217–223.

Masters, W.M. and Markl, H. 1981. Vibration signal transmission in spider orb webs. Science 213: 363–365.

Matsuura, K. 2001. Nestmate recognition mediated by intestinal bacteria in a termite, *Reticulitermes speratus*. Oikos 92: 20–26.

Mayer, A.G. 1900. On the mating instinct in moths. Psyche 9: 15–20.

Mayer, M.S. and McLaughlin, J.R. 1991. Handbook of Insect Pheromones and Sex Attractants. CRC Press, Boca Raton, FL.

Maynard Smith, J. 1991a. Honest signalling: the Philip Sydney game. Animal Behaviour 42: 1034–1035.

Maynard Smith, J. 1991b. Theories of sexual selection. Trends in Ecology and Evolution 6: 146–151.

Maynard Smith, J. 1994. Must reliable signals always be costly? Animal Behaviour 47: 1115–1120.

Maynard Smith, J., Mayr, E. and Williams, G.C. 1999. The 1999 Crafoord Prize Lectures: Introduction. Quarterly Review of Biology 74: 391–393.

McNeil, J.N. and Delisle, J. 1989. Are host plants important in pheromone-mediated mating systems of Lepidoptera? Experientia 45: 236–240.

McVean, A. and Field, L.H. 1996. Communication by substratum vibration in the New Zealand tree weta, *Hemideina femorata* (Stenopelmatidae: Orthoptera). Journal of Zoology 239: 101–122.

Meier, T. and Reichert, H. 1990. Embryonic development and evolutionary origin of the orthopteran auditory organs. Journal of Neurobiology 21: 592–610.

Meinwald, J., Wiemer, D.F. and Eisner, T. 1979. Defense mechanisms of arthropods. 61. Lucibufagins. 2. esters of 12-oxo-2-beta,5-beta,11-alpha-trihydroxybufalin, the major defensive steroids of the firefly *Photinus pyralis* (Coleoptera, Lampyridae). Journal of the American Chemical Society 101: 3055–3060.

Menzel, R., Ventura, D.F., Hertel, H., DeSouza, J.M. and Greggers, U. 1986. Spectral sensitivity of photoreceptors in insect compound eyes: comparison of species and methods. Journal of Comparative Physiology A 158: 165–177.

Meyer, J. and Elsner, N. 1995. How respiration affects auditory sensitivity in the grasshopper *Chorthippus biguttulus* (L). Journal of Comparative Physiology A 176: 563–573.

Meyer, J. and Elsner, N. 1996. How well are frequency sensitivities of grasshopper ears tuned to species specific song spectra? Journal of Experimental Biology 199: 1631–1642.

Meyer, J. and Hedwig, B. 1995. The influence of tracheal pressure changes on the responses of the tympanal membrane and auditory receptors in the locust *Locusta migratoria* L. Journal of Experimental Biology 198: 1327–1339.

Meyer-Rochow, V.B. and Penrose, J.D. 1976. Sound production by western rock lobster *Panulirus longipes* (Milne Edwards). Journal of Experimental Marine Biology and Ecology 23: 191–209.

Meyer-Rochow, V.B., Penrose, J.D., Oldfield, B.P. and Bailey, W.J. 1982. Phonoresponses in the rock lobster *Panulirus longipes* (Milne Edwards). Behavioral and Neural Biology 34: 331–336.

Michel, K. 1974. Tympanic organ of *Gryllus bimaculatus* deGeer (Saltatoria: Gryllidae). Zeitschrift für Morphologie der Tiere 77: 285–315.

Michelsen, A. 1971. The physiology of the locust ear. 1. frequency sensitivity of single cells in the isolated ear. Zeitschrift für Vergleichende Physiologie 71: 49–62.

Michelsen, A. and Elsner, N. 1999. Sound emission and the acoustic far field of a singing acridid grasshopper (*Omocestus viridulus* L.). Journal of Experimental Biology 202: 1571–1577.

Michelsen, A. and Fonseca, P. 2000. Spherical sound radiation patterns of singing grass cicadas, *Tympanistalna gastrica*. Journal of Comparative Physiology A 186: 163–168.

Michelsen, A. and Larsen, O.N. 1985. Hearing and sound. In Comprehensive Insect Physiology, Biochemistry and Pharmacology, Vol. 6 (eds G.A. Kerkut and L.I. Gilbert), pp. 496–556. Pergamon Press, Oxford.

Michelsen, A. and Löhe, G. 1995. Tuned directionality in cricket ears. Nature 375: 639.

Michelsen, A., Fink, F., Gogala, M. and Traue, D. 1982. Plants as transmission channels for insect vibrational songs. Behavioral Ecology and Sociobiology 11: 269–281.

Michelsen, A., Kirchner, W.H. and Lindauer, M. 1986. Sound and vibrational signals in the dance language of the honeybee, *Apis mellifera*. Behavioral Ecology and Sociobiology 18: 207–212.

Michelsen, A., Towne, W.F., Kirchner, W.H. and Kryger, P. 1987. The acoustic near field of a dancing honeybee. Journal of Comparative Physiology A 161: 633–643.

Michelsen, A., Andersen, B.B., Kirchner, W.H. and Lindauer, M. 1989. Honeybees can be recruited by a mechanical model of a dancing bee. Naturwissenschaften 76: 277–280.

Michelsen, A., Heller, K.G., Stumpner, A. and Rohrseitz, K. 1994a. A new biophysical method to determine the gain of the acoustic trachea in bush crickets. Journal of Comparative Physiology A 175: 145–151.

Michelsen, A., Popov, A.V. and Lewis, B. 1994b. Physics of directional hearing in the cricket *Gryllus bimaculatus*. Journal of Comparative Physiology A 175: 153–164.

Miklas, N., Stritih, N., Čokl, A., Virant-Doberlet, M. and Renou, M. 2001. The influence of substrate on male responsiveness to the female calling song in *Nezara viridula*. Journal of Insect Behavior 14: 313–332.

Millar, J.G. 2000. Polyene hydrocarbons and epoxides: a second major class of lepidopteran sex attractant pheromones. Annual Review of Entomology 45: 575–604.

Miller, J.R. and Roelofs, W.L. 1978. Sustained-flight tunnel for measuring insect responses to wind-borne sex pheromones. Journal of Chemical Ecology 4: 187–198.

Miller, J.R. and Strickler, K.L. 1984. Finding and accepting host plants. In Chemical Ecology of Insects (eds W.J. Bell and R.T. Cardé), pp. 127–157. Chapman and Hall, London.

Miller, J.R., Baker, T.C., Cardé, R.T. and Roelofs, W.L. 1976. Reinvestigation of oak leaf roller sex-pheromone components and hypothesis that they vary with diet. Science 192: 140–143.

Miller, L.A. 1977. Directional hearing in the locust *Schistocerca gregaria* Forskl (Acrididae, Orthoptera). Journal of Comparative Physiology A 119: 85–98.

Minckley, R.L. and Greenfield, M.D. 1995. Psychoacoustics of female phonotaxis and the evolution of male signal interactions in Orthoptera. Ethology, Ecology and Evolution 7: 235–243.

Minckley, R.L., Buchmann, S.L. and Wcislo, W.T. 1991. Bioassay evidence for a sex attractant pheromone in the large carpenter bee, *Xylocopa varipuncta* (Anthophoridae, Hymenoptera). Journal of Zoology 224: 285–291.

Minet, J. 1983. Étude morphologique et phylogénétique des organes tympaniques des Pyraloidea. 1. Généralités et homologies. (Lep. Glossata). Annales de la Société Entomologique de France 19: 175–207.

Minet, J. 1985. Étude morphologique et phylogénétique des organes tympaniques des Pyraloidea. 2. Pyralidae; Crambidae, première partie. (Lepidoptera. Glossata). Annales de la Société Entomologique de France 21: 69–86.

Miyashita, T. and Hayashi, H. 1996. Volatile chemical cue elicits mating behavior of cohabiting males of *Nephila clavata* (Araneae: Tetragnathidae). Journal of Arachnology 24: 9–15.

Moglich, M., Maschwitz, U. and Hölldobler, B. 1974. Tandem calling: new kind of signal in ant communication. Science 186: 1046–1047.

Mohl, B. and Miller, L.A. 1976. Ultrasonic clicks produced by the peacock butterfly: a possible bat-repellent mechanism. Journal of Experimental Biology 64: 639–644.

Moiseff, A. and Copeland, J. 1995. Mechanisms of synchrony in the North American firefly *Photinus carolinus* (Coleoptera: Lampyridae). Journal of Insect Behavior 8: 395–407.

Moiseff, A. and Copeland, J. 2000. A new type of synchronized flashing in a North American firefly. Journal of Insect Behavior 13: 597–612.

Moiseff, A., Pollack, G.S. and Hoy, R.R. 1978. Steering responses of flying crickets to sound and ultrasound: mate attraction and predator avoidance. Proceedings of the National Academy of Sciences, U.S.A. 75: 4052–4056.

Møller, A.P. and Alatalo, R.V. 1999. Good-genes effects in sexual selection. Proceedings of the Royal Society of London B 266: 85–91.

Møller, A.P. and Swaddle, J.P. 1997. Asymmetry, Developmental Stability, and Evolution. Oxford University Press, Oxford.

Monge-Najera, J., Hernandez, F., Gonzalez, M.I., Soley, J., Araya, J. and Zolla, S. 1998. Spatial distribution, territoriality and sound production by tropical butterflies (*Hamadryas*, Lepidoptera: Nymphalidae): implications for the "industrial melanism" debate. Revista de Biologia Tropical 46: 297–330.

Moore, A.J. 1994. Genetic evidence for the good genes process of sexual selection. Behavioral Ecology and Sociobiology 35: 235–241.

Moore, A.J. 1997. The evolution of social signal: morphological, functional, and genetic integration of the sex pheromone in *Nauphoeta cinerea*. Evolution 51: 1920–1928.

Moore, A.J. and Moore, P.J. 1999. Balancing sexual selection through opposing mate choice and male competition. Proceedings of the Royal Society of London B 266: 711–716.

Moore, P.A. and Atema, J. 1991. Spatial information in the 3-dimensional fine-structure of an aquatic odor plume. Biological Bulletin 181: 408–418.

Moore, P.J., Reagan-Wallin, N.L., Haynes, K.F. and Moore, A.J. 1997. Odour conveys status on cockroaches. Nature 389: 25.

Morin, J.G. 1986. Firefleas of the sea: luminescent signaling in marine ostracode crustaceans. Florida Entomologist 69: 105–121.

Morris, G.K. 1980. Calling display and mating behavior of *Copiophora rhinoceros* (Orthoptera: Tettigoniidae). Animal Behaviour 28: 42–51.

Morris, G.K. and Gwynne, D.T. 1978. Geographical distribution and biological observations of *Cyphoderris* (Orthoptera: Haglidae) with a description of a new species. Psyche 85: 147–167.

Morris, G.K. and Mason, A.C. 1995. Covert stridulation: novel sound generation by a South American katydid. Naturwissenschaften 82: 96–98.

Morris, G.K., Mason, A.C., Wall, P. and Belwood, J.J. 1994. High ultrasonic and tremulation signals in neotropical katydids (Orthoptera: Tettigoniidae). Journal of Zoology 233: 129–163.

Morris, R.B. 1975. Iridescence from diffraction structures in wing scales of *Callophrys rubi*, green hairstreak. Journal of Entomology A 49: 149–154.

Muller, K.L. 1998. The role of conspecifics in habitat settlement in a territorial grasshopper. Animal Behaviour 56: 479–485.

Murlis, J. and Jones, C.D. 1981. Fine scale structure of odor plumes in relation to insect orientation to distant pheromone and other attractant sources. Physiological Entomology 6: 71–86.

Murlis, J., Elkinton, J.S. and Cardé, R.T. 1992. Odor plumes and how insects use them. Annual Review of Entomology 37: 505–532.

Murphy, C.G. 1994a. Chorus tenure of male barking treefrogs, *Hyla gratiosa*. Animal Behaviour 48: 763–777.

Murphy, C.G. 1994b. Determinants of chorus tenure in barking treefrogs (*Hyla gratiosa*). Behavioral Ecology and Sociobiology 34: 285–294.

Murphy, C.G. 1999. Nightly timing of chorusing by male barking treefrogs (*Hyla gratiosa*): the influence of female arrival and energy. Copeia 333–347.

Mustaparta, H. 1997. Olfactory coding mechanisms for pheromone and interspecific signal information in related moth species. In Insect Pheromone Research: New Directions (eds R.T. Cardé and A.K. Minks), pp. 144–163. Chapman and Hall, New York.

Naguib, M., Klump, G.M., Hillmann, E., Griessmann, B. and Teige, T. 2000. Assessment of auditory distance in a territorial songbird: Accurate feat or rule of thumb? Animal Behaviour 59: 715–721.

Narins, P.M. 1992. Reduction of tympanic membrane displacement during vocalization of the arboreal frog, *Eleutherodactylus coqui*. Journal of the Acoustical Society of America 91: 3551–3557.

Neda, Z., Ravasz, E., Brechet, Y., Vicsek, T. and Barabasi, A.L. 2000. The sound of many hands clapping: tumultuous applause can transform itself into waves of synchronized clapping. Nature 403: 849–850.

Needham, A.E. 1974. The Significance of Zoochromes. Springer-Verlag, Berlin.

Nelson, C.M. and Nolen, T.G. 1997. Courtship song, male agonistic encounters, and female mate choice in the house cricket, *Acheta domesticus* (Orthoptera: Gryllidae). Journal of Insect Behavior 10: 557–570.

Nelson, M.C. and Fraser, J. 1980. Sound production in the cockroach, *Gromphadorina portentosa*: evidence for communication by hissing. Behavioral Ecology and Sociobiology 6: 305–314.

Niassy, A., Torto, B., Njagi, P.G.N., Hassanali, A., Obeng-Ofori, D. and Ayertey, J.N. 1999. Intra- and interspecific aggregation responses of *Locusta migratoria migratorioides* and *Schistocerca gregaria* and a comparison of their pheromone emissions. Journal of Chemical Ecology 25: 1029–1042.

Nickle, D.A. and Carlysle, T.C. 1975. Morphology and function of female sound-producing structures in ensiferan Orthoptera with special emphasis on the Phaneropterinae. International Journal of Insect Morphology and Embryology 4: 159–168.

Nieh, J.C. 1998a. The food recruitment dance of the stingless bee, *Melipona panamica*. Behavioral Ecology and Sociobiology 43: 133–145.

Nieh, J.C. 1998b. The role of a scent beacon in the communication of food location by the stingless bee, *Melipona panamica*. Behavioral Ecology and Sociobiology 43: 47–58.

Nieh, J.C. 1999. Stingless bee communication. American Scientist 87: 428–435.

Nieh, J.C. and Roubik, D.W. 1995. A stingless bee (*Melipona panamica*) indicates food location without using a scent trail. Behavioral Ecology and Sociobiology 37: 63–70.

Nieh, J.C. and Roubik, D.W. 1998. Potential mechanisms for the communication of height and distance by a stingless bee, *Melipona panamica*. Behavioral Ecology and Sociobiology 43: 387–399.

Nieh, J.C. and Tautz, J. 2000. Behaviour-locked signal analysis reveals weak 200–300 Hz comb vibrations during the honeybee waggle dance. Journal of Experimental Biology 203: 1573–1579.

Nieh, J.C., Tautz, J., Spaethe, J. and Bartareau, T. 1999. The communication of food location by a primitive stingless bee, *Trigona carbonaria*. Zoology-Analysis of Complex Systems 102: 238–246.

Nielsen, E.S. 1989. Phylogeny of major lepidopteran groups. Excerpta Medica International Congress Series 824: 281–294.

Nijhout, H.F. 1981. The color patterns of butterflies and moths. Scientific American 245(5): 104–115.

Nijhout, H.F. 1991. The Development and Evolution of Butterfly Wing Patterns. Smithsonian Institution Press, Washington, D.C.

Nilsson, D.E. 1989. Optics and evolution of the compound eye. In Facets of Vision (eds D.G. Stavenga and R.C. Hardie), pp. 30–73. Springer-Verlag, Berlin.

Njagi, P.G.N., Torto, B., ObengOfori, D. and Hassanali, A. 1996. Phase-independent responses to phase-specific aggregation pheromone in adult desert locusts, *Schistocerca gregaria* (Orthoptera: Acrididae). Physiological Entomology 21: 131–137.

Nocke, H. 1971. Biophysik der Schallerzeugung durch die Vorderflügel der Grillen. Zeitschrift für Vergleichende Physiologie 74: 272–314.

Noldus, L.P.J.J. 1988. Response of the egg parasitoid Trichogramma *pretiosum* to the sex pheromone of its host *Heliothis zea*. Entomologia Experimentalis et Applicata 48: 293–300.

Noldus, L.P.J.J., Potting, R.P.J. and Barendregt, H.E. 1991a. Moth sex pheromone adsorption to leaf surface: bridge in time for chemical spies. Physiological Entomology 16: 329–344.

Noldus, L.P.J.J., Vanlenteren, J.C. and Lewis, W.J. 1991b. How *Trichogramma* parasitoids use moth sex pheromones as kairomones: orientation behavior in a wind tunnel. Physiological Entomology 16: 313–327.

Nolen, T.G. and Hoy, R.R. 1986. Phonotaxis in flying crickets.1. attraction to the calling song and avoidance of bat-like ultrasound are discrete behaviors. Journal of Comparative Physiology A 159: 423–439.

Noor, M.A.F. 1999. Reinforcement and other consequences of sympatry. Heredity 83: 503–508.

Novacek, M.J. 1985. Evidence for echolocation in the oldest known bats. Nature 315: 140–141.

Nufio, C.R. and Papaj, D.R. 2001. Host marking behavior in phytophagous insects and parasitoids. Entomologia Experimentalis et Applicata 99: 273–293.

O'Carroll, D. 1993. Feature-detecting neurons in dragonflies. Nature 362: 541–543.

Ocker, W.G. and Hedwig, B. 1996. Interneurones involved in stridulatory pattern generation in the grasshopper *Chorthippus mollis* (Charp). Journal of Experimental Biology 199: 653–662.

Olberg, R.M. 1983. Pheromone-triggered flip-flopping interneurons in the ventral nerve cord of the silkworm moth, *Bombyx mori*. Journal of Comparative Physiology A 152: 297–307.

Oldfield, B.P. 1980. Accuracy of orientation in female crickets, *Teleogryllus oceanicus* (Gryllidae): dependence on the song spectrum. Journal of Comparative Physiology A 141: 93–99.

Oldfield, B.P. 1982. Tonotopic organization of auditory receptors in Tettigoniidae (Orthoptera: Ensifera). Journal of Comparative Physiology A 147: 461–469.

Oldfield, B.P., Kleindienst, H.U. and Huber, F. 1986. Physiology and tonotopic organization of auditory receptors in the cricket *Gryllus bimaculatus* deGeer. Journal of Comparative Physiology A 159: 457–464.

Oliveira, R.F. and Custodio, M.R. 1998. Claw size, waving display and female choice in the European fiddler crab, *Uca tangeri*. Ethology, Ecology and Evolution 10: 241–251.

Olson, H.F. 1967. Music Engineering, 2nd edn. Dover, New York.

Olvido, A.E. and Mousseau, T.A. 1995. Effect of rearing environment on calling-song plasticity in the striped ground cricket. Evolution 49: 1271–1277.

Osorio, D., Marshall, N.J. and Cronin, T.W. 1997. Stomatopod photoreceptor spectral tuning as an adaptation for colour constancy in water. Vision Research 37: 3299–3309.

Otte, D. 1970. A comparative study of communicative behavior in grasshoppers. Miscellaneous Publications of the Museum of Zoology, University of Michigan, Ann Arbor. No. 141: 1–168.

Otte, D. 1992. Evolution of cricket songs. Journal of Orthoptera Research 1: 25–49.

Otte, D. and Alexander, R.D. 1983. The Australian crickets (Orthoptera: Gryllidae). Monographs of the Academy of Natural Sciences of Philadelphia 22: 1–477.

Otte, D. and Smiley, J. 1977. Synchrony in Texas fireflies with a consideration of male interaction models. Biology of Behaviour 2: 143–158.

Parri, S., Alatalo, R.V., Kotiaho, J. and Mappes, J. 1997. Female choice for male drumming in the wolf spider *Hygrolycosa rubrofasciata*. Animal Behaviour 53: 305–312.

Partan, S. and Marler, P. 1999. Behavior: communication goes multimodal. Science 283: 1272–1273.

Partridge, L. 1980. Mate choice increases a component of offspring fitness in fruit flies. Nature 283: 290–291.

Patek, S.N. 2001. Spiny lobsters stick and slip to make sound. Nature 411: 153–154.

Paul, R.C. and Walker, T.J. 1979. Arboreal singing in a burrowing cricket, *Anurogryllus arboreus*. Journal of Comparative Physiology A 132: 217–223.

Pener, M.P. and Yerushalmi, Y. 1998. The physiology of locust phase polymorphism: an update. Journal of Insect Physiology 44: 365–377.

Percy-Cunningham, J.E. and MacDonald, J.A. 1987. Biology and ultrastructure of sex pheromone-producing glands. In Pheromone Biochemistry (eds G.D. Prestwich and G.J. Blomquist), pp. 27–75. Academic Press, Orlando, FL.

Perdeck, A. 1957. The isolating value of specific song patterns in two sibling species of grasshoppers (*Chorthippus brunneus* Thunb. and *C. biguttulus* L.). Behaviour 12: 1–75.

Petrie, M. 1994. Improved growth and survival of offspring of peacocks with more elaborate trains. Nature 371: 598–599.

Pfau, H.K. and Koch, U.T. 1994. The functional morphology of singing in the cricket. Journal of Experimental Biology 195: 147–167.

Phelan, P.L. 1992. Evolution of sex pheromones and the role of asymmetric tracking. In Insect Chemical Ecology (eds B.D. Roitberg and M.B. Isman), pp. 265–314. Chapman and Hall, New York.

Phelan, P.L. and Baker, T.C. 1986. Male size-related courtship success and intersexual selection in the tobacco moth, *Ephestia elutella*. Experientia 42: 1291–1293.

Phelan, P.L. and Baker, T.C. 1987. Evolution of male pheromones in moths: reproductive isolation through sexual selection. Science 235: 205–207.

Phelan, P.L. and Baker, T.C. 1990. Comparative study of courtship in 12 phycitine moths (Lepidoptera: Pyralidae). Journal of Insect Behavior 3: 303–326.

Pickett, J.A., Wadhams, L.J. and Woodcock, C.M. 1997. First steps in the use of aphid sex pheromones. In Insect Pheromone Research: New Directions (eds R.T. Cardé and A.K. Minks), pp. 493–444. Chapman and Hall, New York.

Pickett, J.A., Wadhams, L.J., Woodcock, C.M. and Hardie, J. 1992. The chemical ecology of aphids. Annual Review of Entomology 37: 67–90.

Pires, A. and Hoy, R.R. 1992a. Temperature coupling in cricket acoustic communication. 1. field and laboratory studies of temperature effects on calling song production and recognition in *Gryllus firmus*. Journal of Comparative Physiology A 171: 69–78.

Pires, A. and Hoy, R.R. 1992b. Temperature coupling in cricket acoustic communication. 2. localization of temperature effects on song production and recognition networks in *Gryllus firmus*. Journal of Comparative Physiology A 171: 79–92.

Plettner, E., Slessor, K.N., Winston, M.L. and Oliver, J.E. 1996. Caste-selective pheromone biosynthesis in honeybees. Science 271: 1851–1853.

Plettner, E., Otis, G.W., Wimalaratne, P.D.C., Winston, M.L., Slessor, K.N., Pankiw, T. and Punchihewa, P.W.K. 1997. Species- and caste-determined mandibular gland signals in honeybees (*Apis*). Journal of Chemical Ecology 23: 363–377.

Plettner, E., Slessor, K.N. and Winston, M.L. 1998. Biosynthesis of mandibular acids in honey bees (*Apis mellifera*): de novo synthesis, route of fatty acid hydroxylation and caste selective beta-oxidation. Insect Biochemistry and Molecular Biology 28: 31–42.

Podos, J. 1996. Motor constraints on vocal development in a songbird. Animal Behaviour 51: 1061–1070.

Pollack, G.S. 1982. Sexual differences in cricket calling song recognition. Journal of Comparative Physiology A 146: 217–221.

Pollack, G.S. 1988. Selective attention in an insect auditory neuron. Journal of Neuroscience 8: 2635–2639.

Pollack, G.S. 1998. Neural processing of acoustic signals. In Handbook of Auditory Research, Vol. 10, Comparative Hearing: Insects (eds R.R. Hoy, A.N. Popper and R.R. Fay), pp. 139–196. Springer-Verlag, New York.

Pollack, G.S. and Hoy, R. 1981. Phonotaxis to individual rhythmic components of a complex cricket calling song. Journal of Comparative Physiology A 144: 367–373.

Pollack, G.S. and Hoy, R.R. 1979. Temporal pattern as a cue for species-specific calling song recognition in crickets. Science 204: 429–432.

Pollack, G.S., Huber, F. and Weber, T. 1984. Frequency and temporal pattern dependent phonotaxis of crickets (*Teleogryllus oceanicus*) during tethered flight and compensated walking. Journal of Comparative Physiology A 154: 13–26.

Pomiankowski, A. 1987. Sexual selection: the handicap principle does work - sometimes. Proceedings of the Royal Society of London B 231: 123–145.

Pomiankowski, A. 1988. The evolution of female mate preferences for male genetic quality. Oxford Surveys in Evolutionary Biology 5: 136–184.

Pomiankowski, A. and Møller, A.P. 1995. A resolution of the lek paradox. Proceedings of the Royal Society of London B 260: 21–29.

Popov, A. and Shuvalov, V. 1977. The phonotactic behavior of crickets. Journal of Comparative Physiology A 119: 111–126.

Popper, A.N, Platt, C. and Edds, P.L. 1992. Evolution of the vertebrate inner ear: an overview of ideas. In The Evolutionary Biology of Hearing (eds D.B. Webster, R.R. Fay and A.N. Popper), pp. 49–57. Springer-Verlag, New York.

Popper, A.N., Salmon, M. and Horch, K.W. 2001. Acoustic detection and communication by decapod crustaceans. Journal of Comparative Physiology A 187: 83–89.

Prager, J. and Larsen, O.N. 1981. Asymmetrical hearing in the water bug *Corixa punctata* observed with laser vibrometry. Naturwissenschaften 68: 579–580.

Prager, J. and Streng, R. 1982. The resonance properties of the physical gill of *Corixa punctata* and their significance in sound reception. Journal of Comparative Physiology A 148: 323–335.

Prager, J. and Theiss, J. 1982. The effect of tympanal position on the sound-sensitivity of the metathoracic scolopale organ of *Corixa*. Journal of Insect Physiology 28: 447–452.

Prestwich, K.N. and Walker, T.J. 1981. Energetics of singing in crickets: effect of temperature in 3 trilling species (Orthoptera: Gryllidae). Journal of Comparative Physiology B 143: 199–212.

Prestwich, K.N., Lenihan, K.M. and Martin, D.M. 2000. The control of carrier frequency in cricket calls: a refutation. Journal of Experimental Biology 203: 585–596.

Proctor, H.C. 1991. Courtship in the water mite *Neumania papillator*: males capitalize on female adaptations for predation. Animal Behaviour 42: 589–598.

Proctor, H.C. 1992. Sensory exploitation and the evolution of male mating behavior: a cladistic test using water mites (Acari: Parasitengona). Animal Behaviour 44: 745–752.

Prokopy, R.J. 1972. Evidence for a pheromone deterring repeated oviposition in apple maggot flies. Environmental Entomology 1: 326–332.

Prozesky-Schulze, L., Prozesky, O.P.M., Andersen, F. and Vander Merwe, G.J.J. 1975. Use of a self-made sound baffle by a tree cricket. Nature 255: 142–143.

Prum, R.O., Torres, T.H., Williamson, S. and Dyck, J. 1998. Coherent light scattering by blue feather barbs. Nature 396: 28–29.

Prum, R.O., Torres, R., Kovach, C., Williamson, S. and Goodman, S.M. 1999. Coherent light scattering by nanostructured collagen arrays in the caruncles of the Malagasy asities (Eurylaimidae: Aves). Journal of Experimental Biology 202: 3507–3522.

Pugh, A.R.G. and Ritchie, M.G. 1996. Polygenic control of a mating signal in *Drosophila*. Heredity 77: 378–382.

Radl, R.C. and Linsenmair, K.E. 1991. Maternal behavior and nest recognition in the subsocial earwig *Labidura riparia* Pallas (Dermaptera: Labiduridae). Ethology 89: 287–296.

Rafaeli, A., Soroker, V., Hirsch, J., Kamensky, B. and Raina, A.K. 1993. Influence of photoperiod and age on the competence of pheromone glands and on the distribution of immunoreactive PBAN in *Helicoverpa* spp. Archives of Insect Biochemistry and Physiology 22: 169–180.

Rafaeli, A., Gileadi, C., Yongliang, F. and Cao, M.X. 1997. Physiological mechanisms of pheromonostatic responses: effects of adrenergic agonists and antagonists on moth (*Helicoverpa armigera*) pheromone biosynthesis. Journal of Insect Physiology 43: 261–269.

Raffa, K.F. and Berryman, A.A. 1983. The role of host plant resistance in the colonization behavior and ecology of bark beetles (Coleoptera: Scolytidae). Ecological Monographs 53: 27–49.

Raffa, K.F. and Berryman, A.A. 1987. Interacting selective pressures in conifer-bark beetle systems: a basis for reciprocal adaptations. American Naturalist 129: 234–262.

Raina, A.K. 1993. Neuroendocrine control of sex-pheromone biosynthesis in lepidoptera. Annual Review of Entomology 38: 329–349.

Raina, A.K., Kingan, T.G. and Mattoo, A.K. 1992. Chemical signals from host plants and sexual behavior in a moth. Science 255: 592–594.

Raina, A.K., Kingan, T.G. and Giebultowicz, J.M. 1994. Mating induced loss of sex pheromone and sexual receptivity in insects with emphasis on *Helicoverpa zea* and *Lymantria dispar*. Archives of Insect Biochemistry and Physiology 25: 317–327.

Raina, A.K., Jackson, D.M. and Severson, R.F. 1997. Increased pheromone production in wild tobacco budworm (Lepidoptera: Noctuidae) exposed to host plants and host chemicals. Environmental Entomology 26: 101–105.

Rau, P. 1932. Rhythmic periodicity and synchronous flashing in the firefly, *Photinus pyralis*, with notes on *Photinus pennsylvanicus*. Ecology 13: 7–11.

Rau, P. and Rau, N.L. 1929. The sex attraction and rhythmic periodicity in the giant saturniid moths. Transactions of the Academy of Science of St. Louis 26: 83–271.

Real, L. 1990. Search theory and mate choice. 1. models of single sex discrimination. American Naturalist 136: 376–405.

Real, L.A. 1991. Search theory and mate choice. 2. mutual interaction, assortative mating, and equilibrium variation in male and female fitness. American Naturalist 138: 901–917.

Réaumur, R.-A.F. de 1926. The Natural History of Ants (English translation by W.M. Wheeler). A.A. Knopf, New York.

Reeve, H.K. 1997. Evolutionarily stable communication between kin: a general model. Proceedings of the Royal Society of London B 264: 1037–1040.

Regen, J. 1913. Über die Anlockung des Weibchens von *Gryllus campestris* L. durch telephonisch übertragene der Stridulationslaute des Männchens. Pflügers Archiv für die Gesamte Physiologie des Menschen und der Tiere 155: 193–200.

Reinhard, J. and Kaib, M. 2001. Trail communication during foraging and recruitment in the subterranean termite *Reticulitermes santonensis* DeFeytaud (Isoptera, Rhinotermitidae). Journal of Insect Behavior 14: 157–171.

Reinhold, K. 1998. Sex linkage among genes controlling sexually selected traits. Behavioral Ecology and Sociobiology 44: 1–7.

Reinhold, K. 1999. Evolutionary genetics of sex-limited traits under fluctuating selection. Journal of Evolutionary Biology 12: 897–902.

Reinhold, K., Greenfield, M.D., Jang, Y.W. and Broce, A. 1998. Energetic cost of sexual attractiveness: ultrasonic advertisement in wax moths. Animal Behaviour 55: 905–913.

Reynolds, J.D. and Gross, M.R. 1992. Female mate preferences enhance offspring growth and reproduction in a fish, *Poecilia reticulata*. Proceedings of the Royal Society of London B 250: 57–62.

Rheinlaender, J. and Blätgen, G. 1982. The precision of auditory lateralization in the cricket, *Gryllus bimaculatus*. Physiological Entomology 7: 209–218.

Riddiford, L.M. and Williams, C.M. 1967. Chemical signaling between polyphemus moths and between moths and host plant. Science 156: 541.

Riede, K. 1987. A comparative study of mating behavior in some neotropical grasshoppers (Acridoidea). Ethology 76: 265–296.

Riede, K., Kamper, G. and Höffler, I. 1990. Tympana, auditory thresholds, and projection areas of tympanal nerves in singing and silent grasshoppers (Insecta: Acridoidea). Zoomorphology 109: 223–230.

Ritchie, M.G. 1996. The shape of female mating preferences. Proceedings of the National Academy of Sciences, U.S.A. 93: 14628–14631.

Ritchie, M.G. 2000. The inheritance of female preference functions in a mate recognition system. Proceedings of the Royal Society of London B 267: 327–332.

Ritchie, M.G. and Kyriacou, C.P. 1994. Genetic variability of courtship song in a population of *Drosophila melanogaster*. Animal Behaviour 48: 425–434.

Ritchie, M.G. and Philipps, S.D.F. 1998. The genetics of sexual isolation. In Endless Forms: Species and Speciation (eds D.J. Howard and S.H. Berlocher), pp. 291–308. Oxford University Press, Oxford.

Ritchie, M.G., Yate, V.H. and Kyriacou, C.P. 1994. Genetic variability of the interpulse interval of courtship song among some European populations of *Drosophila melanogaster*. Heredity 72: 459–464.

Ritchie, M.G., Couzin, I.D. and Snedden, W.A. 1995. What's in a song? Female bushcrickets discriminate against the song of older males. Proceedings of the Royal Society of London B 262: 21–27.

Ritchie, M.G., Townhill, R.M. and Hoikkala, A. 1998. Female preference for fly song: playback experiments confirm the targets of sexual selection. Animal Behaviour 56: 713–717.

Ritchie, M.G., Saarikettu, M., Livingstone, S. and Hoikkala, A. 2001. Characterization of female preference functions for *Drosophila montana* courtship song and a test for the temperature coupling hypothesis. Evolution 55: 721–727.

Ritzmann, R. 1973. Snapping behavior of the shrimp *Alpheus californiensis*. Science 181: 459–460.

Ritzmann, R.E. 1974. Mechanisms for snapping behavior of 2 alpheid shrimp, *Alpheus californiensis* and *Alpheus heterochelis*. Journal of Comparative Physiology A 95: 217–236.

Robert, D. and Hoy, R.R. 1998. The evolutionary innovation of tympanal hearing in Diptera. In Handbook of Auditory Research, Vol. 10, Comparative Hearing: Insects (eds R.R. Hoy, A.N. Popper and R.R. Fay), pp. 197–227. Springer-Verlag, New York.

Robert, D., Read, M.P. and Hoy, R.R. 1994. The tympanal hearing organ of the parasitoid fly *Ormia ochracea* (Diptera: Tachinidae: Ormiini). Cell and Tissue Research 275: 63–78.

Robert, D., Miles, R.N. and Hoy, R.R. 1996. Directional hearing by mechanical coupling in the parasitoid fly *Ormia ochracea*. Journal of Comparative Physiology A 179: 29–44.

Robert, D., Miles, R.N. and Hoy, R.R. 1998. Tympanal mechanics in the parasitoid fly *Ormia ochracea*: intertympanal coupling during mechanical vibration. Journal of Comparative Physiology A 183: 443–452.

Robert, D., Miles, R.N. and Hoy, R.R. 1999. Tympanal hearing in the sarcophagid parasitoid fly *Emblemasoma* sp.: the biomechanics of directional hearing. Journal of Experimental Biology 202: 1865–1876.

Roberts, O.F.T. 1923. The theoretical scattering of smoke in a turbulent atmosphere. Proceedings of the Royal Society of London A 104: 640–654.

Robinson, D.J. 1980. Acoustic communication between the sexes of the bush cricket, *Leptophyes Punctatissima*. Physiological Entomology 5: 183–189.

Robinson, D. 1990. Acoustic communication between the sexes in bushcrickets. In The Tettigoniidae: Biology, Systematics and Evolution (eds W.J. Bailey and D.C.F. Rentz), pp. 112–129. Springer-Verlag, Berlin.

Robinson, D., Rheinlaender, J. and Hartley, J. C. 1986. Temporal parameters of male–female sound communication in *Leptophyes punctatissima*. Physiological Entomology 11: 317 323.

Robinson, G.E. 1996. Chemical communication in honeybees. Science 271: 1824–1825.

Robinson, M.H. and Robinson, B. 1978. The evolution of courtship systems in tropical araneid spiders. Symposium of the Zoological Society of London 42: 17–29.

Robinson, M.H. and Robinson, B. 1979. By dawn's early light: matutinal mating and sex attractants in a neotropical mantid. Science 205: 825–827.

Roces, F. and Manrique, G. 1996. Different stridulatory vibrations during sexual behaviour and disturbance in the blood-sucking bug *Triatoma infestans* (Hemiptera: Reduviidae). Journal of Insect Physiology 42: 231–238.

Roeder, K.D. 1967a. Turning tendency of moths exposed to ultrasound while in stationary flight. Journal of Insect Physiology 13: 873–888.

Roeder, K.D. 1967b. Auditory system of noctuid moths. Science 154: 1515–1518.

Roelofs, W.L. 1978. Threshold hypothesis for pheromone perception. Journal of Chemical Ecology 4: 685–699.

Roelofs, W.L. 1984. Electroantennogram assays: rapid and convenient screening procedures for pheromones. In Techniques in Pheromone Research (eds H.E. Hummel and T.A. Miller), pp. 131–159. Springer-Verlag, New York.

Roelofs, W. and Bjostad, L. 1984. Biosynthesis of lepidopteran pheromones. Bioorganic Chemistry 12: 279–298.

Roelofs, W. and Comeau, A. 1968. Sex pheromone perception. Nature 220: 600–601.

Roessingh, P., Bouaichi, A. and Simpson, S.J. 1998. Effects of sensory stimuli on the behavioural phase state of the desert locust, *Schistocerca gregaria*. Journal of Insect Physiology 44: 883–893.

Roff, D.A. 1997. Evolutionary Quantitative Genetics. Chapman and Hall, New York.

Rohrseitz, K. and Tautz, J. 1999. Honey bee dance communication: waggle run direction coded in antennal contacts? Journal of Comparative Physiology A 184: 463–470.

Roitberg, B.D. and Mangel, M. 1988. On the evolutionary ecology of marking pheromones. Evolutionary Ecology 2: 289–315.

Röller, H., Biemann, K., Bjerke, J.S., Norgard, D.W. and McShan, W.H. 1968. Sex pheromones of pyralid moths. I. Isolation and identification of sex-attractant of

Galleria mellonella L. (greater waxmoth). Acta Entomologica Bohemoslovaca 65: 208–211.

Römer, H. 1993. Environmental and biological constraints for the evolution of long-range signaling and hearing in acoustic insects. Philosophical Transactions of the Royal Society of London B 340: 179–185.

Römer, H. 1998. The sensory ecology of acoustic communication in insects. In in Handbook of Auditory Research, Vol. 10, Comparative Hearing: Insects (eds R.R. Hoy, A.N. Popper and R.R. Fay), pp. 63–96. Springer-Verlag, New York.

Römer, H. and Bailey, W.J. 1986. Insect hearing in the field. 2. Male spacing behavior and correlated acoustic cues in the bush cricket *Mygalopsis marki*. Journal of Comparative Physiology A 159: 627–638.

Römer, H. and Krusch, M. 2000. A gain-control mechanism for processing of chorus sounds in the afferent auditory pathway of the bushcricket *Tettigonia viridissima* (Orthoptera: Tettigoniidae). Journal of Comparative Physiology A 186: 181–191.

Römer, H. and Lewald, J. 1992. High frequency sound transmission in natural habitats: implications for the evolution of insect acoustic communication. Behavioral Ecology and Sociobiology 29: 437–444.

Römer, H., Bailey, W.J. and Dadour, I.R. 1989. Insect hearing in the field. 3. Masking by noise. Journal of Comparative Physiology A 164: 609–620.

Römer, H., Hedwig, B. and Ott, S. 1997. Proximate mechanism of female preference for the leader male in synchronizing bushcrickets (*Mecopoda elongata*). In Proceedings of the 25th Göttingen Neurobiology Conference (eds N. Elsner and H. Waessle), p. 322. Thieme, Stuttgart, Germany.

Ronacher, B. and Duft, U. 1996. An image-matching mechanism describes a generalization task in honeybees. Journal of Comparative Physiology A 178: 803–812.

Rossel, S. 1979. Regional differences in photoreceptor performance in the eye of the praying mantis. Journal of Comparative Physiology A 131: 95–112.

Rossel, S. 1989. Polarization sensitivity in compound eyes. In Facets of Vision (eds D.G. Stavenga and R.C. Hardie), pp. 298–316. Springer-Verlag, Berlin.

Rossel, S. and Wehner, R. 1984. How bees analyze the polarization patterns in the sky: experiments and model. Journal of Comparative Physiology A 154: 607–615.

Rotenberry, J.T., Zuk, M., Simmons, L.W. and Hayes, C. 1996. Phonotactic parasitoids and cricket song structure: an evaluation of alternative hypotheses. Evolutionary Ecology 10: 233–243.

Rovner, J.S. 1975. Sound production by Nearctic wolf spiders: a substratum-coupled stridulatory mechanism. Science 190: 1309–1310.

Rowe, L. and Houle, D. 1996. The lek paradox and the capture of genetic variance by condition dependent traits. Proceedings of the Royal Society of London B 263: 1415–1421.

Royer, L. and McNeil, J.N. 1992. Evidence of a male sex pheromone in the European corn borer, *Ostrinia nubilalis* (Hubner) (Lepidoptera: Pyralidae). Canadian Entomologist 124: 113–116.

Russell, D.F., Wilkens, L.A. and Moss, F. 1999. Use of behavioural stochastic resonance by paddle fish for feeding. Nature 402: 291–294.

Rust, J., Stumpner, A. and Gottwald, J. 1999. Singing and hearing in a Tertiary bushcricket. Nature 399: 650.

Rutowski, R.L. 1977a. Chemical communication in courtship of small sulfur butterfly *Eurema lisa* (Lepidoptera, Pieridae). Journal of Comparative Physiology A 115: 75–85.

Rutowski, R.L. 1977b. Use of visual cues in sexual and species discrimination by males of small sulfur butterfly *Eurema lisa* (Lepidoptera, Pieridae). Journal of Comparative Physiology A 115: 61–74.

Rutowski, R.L. 1978. Form and function of ascending flights in *Colias* butterflies. Behavioral Ecology and Sociobiology 3: 163–172.

Rutowski, R.L. 1983. The wing-waving display of *Eurema daira* males (Lepidoptera, Pieridae): its structure and role in successful courtship. Animal Behaviour 31: 985–989.

Rutowski, R.L. 1997. Sexual dimorphism, mating systems and ecology in butterflies. In The Evolution of Mating Systems in Insects and Arachnids (eds J.C. Choe and B.J. Crespi), pp. 257–272. Cambridge University Press, Cambridge, U.K.

Rutowski, R.L. 1998. Mating strategies in butterflies. Scientific American 279(1): 64–69.

Rutowski, R.L. 2002. Visual ecology of adult butterflies. In Ecology and Evolution Taking Flight: Butterflies as Model Study Systems (eds C. Boggs, W. Watt and P. Ehrlich), in press. University of Chicago Press, Chicago, IL.

Ryan, M.J. 1988. Energy, calling, and selection. American Zoologist 28: 885–898.

Ryan, M.J. and Rand, A.S. 1993. Species recognition and sexual selection as a unitary problem in animal communication. Evolution 47: 647–657.

Ryan, M.J., Fox, J., Wilczynski, W. and Rand, A.S. 1990. Sexual selection for sensory exploitation in the frog *Physalaemus pustulosus*. Nature 343: 66–67.

Ryan, M.J., Tuttle, M.D. and Taft, L.K. 1981. The costs and benefits of frog chorusing behavior. Behavioral Ecology and Sociobiology 8: 273–278.

Rydell, J. 1998. Bat defence in lekking ghost swifts (*Hepialus humuli*), a moth without ultrasonic hearing. Proceedings of the Royal Society of London B 265: 1373–1376.

Rydell, J. and Lancaster, W.C. 2000. Flight and thermoregulation in moths were shaped by predation from bats. Oikos 88: 13–18.

Rydell, J., Skals, N., Surlykke, A. and Svensson, M. 1997. Hearing and bat defence in geometrid winter moths. Proceedings of the Royal Society of London B 264: 83–88.

Rydell, J., Roininen, H. and Philip, K.W. 2000. Persistence of bat defence reactions in high Arctic moths (Lepidoptera). Proceedings of the Royal Society of London B 267: 553–557.

Ryder, J.J. and Siva-Jothy, M.T. 2000. Male calling song provides a reliable signal of immune function in a cricket. Proceedings of the Royal Society of London B 267: 1171–1175.

Saini, R.K., Rai, M.M., Hassanali, A., Wawiye, J. and Odongo, H. 1995. Semiochemicals from froth of egg pods attract ovipositing female *Schistocerca gregaria*. Journal of Insect Physiology 41: 711–716.

Sakaluk, S.K. 2000. Sensory exploitation as an evolutionary origin to nuptial food gifts in insects. Proceedings of the Royal Society of London B 267: 339–343.

Sakaluk, S.K. and Belwood, J.J. 1984. Gecko phonotaxis to cricket calling song: a case of satellite predation. Animal Behaviour 32: 659–662.

Salmon, M. 1971. Signal characteristics and acoustic detection by fiddler crabs, *Uca rapax* and *Uca pugilator*. Physiological Zoology 44: 210–224.

Salmon, M. and Hyatt, G.W. 1979. Development of acoustic display in the fiddler crab *Uca pugilator*, and its hybrids with *Uca panacea*. Marine Behaviour and Physiology 6: 197–209.

Sanborn, A.F. and Phillips, P.K. 1995. Scaling of sound pressure level and body size in cicadas (Homoptera: Cicadidae: Tibicinidae). Annals of the Entomological Society of America 88: 479–484.

Sanborn, A.F. and Phillips, P.K. 1999. Analysis of acoustic signals produced by the cicada *Platypedia putnami* variety *lutea* (Homoptera: Tibicinidae). Annals of the Entomological Society of America 92: 451–455.

Sanborn, A.F., Heath, J.E., Heath, M.S. and Noriega, F.G. 1995. Thermoregulation by endogenous heat production in 2 South American grass dwelling cicadas (Homoptera: Cicadidae: *Proarna*). Florida Entomologist 78: 319–328.

Schal, C. 1982. Intraspecific vertical stratification as a mate-finding mechanism in tropical cockroaches. Science 215: 1405–1407.

Schal, C. and Cardé, R.T. 1985. Rhythmic extrusion of pheromone gland elevates pheromone release rate. Experientia 41: 1617–1619.

Schal, C., Charlton, R.E. and Cardé, R.T. 1987. Temporal patterns of sex pheromone titers and release rates in *Holomelina lamae* (Lepidoptera: Arctiidae). Journal of Chemical Ecology 13: 1115–1129.

Schal, C., Sevala, V. and Cardé, R.T. 1998. Novel and highly specific transport of a volatile sex pheromone by hemolymph lipophorin in moths. Naturwissenschaften 85: 339–342.

Schatral, A. 1990a. Interspecific acoustic behaviour among bushcrickets. In The Tettigoniidae: Biology, Systematics and Evolution (eds W.J. Bailey and D.C.F. Rentz), pp. 152–165. Springer-Verlag, Berlin.

Schatral, A. 1990b. Body size, song frequency and mating success of male bush-crickets *Requena verticalis* (Orthoptera: Tettigoniidae: Listrocelidinae) in the field. Animal Behaviour 40: 982–984.

Schatral, A. and Bailey, W.J. 1991. Song variability and the response to conspecific song and to song models of different frequency contents in males of the bushcricket *Requena verticalis* (Orthoptera: Tettigoniidae). Behaviour 116: 163–179.

Schiestl, F.P., Ayasse, M., Paulus, H.F., Löfstedt, C., Hansson, B.S., Ibarra, F. and Francke, W. 1999. Orchid pollination by sexual swindle. Nature 399: 421–422.

Schildberger, K. 1988. Behavioral and neuronal mechanisms of cricket phonotaxis. Experientia 44: 408–415.

Schildberger, K. 1994. The auditory pathway for crickets: adaptations for intraspecific communication. Fortschritte der Zoologie 39: 209–225.

Schildberger, K. and Kleindienst, H.-U. 1989. Sound localization in intact and one-eared crickets: comparison of neuronal properties with open-loop and closed-loop behavior. Journal of Comparative Physiology A 165: 615–626.

Schiolten, P., Larsen, O.N. and Michelsen, A. 1981. Mechanical time resolution in some insect ears. 1. Impulse responses and time constants. Journal of Comparative Physiology A 143: 289–295.

Schlyter, F. and Birgersson, G. 1989. Individual variation in bark beetle and moth pheromones: a comparison and an evolutionary background. Holarctic Ecology 12: 457–465.

Schlyter, F., Birgersson, G. and Leufven, A. 1989. Inhibition of attraction to aggregation pheromone by verbenone and ipsenol: density regulation mechanisms in bark beetle *Ips typographus*. Journal of Chemical Ecology 15: 2263–2277.

Schmidt, J.O. 1998. Mass action in honey bees: alarm, swarming and the role of releaser pheromones. In Pheromone Communication in Social Insects: Ants, Wasps, Bees and Termites (eds R.K. Vander Meer, M.D. Breed, K.E. Espelie and M.L. Winston), pp. 257–290. Westview Press, Boulder, CO.

Schmitt, A., Friedel, T. and Barth, F.G. 1993. Importance of pause between spider courtship vibrations and general problems using synthetic stimuli in behavioral studies. Journal of Comparative Physiology A 172: 707–714.

Schmitt, A., Schuster, M. and Barth, F.G. 1994. Vibratory communication in a wandering spider, *Cupiennius getazi*: female and male preferences for features of the conspecific males releaser. Animal Behaviour 48: 1155–1171.

Schmitz, B., Kleindienst, H.U., Schildberger, K. and Huber, F. 1988. Acoustic orientation in adult, female crickets (*Gryllus bimaculatus* deGeer) after unilateral foreleg amputation in the larva. Journal of Comparative Physiology A 162: 715–728.

Schmitz, H. and Bleckmann, H. 1998. The photomechanic infrared receptor for the detection of forest fires in the beetle *Melanophila acuminata* (Coleoptera: Buprestidae). Journal of Comparative Physiology A 182: 647–657.

Schmitz, H., Bleckmann, H. and Murtz, M. 1997. Infrared detection in a beetle. Nature 386: 773–774.

Schneider, D. 1964. Insect antennae. Annual Review of Entomology 9: 103–122.

Schneider, D. 1971. Molekulare Grundlagen der chemischen Sinne bei Insekten. Naturwissenschaften 58: 194–200.

Schneider, D. 1992. 100 years of pheromone research: an essay on Lepidoptera. Naturwissenschaften 79: 241–250.

Schneider, D., Boppré, M., Schneider, H. Thompson, W.R., Boriack, C.J., Petty, R.L. and Meinwald, J. 1975. Pheromone precursor and its uptake in male *Danaus* butterflies. Journal of Comparative Physiology A 97: 245–256.

Schneider, D., Boppré, M., Zweig, J., Horsley, S.B., Bell, T.W., Meinwald, J., Hansen, K. and Diehl, E.W. 1982. Scent organ development in *Creatonotus* moths: regulation by pyrrolizidine alkaloids. Science 215: 1264–1265.

Schneider, D., Schulz, S., Priesner, E., Ziesmann, J. and Francke, W. 1998a. Autodetection and chemistry of female and male pheromone in both sexes of the tiger moth *Panaxia quadripunctaria*. Journal of Comparative Physiology A 182: 153–161.

Schneider, R.W.S., Lanzen, J. and Moore, P.A. 1998b. Boundary-layer effect on chemical signal movement near the antennae of the sphinx moth, *Manduca sexta*: temporal filters for olfaction. Journal of Comparative Physiology A 182: 287–298.

Schneiderman, A.M., Hildebrand, J.G., Brennan, M.M. and Tumlinson, J.H. 1986. Transsexually grafted antennae alter pheromone-directed behavior in a moth. Nature 323: 801–803.

Schüch, W. and Barth, F.G. 1985. Temporal patterns in the vibratory courtship signals of the wandering spider *Cupiennius salei* Keys. Behavioral Ecology and Sociobiology 16: 263–271.

Schul, J. 1999. Neuronal basis for spectral song discrimination in the bushcricket *Tettigonia cantans*. Journal of Comparative Physiology A 184: 457–461.

Schul, J., von Helversen, D. and Weber, T. 1998. Selective phonotaxis in *Tettigonia cantans* and *T. viridissima* in song recognition and discrimination. Journal of Comparative Physiology A 182: 687–694.

Schultz, T.D. and Bernard, G.D. 1989. Pointillistic mixing of interference colors in cryptic tiger beetles. Nature 337: 72–73.

Schulz, S., Francke, W., Boppré, M., Eisner, T. and Meinwald, J. 1993. Defense-mechanisms of arthropods. 117. insect pheromone biosynthesis: stereochemical pathway of hydroxydanaidal production from alkaloidal precursors in *Creatonotos transiens* (Lepidoptera: Arctiidae). Proceedings of the National Academy of Sciences, U.S.A. 90: 6834–6838.

Schulze, W. and Schul, J. 2001. Ultrasound avoidance behaviour in the bushcricket *Tettigonia viridissima* (Orthoptera: Tettigoniidae). Journal of Experimental Biology 204: 733–740.

Schwartz, J.J. 1991. Why stop calling? A study of unison bout singing in a neotropical treefrog. Animal Behaviour 42: 565–578.

Schwartz, J.J. and Gerhardt, H.C. 1989. Spatially mediated release from auditory masking in an anuran amphibian. Journal of Comparative Physiology A 166: 37–41.

Schwemer, J. 1989. Visual pigments of compound eyes: structure, photochemistry, and regeneration. In Facets of Vision (eds D.G. Stavenga and R.C. Hardie), pp. 122–133. Springer-Verlag, Berlin.

Schwind, R. 1984. Evidence for true polarization vision based on a two-channel analyser system in the eye of the water bug, *Notonecta glauca*. Journal of Comparative Physiology A 154: 53–57.

Schwind, R. 1989. Size and distance perception in compound eyes. In Facets of Vision (eds D.G. Stavenga and R.C. Hardie), pp. 425–444. Springer-Verlag, Berlin.

Scoble, M.J. 1992. The Lepidoptera: Form, Function and Diversity. Oxford University Press, Oxford, U.K.

Searcy, W.A. 1989. Pseudoreplication, external validity and the design of playback experiments. Animal Behaviour 38: 715–717.

Seliger, H.H., Lall, A.B., Lloyd, J.E. and Biggley, W.H. 1982. The colors of firefly bioluminescence. 1. Optimization model. Photochemistry and Photobiology 36: 673–680.

Selverston, A., Kleindienst, H.-U. and Huber, F. 1985. Synaptic connectivity between cricket auditory interneurons as studied by selective photoinactivation. Journal of Neuroscience Research 5: 1283–1292.

Sharov, A.G. 1971. Phylogeny of the Orthopteroidea. Israel Program for Scientific Translations, Jerusalem.

Shaw, K. 1995. Phylogenetic tests of the sensory exploitation model of sexual selection. Trends in Ecology and Evolution 10: 117–120.

Shaw, K.L. 1996. Polygenic inheritance of a behavioral phenotype: interspecific genetics of song in the Hawaiian cricket genus *Laupala*. Evolution 50: 256–266.

Shaw, K.L. 2000. Interspecific genetics of mate recognition: inheritance of female acoustic preference in Hawaiian crickets. Evolution 54: 1303–1312.

Shaw, K.L. and Herlihy, D.P. 2000. Acoustic preference functions and song variability in the Hawaiian cricket *Laupala cerasina*. Proceedings of the Royal Society of London B 267: 577–584.

Shaw, S.R. 1967. Simultaneous recordings from two cells in the locust retina. Zeitschrift für Vergleichende Physiologie. 55: 183–194.

Shaw, S.R. 1984. Early visual processing in insects. Journal of Experimental Biology 112: 225–251.

Shaw, S.R. 1994. Detection of airborne sound by a cockroach vibration detector: a possible missing link in insect auditory evolution. Journal of Experimental Biology 193: 13–47.

Shelly, T.E. and Greenfield, M.D. 1989. Satellites and transients: ecological constraints on alternative mating tactics in male grasshoppers. Behaviour 109: 200–221.

Shelly, T.E. and Greenfield, M.D. 1991. Dominions and desert clickers (Orthoptera, Acrididae): influences of resources and male signaling on female settlement patterns. Behavioral Ecology and Sociobiology 28: 133–140.

Shelly, T.E. and Whittier, T.S. 1997. Lek behavior of insects. In The Evolution of Mating Systems in Insects and Arachnids (eds J.C. Choe and B.J. Crespi), pp. 273–293. Cambridge University Press, Cambridge.

Sherman, P.W. and Holmes, W.G. 1985. Kin recognition: issues and evidence. Fortschritte der Zoologie 31: 437–460.

Shuvalov, V.F. 1999. Effect of environment on the parameters of sound signals in the cricket, *Gryllus bimaculatus*. Journal of Evolutionary Biochemistry and Physiology 35: 55–58.

Silberglied, R.E. 1979. Communication in the ultraviolet. Annual Review of Ecology and Systematics 10: 373–398.

Silberglied, R.E. 1984. Visual communication and sexual selection among butterflies. Symposia of the Royal Entomological Society of London 11: 207–223.

Silberglied, R.E. and Taylor, O.R. Jr. 1978. Ultraviolet reflection and its behavioral role in the courtship of the sulfur butterflies, *Colias eurytheme* and *C. philodice*. Behavioral Ecology and Sociobiology 3: 203–243.

Simmons, L.W. 1988. The calling song of the field cricket, *Gryllus bimaculatus* (deGeer): constraints on transmission and its role in intermale competition and female choice. Animal Behaviour 36: 380–394.

Simmons, L.W. and Bailey, W.J. 1993. Agonistic communication between males of a zaprochiline katydid (Orthoptera: Tettigoniidae). Behavioral Ecology 4: 364–368.

Simmons, L.W. and Ritchie, M.G. 1996. Symmetry in the songs of crickets. Proceedings of the Royal Society of London B 263: 1305–1311.

Simmons, L.W. and Zuk, M. 1992. Variability in call structure and pairing success of male field crickets, *Gryllus bimaculatus*: the effects of age, size and parasite load. Animal Behaviour 44: 1145–1152.

Simmons, L.W., Teale, R.J., Maier, M., Standish, R.J., Bailey, W.J. and Withers, P.C. 1992. Some costs of reproduction for male bushcrickets, *Requena verticalis* (Orthoptera: Tettigoniidae): allocating resources to mate attraction and nuptial feeding. Behavioral Ecology and Sociobiology 31: 57–62.

Simon, T. and Hefetz, A. 1991. Trail following responses of *Tapinoma simrothi* (Formicidae: Dolichoderinae) to pygidial gland extracts. Insectes Sociaux 38: 17–25.

Simpson, S.J., McCaffery, A.R. and Hagele, B.F. 1999. A behavioural analysis of phase change in the desert locust. Biological Reviews of the Cambridge Philosophical Society 74: 461–480.

Sismondo, E. 1993. Ultrasubharmonic resonance and nonlinear dynamics in the song of *Oecanthus nigricornis* F. Walker (Orthoptera: Gryllidae). International Journal of Insect Morphology and Embryology 22: 217–231.

Sivinski, J.M. 1998. Phototropism, bioluminescence, and the Diptera. Florida Entomologist 81: 282–292.

Skovmand, O. and Pedersen, S.B. 1983. Song recognition and song pattern in a shorthorned grasshopper. Journal of Comparative Physiology A 153: 393–401.

Slatkin, M. 1978. Spatial patterns in the distribution of polygenic characters. Journal of Theoretical Biology 70: 213–228.

Smith, B.H. and Breed, M.D. 1995. The chemical basis for nestmate recognition and mate discrimination in social insects. In Chemical Ecology of Insects, Vol. 2 (eds R.T. Cardé and W.J. Bell), pp. 287–317. Chapman and Hall, New York.

Snedden, W.A. 1996. Lifetime mating success in male sagebrush crickets: sexual selection constrained by a virgin male mating advantage. Animal Behaviour 51: 1119–1125.

Snedden, W.A. and Greenfield, M.D. 1998. Females prefer leading males: relative call timing and sexual selection in katydid choruses. Animal Behaviour 56: 1091–1098.

Snedden, W.A., Greenfield, M.D. and Jang, Y. 1998. Mechanisms of selective attention in grasshopper choruses: who listens to whom? Behavioral Ecology and Sociobiology 43: 59–66.

Snedden, W.A., Tosh, C.R. and Ritchie, M.G. 1994. The ultrasonic mating signal of the male lesser wax moth. Physiological Entomology 19: 367–372.

Sobel, E.C. and Tank, D.W. 1994. In vivo Ca^{2+} dynamics in a cricket auditory neuron: an example of chemical computation. Science 263: 823–826.

Solis, M.A. and Mitter, C. 1992. Review and preliminary phylogenetic analysis of the subfamilies of the Pyralidae (sensu stricto) (Lepidoptera: Pyraloidea). Systematic Entomology 17: 79–90.

Souroukis, K., Cade, W.H. and Rowell, G. 1992. Factors that possibly influence variation in the calling song of field crickets: temperature, time, and male size, age, and wing morphology. Canadian Journal of Zoology 70: 950–955.

Sower, L.L. and Fish, J.C. 1975. Rate of release of the sex pheromone of the female Indian meal moth. Environmental Entomology 4: 168–169.

Spangler, H.G. 1987. Ultrasonic communication in *Corcyra cephalonica* (Stainton) (Lepidoptera, Pyralidae). Journal of Stored Products Research 23: 203–211.

Spangler, H.G. 1988a. Moth hearing, defense, and communication. Annual Review of Entomology 33: 59–81.

Spangler, H.G. 1988b. Hearing in tiger beetles (Cicindelidae). Physiological Entomology 13: 447–452.

Spangler, H.G. 1991. Do honey bees encode distance information into the wing vibrations of the waggle dance? Journal of Insect Behavior 4: 15–20.

Spangler, H.G., Greenfield, M.D. and Takessian, A. 1984. Ultrasonic mate calling in the lesser wax moth. Physiological Entomology 9: 87–95.

Spurgeon, D.W., Lingren, P.D., Raulston, J.R. and Shaver, T.N. 1995. Age-specific mating activities of Mexican rice borers (Lepidoptera: Pyralidae). Environmental Entomology 24: 105–109.

Srinivasan, M.V. 1994. Pattern recognition in the honeybee: recent progress. Journal of Insect Physiology. 40: 183–194.

Srinivasan, M.V., Zhang, S.W. and Rolfe, B. 1993. Is pattern vision in insects mediated by cortical processing. Nature 362: 539–540.

Srinivasan, M.V., Zhang, S.W. and Witney, K. 1994. Visual discrimination of pattern orientation by honeybees: performance and implications for cortical processing. Philosophical Transactions of the Royal Society of London B 343: 199–210.

Srinivasan, M.V., Zhang, S.W., Altwein, M. and Tautz, J. 2000. Honeybee navigation: nature and calibration of the "odometer." Science 287: 851–853.

Srinivasarao, M. 1999. Nano-optics in the biological world: beetles, butterflies, birds, and moths. Chemical Reviews 99: 1935–1961.

Stamps, J.A. 1987. Conspecifics as cues to territory quality: a preference of juvenile lizards (*Anolis aeneus*) for previously used territories. American Naturalist 129: 629–642.

Stavenga, D.G. 1992. Eye regionalization and spectral tuning of retinal pigments in insects. Trends in Neurosciences 15: 213–218.

Stephen, R.O. and Bennet-Clark, H.C. 1982. The anatomical and mechanical basis of stimulation and frequency analysis in the locust ear. Journal of Experimental Biology 99: 279–314.

Stephen, R.O. and Hartley, J.C. 1995. Sound production in crickets. Journal of Experimental Biology 198: 2139–2152.

Stephens, K. 1986. Pheromones among the procaryotes. Critical Reviews in Microbiology. 13: 309–334.

Stern, D.L. and Foster, W.A. 1996. The evolution of soldiers in aphids. Biological Reviews of the Cambridge Philosophical Society 71: 27–79.

Stern, D.L. and Foster, W.A. 1997. The evolution of sociality in aphids: a clone's-eye view. In The Evolution of Social Behavior in Insects and Arachnids (eds J.C. Choe and B.J. Crespi), pp. 150–165. Cambridge University Press, Cambridge, U.K.

Stevens, E. and Josephson, R.K. 1977. Metabolic rate and body temperature in singing katydids. Physiological Zoology 50: 31–42.

Stewart, K.W., Bottorff, R.L., Knight, A.W. and Moring, J.B. 1991. Drumming of 4 North American euholognathan stonefly species and a new complex signal pattern in *Nemoura spiniloba* Jewett (Plecoptera: Nemouridae). Annals of the Entomological Society of America 84: 201–206.

Stickland, T.R., Tofts, C. and Franks, N.R. 1993. Algorithms for ant foraging. Naturwissenschaften 80: 427–430.

Stickland, T.R., Britton, N.F. and Franks, N.R. 1995. Complex trails and simple algorithms in ant foraging. Proceedings of the Royal Society of London B 260: 53–58.

Stirling, D. and Hamilton, P.V. 1986. Observations on the mechanism of detecting mucous trail polarity in the snail *Littorina irrorata*. Veliger 29: 31–37.

Stout, J., Atkins, G. and Zacharias, D. 1991. Regulation of cricket phonotaxis through hormonal control of the threshold of an identified auditory neuron. Journal of Comparative Physiology A 169: 765–772.

Stowe, M.K., Tumlinson, J.H. and Heath, R.R. 1987. Chemical mimicry: bolas spiders emit components of moth prey species sex pheromones. Science 236: 964–967.

Stowe, M.K., Turlings, T.C.J., Loughrin, J.H., Lewis, W.J. and Tumlinson, J.H. 1995. The chemistry of eavesdropping, alarm, and deceit. Proceedings of the National Academy of Sciences, U.S.A. 92: 23–28.

Strausfeld, N.J. 1976. Mosaic organizations, layers and visual pathways in the insect brain. In Neural Principles in Vision (eds F. Zettler and R. Weiler), pp. 245–279. Springer-Verlag, Berlin.

Strausfeld, N.J. and Barth, F.G. 1993. 2 visual systems in one brain: neuropils serving the secondary eyes of the spider *Cupiennius salei*. Journal of Comparative Neurology 328: 43–62.

Strausfeld, N.J., Weltzien, P. and Barth, F.G. 1993. Two visual systems in one brain: neuropils serving the principal eyes of the spider *Cupiennius salei*. Journal of Comparative Neurology 328: 63–75.

Strogatz, S.H. and Stewart, I. 1993. Coupled oscillators and biological synchronization. Scientific American 269(6): 102–109.

Stumpner, A. and von Helversen, D. 1992. Recognition of a two-element song in the grasshopper *Chorthippus dorsatus* (Orthoptera: Gomphocerinae). Journal of Comparative Physiology A 171: 405–412.

Suckling, D.M., Green, S.R., Gibb, A.R. and Karg, G. 1999a. Predicting atmospheric concentration of pheromone in treated apple orchards. Journal of Chemical Ecology 25: 117–139.

Suckling, D.M., Karg, G., Green, S. and Gibb, A.R. 1999b. The effect of atmospheric pheromone concentrations on behavior of lightbrown apple moth in an apple orchard. Journal of Chemical Ecology 25: 2011–2025.

Surlykke, A., Larsen, O.N. and Michelsen, A. 1988. Temporal coding in the auditory receptor of the moth ear. Journal of Comparative Physiology A 162: 367–374.

Surlykke, A., Skals, N., Rydell, J. and Svensson, M. 1998. Sonic hearing in a diurnal geometrid moth, *Archiearis parthenias*, temporally isolated from bats. Naturwissenschaften 85: 36–37.

Surlykke, A., Filskov, M., Fullard, J.H. and Forrest, E. 1999. Auditory relationships to size in noctuid moths: bigger is better. Naturwissenschaften 86: 238–241.

Sutton, O.G. 1953. Micrometeorology. McGraw-Hill, New York.

Swaddle, J.P. 1999. Visual signalling by asymmetry: a review of perceptual processes. Philosophical Transactions of the Royal Society of London B 354: 1383–1393.

Tang, J.D., Wolf, W.A., Roelofs, W.L. and Knipple, D.C. 1991. Development of functionally competent cabbage looper moth sex pheromone glands. Insect Biochemistry 21: 573–581.

Tang, J.D., Charlton, R.E., Cardé, R.T. and Yin, C.M. 1992. Diel periodicity and influence of age and mating on sex pheromone titer in gypsy moth, *Lymantria dispar* (L.). Journal of Chemical Ecology 18: 749–760.

Tauber, E., Cohen, D., Greenfield, M.D. and Pener, M.P. 2001. Duet singing and female choice in the bushcricket *Phaneroptera nana*. Behaviour 138: 411–430.

Tautz, J. and Markl, H. 1978. Caterpillars detect flying wasps by hairs sensitive to airborne vibration. Behavioral Ecology and Sociobiology 4: 101–110.

Taylor, P.W., Hasson, O. and Clark, D.L. 2000. Body postures and patterns as amplifiers of physical condition. Proceedings of the Royal Society of London B 267: 917–922.

Theiss, J. 1982. Generation and radiation of sound by stridulating water insects as exemplified by the corixids. Behavioral Ecology and Sociobiology 10: 225–235.

Theiss, J. 1983. An acoustic duet is necessary for successful mating in *Corixa dentipes*. Naturwissenschaften 70: 467–468.

Theiss, J. and Prager, J. 1984. Range of corixid sound signals in the biotope. Physiological Entomology 9: 107–114.

Theiss, J., Prager, J. and Streng, R. 1983. Underwater stridulation by corixids: stridulatory signals and sound producing mechanism in *Corixa dentipes* and *Corixa punctata*. Journal of Insect Physiology 29: 761–771.

Thiele, D. and Bailey, W.J. 1980. The function of sound in male spacing behavior in bush crickets (Tettigoniidae: Orthoptera). Australian Journal of Ecology 5: 275–286.

Thornhill, R. and Alcock, J. 1983. The Evolution of Insect Mating Systems. Harvard University Press, Cambridge, MA.

Tillman, J.A., Seybold, S.J., Jurenka, R.A. and Blomquist, G.J. 1999. Insect pheromones: an overview of biosynthesis and endocrine regulation. Insect Biochemistry and Molecular Biology 29: 481–514.

Todd, J.L. and Baker, T.C. 1993. Response of single antennal neurons of female cabbage loopers to behaviorally active attractants. Naturwissenschaften 80: 183–186.

Todd, J.L. and Baker, T.C. 1999. Function of peripheral olfactory organs. In Insect Olfaction (ed. B.S. Hansson), pp. 67–96. Springer-Verlag, Berlin.

Todd, J.L., Haynes, K.F. and Baker, T.C. 1992. Antennal neurons specific for redundant pheromone components in normal and mutant *Trichoplusia ni* males. Physiological Entomology 17: 183–192.

Todd, J.L., Anton, S., Hansson, B.S. and Baker, T.C. 1995. Functional organization of the macroglomerular complex related to behaviourally expressed olfactory redundancy in male cabbage looper moths. Physiological Entomology 20: 349–361.

Torto, B., Obengofori, D., Njagi, P.G.N., Hassanali, A. and Amiani, H. 1994. Aggregation pheromone system of adult gregarious desert locust *Schistocerca gregaria* (Forskal). Journal of Chemical Ecology 20: 1749–1762.

Torto, B., Njagi, P.G.N., Hassanali, A. and Amiani, H. 1996. Aggregation pheromone system of nymphal gregarious desert locust, *Schistocerca gregaria* (Forskal). Journal of Chemical Ecology 22: 2273–2281.

Towne, W.F. 1985. Acoustic and visual cues in the dances of 4 honey bee species. Behavioral Ecology and Sociobiology 16: 187.

Towne, W.F. 1995. Frequency discrimination in the hearing of honey bees (Hymenoptera: Apidae). Journal of Insect Behavior 8: 281–286.

Towne, W.F. and Kirchner, W.H. 1989. Hearing in honey bees: detection of air particle oscillations. Science 244: 686–688.

Traniello, J.F.A. and Robson, S.K. 1995. Trail and territorial communication in social insects. In Chemical Ecology of Insects, Vol. 2 (eds R.T. Cardé and W.J. Bell), pp. 241–286. Chapman and Hall, New York.

Travassos, M.A. and Pierce, N.E. 2000. Acoustics, context and function of vibrational signalling in a lycaenid butterfly-ant mutualism. Animal Behaviour 60: 13–26.

Trimmer, B.A., Aprille, J.R., Dudzinski, D.M., Lagace, C.J., Lewis, S.M., Michel, T., Qazi, S. and Zayas, R.M. 2001. Nitric oxide and the control of firefly flashing. Science 292: 2486–2488.

Tuckerman, J.F., Gwynne, D.T. and Morris, G.K. 1993. Reliable acoustic cues for female mate preference in a katydid (*Scudderia curvicauda*, Orthoptera: Tettigoniidae). Behavioral Ecology 4: 106–113.

Uetz, G.W. and Smith, E.I. 1999. Asymmetry in a visual signaling character and sexual selection in a wolf spider. Behavioral Ecology and Sociobiology 45: 87–93.

Ulagaraj, S.M. 1976. Sound production in mole crickets (Orthoptera: Gryllotalpidae: *Scapteriscus*). Annals of the Entomological Society of America 69: 299–306.

Ulagaraj, S.M. and Walker, T.J. 1973. Phonotaxis of crickets in flight: attraction of male and female crickets to male calling songs. Science 182: 1278–1279.

Ulagaraj, S.M. and Walker, T.J. 1975. Response of flying mole crickets to 3 parameters of synthetic songs broadcast outdoors. Nature 253: 530–532.

Uvarov, B.P. 1966. Grasshoppers and Locusts, Vol. 1. Cambridge University Press, Cambridge, U.K.

Van Dongen, S., Matthysen, E., Sprengers, E. and Dhondt, A.A. 1998. Mate selection by male winter moths *Operophtera brumata* (Lepidoptera: Geometridae): Adaptive male choice or female control? Behaviour 135: 29–42.

van Staaden, M.J. and Römer, H. 1997. Sexual signalling in bladder grasshoppers: tactical design for maximizing calling range. Journal of Experimental Biology 200: 2597–2608.

van Staaden, M.J. and Römer, H. 1998. Evolutionary transition from stretch to hearing organs in ancient grasshoppers. Nature 394: 773–776.

Vander Meer, R.K. and Alonso, L.E. 1998. Pheromone directed behavior in ants. In Pheromone Communication in Social Insects: Ants, Wasps, Bees and Termites (eds R.K. Vander Meer, M.D. Breed, K.E. Espelie and M.L. Winston), pp. 159–192. Westview Press, Boulder, CO.

Vander Meer, R.K. and Morel, L. 1998. Nestmate recognition in ants. In Pheromone Communication in Social Insects: Ants, Wasps, Bees and Termites (eds R.K. Vander Meer, M.D. Breed, K.E. Espelie and M.L. Winston), pp. 79–103. Westview Press, Boulder, CO.

Vander Meer, R.K., Breed, M.D., Espelie, K.E. and Winston, M.L.. 1998. Pheromone Communication in Social Insects: Ants, Wasps, Bees and Termites. Westview Press, Boulder, CO.

Vane-Wright, R.I. and Boppré, M. 1993. Visual and chemical signaling in butterflies: functional and phylogenetic perspectives. Philosophical Transactions of the Royal Society of London B 340: 197–205.

Vargo, E.L. 1998. Primer pheromones in ants. In Pheromone Communication in Social Insects: Ants, Wasps, Bees and Termites (eds R.K. Vander Meer, M.D. Breed, K.E. Espelie and M.L. Winston), pp. 293–313. Westview Press, Boulder, CO.

Vencl, F.V. and Carlson, A.D. 1998. Proximate mechanisms of sexual selection in the firefly *Photinus pyralis* (Coleoptera : Lampyridae). Journal of Insect Behavior 11: 191–207.

Vencl, F.V., Blasko, B.J. and Carlson, A.D. 1994. Flash behavior of female *Photuris versicolor* fireflies (Coleoptera, Lampyridae) in simulated courtship and predatory dialogs. Journal of Insect Behavior 7: 843–858.

Versluis, M., Schmitz, B., von der Heydt, A. and Lohse, D. 2000. How snapping shrimp snap: through cavitating bubbles. Science 289: 2114–2117.

Vickers, N.J. 2000. Mechanisms of animal navigation in odor plumes. Biological Bulletin 198: 203–212.

Vickers, N.J. and Baker, T.C. 1994. Reiterative responses to single strands of odor promote sustained upwind flight and odor source location by moths. Proceedings of the National Academy of Sciences, U.S.A. 91: 5756–5760.

Vickers, N.J. and Baker, T.C. 1997. Flight of *Heliothis virescens* males in the field in response to sex pheromone. Physiological Entomology 22: 277–285.

Vickers, N.J., Christensen, T.A., Baker, T.C. and Hildebrand, J.G. 2001. Odour-plume dynamics influence the brain's olfactory code. Nature 410: 466–470.

Vité, J.P. and Williamson, D.L. 1970. *Thanasimus dubius*: prey perception. Journal of Insect Physiology 16: 233–239.

Viviani, V.R. and Bechara, E.J.H. 1997. Bioluminescence and biological aspects of Brazilian railroad worms (Coleoptera: Phengodidae). Annals of the Entomological Society of America 90: 389–398.

Vogel, S. 1983. How much air passes through a silkmoth's antenna? Journal of Insect Physiology 29: 597–602.

Vogt, R.G. and Riddiford, L.M. 1981. Pheromone binding and inactivation by moth antennae. Nature 293: 161–163.

Vogt, R.G. and Riddiford, L.M. 1986. Scale esterase: a pheromone degrading enzyme from scales of silk moth *Antheraea polyphemus*. Journal of Chemical Ecology 12: 469–482.

von Frisch, K. 1914. Der Farbensinn und Formensinn der Biene. Zool. J. Physiol. 37: 1–238.

von Frisch, K. 1949. Die Polarisation des Himmelslichtes als orientierender Faktor bei den Tänzen Bienen. Experientia 5: 142–148.

von Frisch, K. 1967. The Dance Language and Orientation of Bees. Belknap Press of Harvard University Press, Cambridge, Mass.

von Helversen, D. 1984. Parallel processing in auditory pattern recognition and directional analysis by the grasshopper *Chorthippus biguttulus* L. (Acrididae). Journal of Comparative Physiology A 154: 837–846.

von Helversen, D. 1993. "Absolute steepness" of ramps as an essential cue for auditory pattern recognition by a grasshopper (Orthoptera: Acrididae; *Chorthippus bigutullus* L.). Journal of Comparative Physiology A 172: 633–639.

von Helversen, D. 1998. Is the ramped shape of pulses in the song of grasshoppers adaptive for directional hearing? Naturwissenschaften 85: 186–188.

von Helversen, D. and von Helversen, O. 1981. Korrespondenz zwischen Gesang und auslösendem Schema bei Feldheuschrecken. Nova Acta Leopold NF 54 Nr. 245: 449–462.

von Helversen, D. and von Helversen, O. 1983. Species recognition and acoustic localization in acridid grasshoppers: a behavioral approach. In Neuroethology and Behavioral Physiology (eds F. Huber and H. Markl), pp. 95–107. Springer-Verlag, Berlin.

von Helversen, D. and von Helversen, O. 1995. Acoustic pattern recognition and orientation in orthopteran insects: parallel or serial processing? Journal of Comparative Physiology A 177: 767–774.

von Helversen, D. and von Helversen, O. 1997. Recognition of sex in the acoustic communication of the grasshopper *Chorthippus biguttulus* (Orthoptera, Acrididae). Journal of Comparative Physiology A 180: 373–386.

von Helversen, D. and von Helversen, O. 1998. Acoustic pattern recognition in a grasshopper: processing in the time or frequency domain? Biological Cybernetics 79: 467–476.

von Helversen, O. 1972. Zur spektralen Unterschiedsempfindlichkeit der Honigbiene. Journal of Comparative Physiology 80: 439–472.

von Helversen, O. and Edrich, W. 1974. Der polarisationsempfänger im Bienenauge: ein Ultraviolettrezeptor. Journal of Comparative Physiology A 94: 33–47.

von Helversen, O. and Elsner, N. 1977. Stridulatory movements of acridid grasshoppers recorded with an opto-electronic device. Journal of Comparative Physiology A 122: 53–64.

von Helversen, O. and von Helversen, D. 1994. Forces driving coevolution of song and song recognition in grasshoppers. Fortschritte der Zoologie 39: 253–284.

von Nickisch-Rosenegk, E. and Wink, M. 1993. Sequestration of pyrollizidine alkaloids in several arctiid moths (Lepidoptera, Arctiidae). Journal of Chemical Ecology 19: 1889–1903.

Vosshall, L.B., Amrein, H., Morozov, P.S., Rzhetsky, A. and Axel, R. 1999. A spatial map of olfactory receptor expression in the *Drosophila* antenna. Cell 96: 725–736.

Vukusic, P., Sambles, J.R., Lawrence, C.R. and Wootton, R.J. 1999. Quantified interference and diffraction in single *Morpho* butterfly scales. Proceedings of the Royal Society of London B 266: 1403–1411.

Vukusic, P., Sambles, J.R. and Lawrence, C.R. 2000. Structural colour: colour mixing in wing scales of a butterfly. Nature 404: 457.

Vukosic, P., Sambles, S.R., Lawrence, C.R. and Wootton, R.J. 2001. Now you see it—now you don't. Nature 410: 36.

Waddington, K.D. and Kirchner, W.H. 1992. Acoustical and behavioral correlates of profitability of food sources in honey bee round dances. Ethology 92: 1–6.

Wagner, D.L. and Rosovsky, J. 1991. Mating systems in primitive Lepidoptera, with emphasis on the reproductive behavior of *Korscheltellus gracilis* (Hepialidae). Zoological Journal of the Linnean Society 102: 277–303.

Wagner, W.E. Jr. 1996. Convergent song preferences between female field crickets and acoustically orienting parasitoid flies. Behavioral Ecology 7: 279–285.

Wagner, W.E. Jr. 1998. Measuring female mating preferences. Animal Behaviour 55: 1029–1042.

Wagner, W.E. Jr. and Hoback, W.W. 1999. Nutritional effects on male calling behaviour in the variable field cricket. Animal Behaviour 57: 89–95.

Wagner, W.E. Jr. and Reiser, M.G. 2000. The importance of calling song and courtship song in female mate choice in the variable field cricket. Animal Behaviour 59: 1219–1226.

Wagner, W.E. Jr., Murray, A.M. and Cade, W.H. 1995. Phenotypic variation in the mating preferences of female field crickets, *Gryllus integer*. Animal Behaviour 49: 1269–1281.

Walker, T.J. 1957. Specificity in the response of female tree crickets (Orthoptera, Gryllidae, Oecanthinae) to calling songs of the males. Annals of the Entomological Society of America 50: 626–636.

Walker, T.J. 1962. Factors responsible for intraspecific variation in the calling songs of crickets. Evolution 16: 407–428.

Walker, T.J. 1964. Cryptic species among sound producing ensiferan Orthoptera (Gryllidae and Tettigoniidae). Quarterly Review of Biology 39: 345–355.

Walker, T.J. 1969. Acoustic synchrony: two mechanisms in the snowy tree cricket. Science 166: 891–894.

Walker, T.J. 1974. *Gryllus ovisopis* n. sp.: a taciturn cricket with a life cycle suggesting allochronic speciation. Florida Entomologist 57: 13–22.

Walker, T.J. 1975a. Stridulatory movements in 8 species of *Neoconocephalus* (Tettigoniidae). Journal of Insect Physiology 21: 595–603.

Walker, T.J. 1975b. Effects of temperature, humidity, and age on stridulatory rates in *Atlanticus* spp. (Orthoptera: Tettigoniidae: Decticinae). Annals of the Entomological Society of America 68: 607–611.

Walker, T.J. 1975c. Effects of temperature on rates in poikilotherm nervous systems: evidence from calling songs of meadow katydids (Orthoptera: Tettigoniidae: *Orchelimum*) and reanalysis of published data. Journal of Comparative Physiology A 101: 57–69.

Walker, T.J. 1977. *Hapithus melodius* and *H. brevipennis*: musical and mute sister species in Florida (Orthoptera: Gryllidae). Annals of the Entomological Society of America 70: 249–252.

Walker, T.J. 1983. Diel patterns of calling in nocturnal Orthoptera. In Orthopteran Mating Systems: Sexual Competition in a Diverse Group of Insects (eds D.T. Gwynne and G.K. Morris), pp. 45–72. Westview Press, Boulder, CO.

Walker, T.J. and Dew, D. 1972. Wing movements of calling katydids: fiddling finesse. Science 178: 174–176.

Walker, T.J. and Forrest, T.G. 1989. Mole cricket phonotaxis: effects of intensity of synthetic calling song (Orthoptera: Gryllotalpidae, *Scapteriscus acletus*). Florida Entomologist 72: 655–659.

Walker, T.J. and Masaki, S. 1989. Natural history. In Cricket Behavior and Neurobiology (eds F. Huber, T.E. Moore, and W. Loher), pp. 1–42. Cornell University Press, Ithaca, NY.

Walker, T.J., Brandt, J.F. and Dew, D. 1970. Sound-synchronized, ultra-high speed photography: a method for studying stridulation in crickets and katydids (Orthoptera). Annals of the Entomological Society of America 63: 910–912.

Wallach, H., Newman, E.B. and Rosenzweig, M.R. 1949. The precedence effect in sound localization. American Journal of Psychology 62: 315–336.

Wang, G.Y., Greenfield, M.D. and Shelly, T.E. 1990. Intermale competition for high-quality host-plants: the evolution of protandry in a territorial grasshopper. Behavioral Ecology and Sociobiology 27: 191–198.

Warzecha, A.-K. and Egelhauf, E. 1995. Visual pattern discrimination in a butterfly. A behavioral study on the Australian lurcher, *Yoma sabina*. Naturwissenschaften 82: 567–569.

Wcislo, W.T. 1997. Are behavioral classifications blinders to studying natural variation? In The Evolution of Social Behavior in Insects and Arachnids (eds J.C. Choe and B.J. Crespi), pp. 8–13. Cambridge University Press, Cambridge, U.K.

Webb, B. 2000. What does robotics offer animal behaviour? Animal Behaviour 60: 545–558.

Webb, K.L. and Roff, D.A. 1992. The quantitative genetics of sound production in *Gryllus firmus*. Animal Behaviour 44: 823–832.

Wedell, N. 1994. Dual function of the bushcricket spermatophore. Proceedings of the Royal Society of London B 258: 181–185.

Wedell, N. and Sandberg, T. 1995. Female preference for large males in the bush-cricket *Requena* sp. 5 (Orthoptera: Tettigoniidae). Journal of Insect Behavior 8: 513–522.

Wehner, R. and Bernard, G.D. 1993. Photoreceptor twist: a solution to the false-color problem. Proceedings of the National Academy of Sciences, U.S.A. 90: 4132–4135.

Wehner, R. and Srinivasan, M.V. 1984. The world as the insect sees it. In Insect Communication (ed. T. Lewis), pp. 29–47. Academic Press, London.

Weissburg, M.J. 1997. Chemo- and mechanosensory orientation by crustaceans in laminar and turbulent flows: from odor trails to vortex streets. In Orientation and Communication in Arthropods (ed. M. Lehrer), pp. 215–246. Birkhäuser Verlag, Basel.

Weissburg, M.J. 2000. The fluid dynamical context of chemosensory behavior. Biological Bulletin 198: 188–202.

Weissburg, M.J. and Zimmer-Faust, R.K. 1994. Odor plumes and how blue crabs use them in finding prey. Journal of Experimental Biology 197: 349–375.

Welch, A.M., Semlitsch, R.D. and Gerhardt, H.C. 1998. Call duration as an indicator of genetic quality in male gray tree frogs. Science 280: 1928–1930.

Weller, S.J., Jacobson, N.L. and Conner, W.E. 1999. The evolution of chemical defences and mating systems in tiger moths (Lepidoptera: Arctiidae). Biological Journal of the Linnean Society 68: 557–578.

Wellington, W.G. 1957. Individual differences as a factor in population dynamics: the development of a problem. Canadian Journal of Zoology 35: 293–323.

Wellington, W.G. 1974. Bumblebee ocelli and navigation at dusk. Science 183: 550–551.

Wells, M.M. and Henry, C.S. 1992. Behavioral responses of green lacewings (Neuroptera, Chrysopidae, *Chrysoperla*) to synthetic mating songs. Animal Behaviour 44: 641–652.

Wenner, A.M. 1990. Anatomy of a Controversy: The Question of a "Language" among Bees. Columbia University Press, New York.

West-Eberhard, M.J. 1984. Sexual selection, competitive communication, and species-specific signals in insects. In Insect Communication (ed. T. Lewis), pp. 283–324. Academic Press, London.

White, P.R., Birch, M.C., Church, S., Jay, C., Rowe, E. and Keenlyside, J.J. 1993. Intraspecific variability in the tapping behavior of the deathwatch beetle, *Xestobium rufovillosum* (Coleoptera: Anobiidae). Journal of Insect Behavior 6: 549–562.

Whitesell, J.J. and Walker, T.J. 1978. Photoperiodically determined dimorphic calling songs in a katydid. Nature 274: 887–888.

Wickler, W. 1968. Mimicry in Plants and Animals. McGraw-Hill, New York.

Wiegmann, D.D. 1999. Search behaviour and mate choice by female field crickets, *Gryllus integer*. Animal Behaviour 58: 1293–1298.

Wiegmann, D.D., Real, L.A., Capone, T.A. and Ellner, S. 1996. Some distinguishing features of models of search behavior and mate choice. American Naturalist 147: 188–204.

Wiernasz, D.C. 1989. Female choice and sexual selection of male wing melanin pattern in *Pieris occidentalis* (Lepidoptera). Evolution 43: 1672–1682.

Wilcox, R.S. 1972. Communication by surface waves: mating behavior of a water strider (Gerridae). Journal of Comparative Physiology 80: 255–266.

Wilcox, R.S. 1979. Sex discrimination in *Gerris remigis*: role of a surface wave signal. Science 206: 1325–1327.

Wilcox, R.S. 1995. Ripple communication in aquatic and semiaquatic insects. Ecoscience 2: 109–115.

Wilkinson, G.S., Presgraves, D.C. and Crymes, L. 1998. Male eye span in stalk-eyed flies indicates genetic quality by meiotic drive suppression. Nature 391: 276–279.

Williams, D.S. and McIntyre, P. 1980. The principal eyes of a jumping spider have a telephoto component. Nature 288: 578–580.

Williams, G.C. 1992. Natural Selection: Domains, Levels and Challenges. Oxford University Press, Oxford.

Willis, M.A. and Arbas, E.A. 1991. Odor-modulated upwind flight of the sphinx moth, *Manduca sexta* L. Journal of Comparative Physiology A 169: 427–440.

Willis, M.A. and Arbas, E.A. 1997. Active behavior and reflexive responses: another perspective on odor-modulated locomotion. In Insect Pheromone Research: New Directions (eds R.T. Cardé and A.K. Minks), pp. 304–319. Chapman and Hall, New York.

Willis, M.A. and Birch, M.C. 1982. Male lek formation and female calling in a population of the arctiid moth *Estigmene acrea*. Science 218: 168–170.

Wilson, E.O., Bossert, W.H. and Regnier, F.E. 1969. A general method for estimating threshold concentrations of odorant molecules. Journal of Insect Physiology 15: 597–610.

Wiltschko, R. and Wiltschko, W. 1995. Magnetic Orientation in Animals. Springer-Verlag, Berlin.

Winston, M.L. and Slessor, K.N. 1992. The essence of royalty: honey bee queen pheromone. American Scientist 80: 374–385.

Winston, M.L. and Slessor, K.N. 1998. Honey bee primer pheromones and colony organization: gaps in our knowledge. Apidologie 29: 81–95.

Witzgall, P. 1997. Modulation of pheromone-mediated flight in male moths. In Insect Pheromone Research: New Directions (eds R.T. Cardé and A.K. Minks), pp. 265–274. Chapman and Hall, New York.

Wolf, H. and von Helversen, O. 1986. "Switching off" of an auditory interneuron during stridulation in the acridid grasshopper *Chorthippus biguttulus* L. Journal of Comparative Physiology A 158: 861–871.

Wood, K.V. 1996. The chemical mechanism and evolutionary development of beetle bioluminescence. Photochemistry and Photobiology 62: 662–673.

Wyatt, T.D. 1997. Putting pheromones to work: paths forward for direct control. In Insect Pheromone Research: New Directions (eds R.T. Cardé and A.K. Minks), pp. 445–459. Chapman and Hall, New York.

Wyttenbach, R.A. and Hoy, R.R. 1993. Demonstration of the precedence effect in an insect. Journal of the Acoustical Society of America 94: 777–784.

Wyttenbach, R.A. and Hoy, R.R. 1997. Spatial acuity of ultrasound hearing in flying crickets Journal of Experimental Biology 200: 1999–2006.

Wyttenbach, R.A., May, M.L. and Hoy, R.R. 1996. Categorical perception of sound frequency by crickets. Science 273: 1542–1544.

Yack, J.E. and Fullard, J.H. 1990. The mechanoreceptive origin of insect tympanal organs: a comparative study of similar nerves in tympanate and atympanate moths. Journal of Comparative Neurology 300: 523–534.

Yack, J.E. and Fullard, J.H. 2000. Ultrasonic hearing in nocturnal butterflies. Nature 403: 265–266.

Yager, D.D. 1996a. Serially homologous ears perform frequency range fractionation in the praying mantis, *Creobroter* (Mantodea: Hymenopodidae). Journal of Comparative Physiology A 178: 463–475.

Yager, D.D. 1996b. Nymphal development of the auditory system in the praying mantis *Hierodula membranacea* Burmeister (Dictyoptera: Mantidae). Journal of Comparative Neurology 364: 199–210.

Yager, D.D. 1999. Structure, development, and evolution of insect auditory systems. Microscopy Research and Technique 47: 380–400.

Yager, D.D. and Hoy, R.R. 1986. The cyclopean ear: a new sense for the praying mantis. Science 231: 727–729.

Yager, D.D. and Spangler, H.G. 1995. Characterization of auditory afferents in the tiger beetle, *Cicindela marutha* Dow. Journal of Comparative Physiology A 176: 587–599.

Yeargan, K.V. 1994. Biology of bolas spiders. Annual Review of Entomology 39: 81–99.

Yen, J., Weissburg, M.J. and Doall, M.H. 1998. The fluid physics of signal perception by mate-tracking copepods. Philosophical Transactions of the Royal Society of London B 353: 787–804.

Young, D. 1990. Do cicadas radiate sound through their eardrums? Journal of Experimental Biology 151: 41–56.

Young, D. and Bennet-Clark, H.C. 1995. The role of the timbal in cicada sound production. Journal of Experimental Biology 198: 1001–1019.

Zahavi, A. 1977. Reliability in communication systems and the evolution of altruism. In Evolutionary Ecology (eds B. Stonehouse and C.M. Perrins), pp. 253–259. Macmillan, London.

Zimmer, R.K. and Butman, C.A. 2000. Chemical signaling processes in the marine environment. Biological Bulletin 198: 168–187.

Zimmer-Faust, R.K., Finelli, C.M., Pentcheff, N.D. and Wethey, D.S. 1995. Odor plumes and animal navigation in turbulent water flow: a field-study. Biological Bulletin 188: 111–116.

Zimmermann, U., Rheinlaender, J. and Robinson, D. 1989. Cues for male phonotaxis in the duetting bushcricket *Leptophyes punctatissima*. Journal of Comparative Physiology A 164: 621–628.

Zuk, M. and Kolluru, G. 1998. Exploitation of sexual signals by predators and parasitoids. Quarterly Review of Biology 73: 415–438.

Zuk, M., Simmons, L.W. and Cupp, L. 1993. Calling characteristics of parasitized and unparasitized populations of the field cricket *Teleogryllus oceanicus*. Behavioral Ecology and Sociobiology 33: 339–343.

Zuk, M., Rotenberry, J.T. and Simmons, L.W. 1998. Calling songs of field crickets (*Teleogryllus oceanicus*) with and without phonotactic parasitoid infection. Evolution 52: 166–171.

Zurek, P.M. 1987. The precedence effect. In Directional Hearing (eds W.A. Yost and G. Gourevitch), pp. 85–105. Springer-Verlag, New York.

Glossary

Across-fiber pattern Process of stimulus evaluation in which inputs from the various receptor categories are combined at the peripheral level, before ascending to higher neural centers.

Active space Volume of air or water surrounding a pheromone source and within which pheromone concentration exceeds a receiver's threshold for detection or behavioral response.

Aggressive mimicry An actor's resemblance of a stimulus attractive to or otherwise judged as positive by a heterospecific individual. By virtue of the resemblance, the actor (aggressive mimic) benefits while the heterospecific receiver (dupe) is harmed.

Aliphatic (compound) Pertaining to organic compounds derived from fats.

Amplifier A morphological device or behavior that increases the conspicuousness of a signal to a receiver.

Anemotaxis Movement oriented with respect to wind velocity; rheotaxis normally refers to movement oriented with respect to water current.

Antagonist A compound, typically one serving as a pheromone component of another species, that inhibits behavioral response to conspecific pheromone.

Apomorphic Derived or advanced character state in phylogeny.

Apposition eye A compound eye in which only light entering the lens of a given ommatidial unit (facet) reaches the receptors in the rhabdom of that unit.

Aromatic (compound) Pertaining to organic compounds bearing a benzene-ring structure(s).

Best frequency Typically used in reference to that carrier frequency of sound for which sensitivity is maximum.

Boundary layer Region very close to the surface of a solid object moving relative to the surrounding fluid; relative fluid velocity in the boundary layer is reduced and turbulent in comparison with the more distant mainstream flow.

Chordotonal organ Internal mechanoreceptor, generally involved in proprioception and lacking associated cuticular structures such as trichoid sensilla.

Chorus Acoustic signaling produced collectively by a group of individuals whose activity is clustered in both space and time; choruses may be temporally structured in alternating or synchronous formats.

Contralateral Of a sensory organ or body surface situated opposite to a reference point such as an external stimulus.

Decibel, absolute A logarithmic measure of the ratio of a physical parameter to a reference value for that parameter. In acoustics, decibels are normally a measure of the ratio of sound pressure (Δp) to a reference pressure value (Δp_{ref}). Decibel values $= 20 \times \log(\Delta p/\Delta p_{ref})$; in air, $\Delta p_{ref} = 20$ µPa; in water, $\Delta p_{ref} = 1$ µPa.

Less commonly, decibels are a measure of the ratio of sound intensity (I) to a reference intensity value (I_{ref}). Decibel values = $10 \times \log(I/I_{ref})$; in air, $I_{ref} = 10^{-12}\,W{\cdot}m^{-2}$.

Decibel, relative A logarithmic measure of the ratio of physical parameter values. In acoustics, a sound whose pressure is twice as great (on a linear scale) as another is 6 dB higher; one that that is 10 times as great is 20 dB higher.

Diastereomer Stereoisomers whose three-dimensional configurations are not mirror images of each other. Also known as geometric isomers, diastereomers of a given compound differ in the way one or more double bonds constrain rotation of the molecule.

Diffraction Redirection of a wave by the edge of a solid (opaque) object. In general, sound waves flow around (are diffracted by) objects $< 1\lambda$ in circumference but are impeded (reflected) by larger objects; light waves are diffracted significantly when passing through an aperture $\lesssim 1\lambda$.

Duetting Antiphonal (alternating) interaction between two acoustic signalers.

Dynamic range Lower and upper limits of stimulus intensity between which a receiver exhibits an increasingly greater neural response as intensity rises.

Duty cycle Signal length divided by signal period; generally used for rhythmic mechanical signals.

Electroantennogram (EAG) Measurement of the neural response in an excised antenna exposed to a chemical stimulus.

Enantiomers Stereoisomers whose three-dimensional configurations are mirror images of each other; i.e., they are not superimposable. The enantiomers of a given compound differ in optical actvity, the direction in which they rotate the plane of polarized light.

Envelope Temporal pattern of the rise and fall in intensity during an acoustic or vibrational stimulus.

Eusocial Traditionally used in reference to social behavior distinguished by (1) overlapping of generations, (2) reproductive altruism (care provided to individuals other than one's offspring), and (3) sterile castes.

Excitatory post-synaptic potential (EPSP) Positive (electric) potential elicited in an interneuron by stimulation from one or more neural inputs.

Far field Region, generally $\gtrsim 1$ wavelength distant from the source of an acoustic stimulus, within which acoustic particle velocity is negligible but relatively high sound pressure levels may be established.

Filter, band-pass A biological or electronic device that selectively removes from incoming stimuli those events whose frequencies (repetition rates) fall below or above specific cutoff values.

Flicker fusion frequency Maximum flash repetition rate that can be temporally resolved by a visual receiver and distinguished from continuous light; acoustic fusion frequency is the analogous value for temporal resolution of sound pulses by an auditory receiver.

Frequency, carrier Repetition rate of sound and other vibrational waves; distinguished from pulse repetition rate, wherein pulses are discrete packets comprised of multiple waves.

Frequency, dominant For sound and other vibrational waves, the carrier frequency at which amplitude is maximum.

Frequency, fundamental For sound and other vibrational waves, the carrier frequency equal to the lowest natural frequency at which the wave source vibrates following an impulse. Higher frequencies, created by more complex modes of vibration, are termed harmonics.

Group velocity Velocity at which the leading edge of a group of surface or bending waves propagates away from the source. In general, group velocity does not equal the velocity of an individual wave within the group.

Hemolymph Fluid which comprises an arthropod's open circulatory system.

Impedance Resistance to the passage or transfer of energy. Specific acoustic impedance, calculated as the velocity of sound in a fluid medium times the density of that medium, is the force needed to move a unit volume of fluid per unit time.

Inhibitory post-synaptic potential (IPSP) Negative (electric) potential elicited in an interneuron by stimulation from one or more neural inputs.

Ipsilateral Of a sensory organ or body surface situated on the same side as a reference point such as an external stimulus.

Johnston's organ A chordotonal organ in the second antennal segment (pedicel).

Juvenile hormone (JH) Sesquiterpenoid substances secreted from the corpora allata and whose titers regulate the form of metamorphosis in arthropods; JH may also control reproductive maturation.

Kairomone An odor, such as a pheromone, that is released externally and inadvertently attracts a species' natural enemies.

Labeled line Process of stimulus evaluation in which inputs from the various receptor categories ascend via independent pathways to separate centers in the brain.

Lateral inhibition Processing of sensory input in which the excitatory potential of a given receptor unit (or organ) is inhibited, via interneuronal messages, by an amount proportional to the excitation of neighboring receptor units (or the opposite organ). In general, lateral inhibition enhances the detection of contrasts in stimulus intensity over space or time.

Macroglomerular complex (MGC) Processing center in the antennal lobes (deutocerebrum) of the brain comprising separate glomeruli that each receive sensory input from different categories of olfactory receptor neurons.

Near field Region, close to the source of an acoustic stimulus, within which acoustic particle velocity is relatively high but sound pressure levels remain low.

Ocellus, dorsal Single-lens eye situated on the vertex of the head; distinguished from the lateral ocellus, found in endopterygote larvae.

Olfactory receptor neuron (ORN) Sensory neuron, as found in the antenna or other chemoreceptive structures; may bear specialist binding sites that only accept conspecific pheromone and antagonists or generalist sites that accept a wide range of host-plant volatiles and other compounds.

Ommatidium Individual unit, a hexagonal facet (= lens + receptor cells), of a compound eye.

Optomotor response Adjustment of orientation velocity with respect to visual input.

Oscillogram Graph of signal energy versus time.

Parallel processing Simultaneous evaluation, by a receiver, of a signal's various components or its characters and source location.

Particle velocity In a sound field, the velocity of fluid particles infinitesimally displaced to and fro as sound waves pass by; the movements are aligned with the direction of wave propagation and in phase with pressure deviations.

Pheromone-binding protein (PBP) A compound that ferries pheromone molecules through the aqueous antennal lymph to the binding sites on the olfactory receptor neurons (ORNs).

Pheromone Substance released to the outside of an organism's body and influencing the behavior (releaser function of pheromone) or physiology (primer function) of a conspecific individual.

Photon Quantal unit of light.

Plesiomorphic Ancestral character state in phylogeny.

Protandry In arthropods, generally used in reference to the seasonal maturation of adult males prior to adult females.

Pulse A brief packet of sound or vibrational waves that generally corresponds with a single repetitive action, such as a wingstroke or legstroke.

Pyrollizidine alkaloid (PA) Secondary plant substance bearing alkaline properties and a basic pyrollizidine structure. PAs are the precursors of male courtship pheromones in various Lepidoptera.

Refraction Redirection of a wave as it passes into a medium in which its velocity of propagation is increased or decreased.

Retinula cell Individual photoreceptor unit, whose lateral projections form a segment (rhabdomere) of the rhabdom of an ommatidium.

Rhabdom Composite of lateral projections of the eight or nine photoreceptor units (tightly packed retinula cells) in the ommatidium; photoactive region of the ommatidium.

Scattering For light, random dispersion of waves by small particles (diameter $<\lambda/10$) in the medium; for sound, random dispersion of waves by acoustically reflecting objects.

Scolopidium Auditory receptor unit in a tympanal organ; generally comprised of a sensory cell surrounded by a sheath cell and a cap cell.

Serial processing Sequential evaluation, by a receiver, of a signal's various components or its characters and source location.

Sonagram Graph of a signal's carrier frequency versus time, with frequencies bearing more energy distinguished by shading or color.

Sound intensity Acoustic power per unit area.

Sound pressure level Pressure deviation between the value found at full compression (or rarefaction) of a sound wave and the ambient value found in the absence of sound.

Spectrogram Graph of signal energy versus carrier frequency; generally determined with the aid of a Fast Fourier Transform (FFT) algorithm.

Spherical spreading Simple attenuation of signal energy over a radial distance from the source such that energy is proportional to the reciprocal of the radial distance squared.

Stereoisomers Forms of the same compound (molecules identical in their molecular formulae and in which atoms are connected to which) that differ in three-dimensional configuration.

Superposition eye A compound eye in which light entering the lens of a given ommatidial unit (facet) as well as those of neighboring units reaches the receptors in the rhabdom of that given unit; an optical design which increases sensitivity at the expense of spatial resolution.

Superposition principle The amplitude of two (or more) intersecting waves equals, at a point of intersection, the simple linear sum of the individual wave amplitudes.

Trichoid chemosensillum Thin-walled, multiporous hair structure bearing olfactor receptors.

Wasmannian mimicry Resemblance of a host, typically a eusocial insect, by a parasite. The resemblance is normally chemical and tactile, and the Wasmannian mimic is accepted by its host as a colony member.

Taxonomic Index

Bold numbers denote references to figures; numbers in parentheses refer to notes.

Subject Index

Bold numbers denote references to figures; numbers in parentheses refer to notes.